Communication Technology Update

9th Edition

Communication Technology Update

9th Edition

Editors

August E. Grant

Jennifer H. Meadows

In association with Technology Futures, Inc.

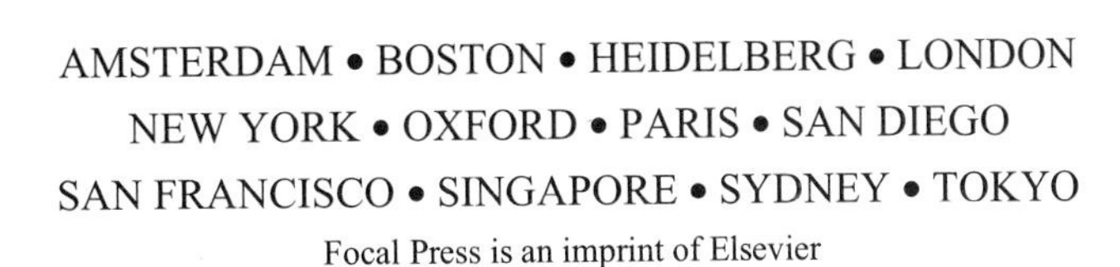

AMSTERDAM • BOSTON • HEIDELBERG • LONDON
NEW YORK • OXFORD • PARIS • SAN DIEGO
SAN FRANCISCO • SINGAPORE • SYDNEY • TOKYO
Focal Press is an imprint of Elsevier

Editors
August E. Grant
Jennifer H. Meadows
Technology Futures, Inc.
Production Editor Debra R. Robison
Art Director Helen Mary V. Marek

Focal Press is an imprint of Elsevier
200 Wheeler Road, Burlington, MA 01803, USA
Linacre House, Jordan Hill, Oxford OX2 8DP, UK

Library of Congress Cataloging-in-Publication Data
Application submitted.

British Library Cataloguing-in-Publication Data
A catalogue record for this book is available from the British Library.

ISBN: 0-240-80640-9

For information on all Focal Press publications
visit our website at www.focalpress.com

04 05 06 07 08 09 10 9 8 7 6 5 4 3 2 1

Printed in the United States of America

Table of Contents

Updates can be found on the
Communication Technology Update Home Page
http://www.tfi.com/ctu/

Preface

This edition of the *Communication Technology Update* represents a milestone of sorts. It is not notable for the edition number (the ninth) or for any particular anniversary (the first edition was published in 1992). Rather, more than any other edition, this one is routine.

What's special is what has become routinized. As with every previous edition, almost every chapter has been completely rewritten to reflect the latest developments in the technologies discussed. The graphics have been sharpened, terms have been added to the glossary, and new chapters have been added, covering new topics that have emerged since publication of the last edition in 2002. The fact that the authors were not allowed to begin their updates until the end of 2003, with text being revised as late as the second week of May 2004 is not exceptional any more—that is the routine of the publication.

We therefore want to thank the people who, through repetition, learning, and application of the technologies discussed in this book, have enabled the production of this one-of-a kind update to become routine: the authors. The contributors to this volume range from senior faculty at major universities to graduate students producing their first publications. They all share a common trait—the desire to understand these emerging communication technologies and to share that understanding with you through their contributions.

Our production staff has also played a major role in distilling 27 diverse contributions into a coherent whole. Deb Robison continues her outstanding work as the copyeditor and production editor, and Helen Mary Marek contributed outstanding graphics up to the day the final text was shipped to the printer. Amy Jollymore of Focal Press facilitated the process on the publication side, making sure that the process flowed as smoothly in printing and distribution as it did in production of the text. We are grateful to them all.

For you, the most important part of our routine may be the home page for the *Communication Technology Update* (www.tfi.com/ctu), which will be periodically updated to supplement the text with updated information and links to a wide variety of information available over the Internet. And, in summer 2005, we will begin planning our next edition—and looking for input from you on how the *Update* can best serve your needs.

As always, we encourage you to suggest new topics, glossary additions, and possible authors for the next edition of this book by communicating directly with us via e-mail, snail mail, or voice. Thank you!

Augie Grant and Jennifer Meadows
May 14, 2004

Augie Grant
College of Mass Communications & Information Studies
University of South Carolina
Columbia, SC 29208
Phone: 803.777.4464
augie@sc.edu

Jennifer H. Meadows
Department of Communication Design
California State University, Chico
Chico, CA 95929-0504
Phone: 530.898.4775
jmeadows@csuchico.edu

1

The Umbrella Perspective on Communication Technology

August E. Grant, Ph.D.*

Communication technologies are the nervous system of contemporary society, transmitting and distributing sensory and control information and interconnecting a myriad of interdependent units. Because these technologies are vital to commerce, control, and even interpersonal relationships, any change in communication technologies has the potential for profound impacts on virtually every area of society.

One of the hallmarks of the industrial revolution was the introduction of new communication technologies as mechanisms of control that played an important role in almost every area of the production and distribution of manufactured goods (Beniger, 1986). These communication technologies have evolved throughout the past two centuries at an increasingly rapid rate. This evolution shows no signs of slowing, so an understanding of this evolution is vital for any individual wishing to attain or retain a position in business, government, or education.

The economic and political challenges faced by the United States and other countries since the beginning of the new millennium clearly illustrate the central role these communication systems play in our society. Just as the prosperity of the 1990s was credited to advances in technology, the economic challenges that followed were linked as well to a major downturn in the technology sector. The aftermath of the September 11 tragedy led many to propose security measures, including control and

---

* Associate Professor, College of Mass Communications and Information Studies, University of South Carolina (Columbia, South Carolina).

monitoring of communication technologies that make extensive use of the technologies discussed in this book.

This text provides a snapshot of the process of technological evolution. The individual chapter authors have compiled facts and figures from hundreds of sources to provide the latest information on more than two dozen communication technologies. Each discussion explains the roots and evolution, the recent developments, and the current status of the technology as of mid-2004. In discussing each technology, we will deal not only with the hardware, but also with the software, organizational structure, political and economic influences, and individual users.

Although the focus throughout the book is on individual technologies, these individual snapshots comprise a larger mosaic representing the communication networks that bind individuals together and enable us to function as a society. No single technology can be understood without understanding the competing and complementary technologies and the larger social environment within which these technologies exist. As discussed in the following section, all of these factors (and others) have been considered in preparing each chapter through application of the "umbrella perspective." Following this discussion, an overview of the remainder of the book is presented.

Defining Communication Technology

The most obvious aspect of communication technology is the hardware—the physical equipment related to the technology. The hardware is the most tangible part of a technology system, and new technologies typically spring from developments in hardware. However, understanding communication technology requires more than just studying the hardware. It is just as important to understand the messages communicated through the technology system. These messages will be referred to in this text as the "software." It must be noted that this definition of "software" is much broader than the definition used in computer programming. For example, our definition of computer software would include information manipulated by the computer (such as this text, a spreadsheet, or any other stream of data manipulated or stored by the computer), as well as the instructions used by the computer to manipulate the data.

The hardware and software must also be studied within a larger context. Rogers' (1986) definition of "communication technology" includes some of these contextual factors, defining it as "the hardware equipment, organizational structures, and social values by which individuals collect, process, and exchange information with other individuals" (p. 2). An even broader range of factors is suggested by Ball-Rokeach (1985) in her media system dependency theory, which suggests that communication media can be understood by analyzing dependency relations within and across levels of analysis, including the individual, organizational, and system levels. Within the system level, Ball-Rokeach (1985) identifies three systems for analysis: the media system, the political system, and the economic system.

These two approaches have been synthesized into the "Umbrella Perspective on Communication Technology" illustrated in Figure 1.1. The bottom level of the umbrella consists of the hardware and software of the technology (as previously defined). The next level is the organizational infrastructure: the group of organizations involved in the production and distribution of the technology. The top level is the system level, including the political, economic, and media systems, as well as other groups of individuals or organizations serving a common set of functions in society. Finally, the "handle" for the umbrella is the individual user, implying that the relationship between the user and a technology must be examined in order to get a "handle" on the technology. The basic premise of the umbrella perspective is that all five areas of the umbrella must be examined in order to understand a technology.

(The use of an "umbrella" to illustrate these five factors is the result of the manner in which they were drawn on a chalkboard during a lecture in 1988. The arrangement of the five attributes resembled an umbrella, and the name stuck. Although other diagrams have since been used to illustrate these five factors, the umbrella remains the most memorable of the lot.)

Figure 1.1
The Umbrella Perspective on Communication Technology

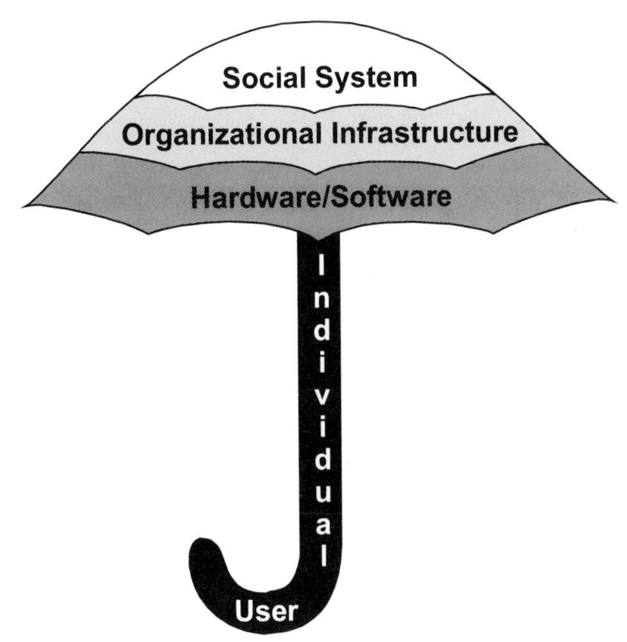

Source: A. E. Grant

Factors within each level of the umbrella may be identified as "enabling," "limiting," "motivating," and "inhibiting." *Enabling factors* are those that make an application possible. For example, the fact that coaxial cable can carry dozens of channels is an enabling factor at the hardware level, and the decision of policy makers to allocate a portion of the spectrum for cellular telephony is an enabling factor at the system level (political system).

Limiting factors are the opposite of enabling factors. Although coaxial cable increased the number of television programs that could be delivered to a home, most analog coaxial networks cannot transmit more than 100 channels of programming. To the viewer, 100 channels might seem to be more than is needed, but to the programmer of a new cable television channel unable to get space on a filled-up cable system, this hardware factor represents a definite limitation. Similarly, the fact that the policy makers discussed above initially permitted only two companies to offer cellular telephone service in each market was a system-level limitation on that technology.

Motivating factors are those that provide a reason for the adoption of a technology. Technologies are not adopted just because they exist. Rather, individuals, organizations, and social systems must have a reason to take advantage of a technology. The desire of local telephone companies for increased profits, combined with the fact that growth in providing local telephone service is limited, is an organizational factor motivating the telcos to enter the markets for new communication technologies. Individual users who desire information more quickly can be motivated to adopt electronic information technologies.

Inhibiting factors are the opposite of motivating ones, providing a disincentive for adoption or use of a communication technology. An example of an inhibiting factor at the software level might be a new electronic information technology that has the capability to update information more quickly than existing technologies, but provides only "old" content that consumers have already received from other sources. One of the most important inhibiting factors for most new technologies is the cost to individual users. Each potential user must decide whether the cost is worth the service, considering his or her budget and the number of competing technologies.

All four types of factors—enabling, limiting, motivating, and inhibiting—can be identified at the system, organizational, software, and individual user levels. However, hardware can only be enabling or limiting; by itself, hardware does not provide any motivating factors. The motivating factors must always come from the messages transmitted (software) or one of the other levels of the umbrella.

The final dimension of the umbrella perspective relates to the environment within which communication technologies are introduced and operate. These factors can be termed "external" factors, while ones relating to the technology itself are "internal" factors. In order to understand a communication technology or be able to predict the manner in which a technology will diffuse, both internal and external factors must be studied and compared.

Each communication technology discussed in this book has been analyzed using the umbrella perspective to ensure that all relevant factors have been included in the discussions. As you will see, in most cases, organizational and system-level factors (especially political factors) are more important in the development and adoption of communication technologies than the hardware itself. For example, political forces have, to date, prevented the establishment of a world standard for high-definition television (HDTV) production and transmission. As individual standards are selected in countries and regions, the standard selected is as likely to be the product of political and economic factors as of technical attributes of the system.

Organizational factors can have similar powerful effects. For example, the entry of a single company, IBM, into the personal computer business in the early 1980s resulted in fundamental changes in the entire industry. Finally, the individuals who adopt (or choose not to adopt) a technology, along with their motivations and the manner in which they use the technology, have profound impacts on the development and success of a technology following its initial introduction.

Perhaps the best indication of the relative importance of organizational and system-level factors is the number of changes individual authors made to the chapters in this book between the time of the initial chapter submission in March 2004 and production of the final, camera-ready text in May 2004. Virtually no new information was added regarding hardware, but numerous changes were made due to developments in the organizational and system levels.

To facilitate your understanding of all of the elements related to the technologies explored, each chapter in this book has been written from the umbrella perspective. The individual writers have endeavored to update developments in each area to the extent possible in the brief summaries provided. Obviously, not every technology experienced developments in each of the five areas, so each report is limited to areas in which relatively recent developments have taken place.

Overview of Book

The next chapter, "Communication Technology Timeline," provides a broad overview of most of the technologies discussed later in the book, allowing a comparison along a number of dimensions: the year each was first introduced, growth rate, number of current users, etc. This chapter co-anchors the book to highlight commonalties in the evolution of individual technologies, as well as present the "big picture" before we delve into the details. By focusing on the number of users over time, this chapter also provides the most useful basis of comparison across technologies.

The technologies discussed in this book have been organized into three sections: electronic mass media, computers and consumer electronics, and telephony and satellite technologies. These three are not necessarily exclusive; for example, direct broadcast satellites (DBS) could be classified as either an electronic mass medium or a satellite technology. The ultimate decision regarding where to put each technology was made by determining which set of current technologies most closely resembled the technology from the user's perspective. Thus, DBS was classified with electronic mass media. This process also locates the discussion of a cable television technology—cable modems—in the "Broadband Networks" chapter in the telephony section.

Each chapter is followed by a brief bibliography. These reference lists represent a broad overview of literally thousands of books and articles that provide details about these technologies. It is hoped that the reader will not only use these references, but will examine the list of source material to determine the best places to find newer information since the publication of this *Update*.

Most of the technologies discussed in this book are continually evolving. As this book was completed, many technological developments were announced but not released, corporate mergers were under discussion, and regulations had been proposed but not passed. Our goal is for the chapters in this book to establish a basic understanding of the structure, functions, and background for each technology, and for the supplementary Internet home page to provide brief synopses of the latest developments for each technology. (The address for the home page is http://www.tfi.com/ctu.)

The final two chapters attempt to draw larger conclusions from the preceding discussions. The first of these two chapters discusses the trend of convergence in media technologies that is blurring the lines among the technologies explored throughout the book. The final chapter then attempts to place these discussions in a larger context, noting commonalties among the technologies and trends over time. It is impossible for any text such as this one to ever be fully comprehensive, but it is hoped that this text will provide you with a broad overview of the current developments in communication technology.

Bibliography

Ball-Rokeach, S. J. (1985). The origins of media system dependency: A sociological perspective. *Communication Research, 12* (4), 485-510.

Beniger, J. (1986). *The control revolution.* Cambridge, MA: Harvard University Press.

Rogers, E. M. (1986). *Communication technology: The new media in society.* New York: Free Press.

2

Communication Technology Timeline

Dan Brown, Ph.D.*

This chapter follows patterns adopted in previous summaries of trends in U.S. communications media (Brown & Bryant, 1989; Brown, 2002). To aid in understanding rates of adoption and use, the premise for this chapter is that non-monetary measures are a more consistent measure of a technology's impact than the dollar value of sales. More meaningful media consumption trends emerge from examining changes in non-monetary media units and penetration (i.e., percentage of marketplace use, such as households) rather than on dollar expenditures. Box office receipts from motion pictures offers a notable exception, with unit attendance figures reported along with the dollar amounts.

Another premise is that government sources should provide as much of the data as possible. Researching the growth or sales figures of various media over time quickly reveals conflict in both dollar figures and units shipped or consumed. Government sources sometimes display the same variability seen in private sources, as the changes in the types of data used for reporting annual publishing of book titles will show. Government sources, although frequently based on private reports, provide some consistency to the reports. Readers should use caution in interpreting data for individual years and instead emphasize the trends over several years.

Figure 2.1 illustrates relative startups of various media types and the increase in the pace of introduction of new media technologies. This rapid increase in development is the logical consequence of

---

* Associate Dean of Arts & Sciences, East Tennessee State University (Johnson City, Tennessee).

the relative degree of permeation of technology in recent years versus the lack of technological sophistication of earlier eras. This figure and this chapter exclude several media that the marketplace abandoned, such as quadraphonic sound, 3D television, CB radios, 8-track audiotapes, and 8mm film cameras. Other media that receive mention have already and may yet suffer this fate of rejection. For example, long-playing vinyl audio recordings, audiocassettes, and compact discs seem doomed in the face of rapid adoption of newer forms of digital audio recordings. This chapter traces trends that reveal clues about what has happened and what may happen in the use of respective media forms.

Figure 2.1
Communication Technology Timeline

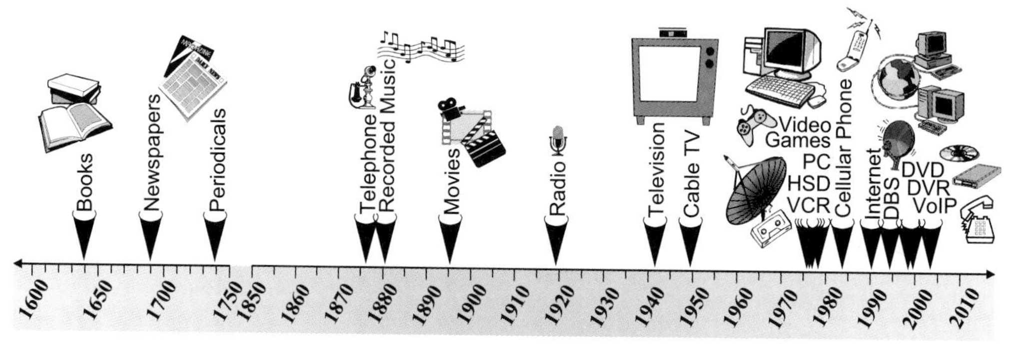

Source: Technology Futures, Inc.

Print Media

The U.S. printing industry is the largest such industry among the printing countries of the world. Although competition and consolidation of ownership reduced the number of U.S. printing firms to 62,000 in 1999, 4.6% fewer than the figure for a decade earlier, the printing industry still enjoyed "unparalleled demand for its products" (U.S. Department of Commerce/International Trade Association, 2000, p. 25-1). Foreign investment in printing is bringing a more global characteristic to the industry, with half of the 20 largest U.S. book publishers having foreign ownership (U.S. Department of Commerce/International Trade Association, 1999).

Newspapers

Publick Occurrences, Both Foreign and Domestick was the first newspaper produced in North America, appearing in 1690 (Lee, 1917). As illustrated in Table 2.1 and Figure 2.2, U.S. newspaper firms and newspaper circulation had extremely slow growth until the 1800s. Early growth suffered from relatively low literacy rates and the lack of discretionary cash among the bulk of the population. The progress of the industrial revolution brought money for workers and improved mechanized printing processes. Lower newspaper prices and the practice of deriving revenue from advertisers encouraged significant growth beginning in the 1830s. Newspapers made the transition from the realm of the educated and wealthy elite to a mass medium serving a wider range of people from this period through the Civil War era (Huntzicker, 1999).

Table 2.1
Newspaper Firms & Daily Newspaper Circulation, 1704-2002

Year	Firms	Circulation*	Year	Firms	Circulation*
1704	1		1935	8,266	40.9
1710	1		1937	8,826	43.3
1720	3		1939	9,173	43.0
1730	7		1947	10,282	53.3
1740	12		1950	12,115	53.8
1750	14		1960	11,315	58.9
1760	18		1965	11,383	60.4
1770	30		1970	11,383	62.1
1780	39		1975	11,400	60.7
1790	92		1980	9,620	62.2
1800	235		1981	9,676	61.4
1810	371		1982	9,183	62.5
1820	512		1983	9,205	62.6
1830	715		1984	9,151	63.1
1840	1,404		1985	9,134	62.8
1850	2,302	0.8	1986	9,144	62.5
1860	3,725	1.5	1987	9,031	62.8
1870	5,091	2.6	1988	10,088	62.7
1880	9,810	3.6	1989	10,457	62.6
1890	12,652	8.4	1990	11,471	62.0
1900	15,904	15.1	1991	11,689	60.0
1904	16,459	19.6	1992	11,339	60.0
1909	17,023	24.2	1993	12,597	60.0
1914	16,944	28.8	1994	12,513	59.0
1919	15,697	33.0	1995	12,246	57.0
1921	9,419	33.7	1996	10,466	57.0
1923	9,248	35.5	1997	10,042	57.0
1925	9,569	37.4	1998	10,504	56.2
1927	9,693	41.4	1999	10,530	56.0
1929	10,176	42.0	2000	10,696	55.8
1931	9,299	41.3	2001	10,739	55.6
1933	6,884	40.9	2002	10,855	55.2

* In millions

Note: The data from 1704 through 1900 are from Lee (1973). The data from 1904 through 1947 are from U.S. Bureau of the Census (1976). The number data between 1947 and 1986 are from U.S. Bureau of the Census (1986). The data from 1987 through 1988 are from U.S. Bureau of the Census (1995). The data for 1988 and 1989 are from U.S. Bureau of the Census (1997). Data from 1989 through 1997 are from U.S. Bureau of the Census (1999). Data from 1997 through 2000 are from U.S. Bureau of the Census (2001). Data after 2000 are from U.S. Bureau of the Census (2003).

Figure 2.2
Newspaper Firms & Daily Newspaper Circulation

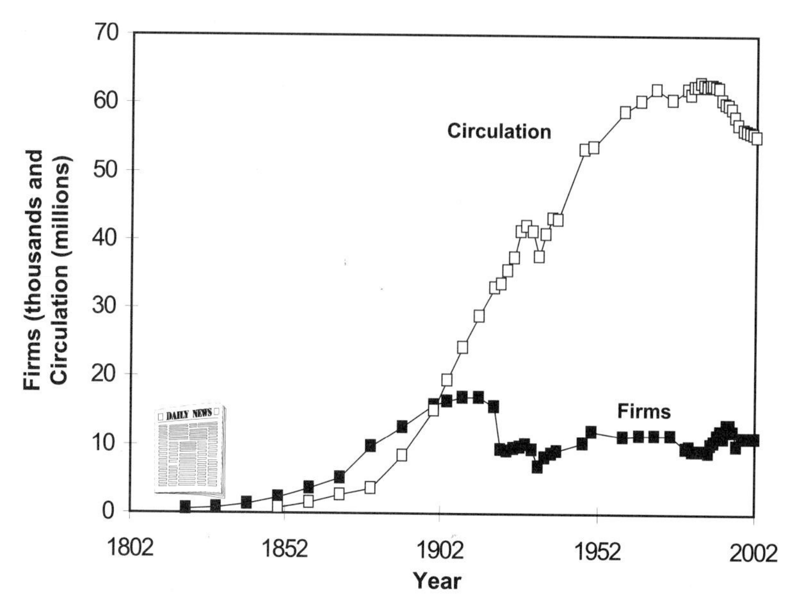

The Mexican and Civil Wars stimulated public demand for news by the middle 1800s, and modern journalism practices, such as assigning reporters to cover specific stories and topics, began to emerge. Circulation wars among big city newspapers in the 1880s featured sensational writing about outrageous stories. Both the number of newspaper firms and newspaper circulation began to soar. Although the number of firms would level off in the 20th century, circulation continued to rise.

The number of morning newspapers more than doubled after 1950, despite a 16% drop in the number of daily newspapers over that period. Circulation remained higher at the start of the new millennium than in 1950, although it inched downward throughout the 1990s. Although circulation actually increased in many developing nations, both U.S. newspaper circulation and the number of U.S. newspaper firms remain lower than the respective figures posted in the early 1990s. Total newspaper circulation in 2002 declined for the fifth year in a row to 55.2 million. The average hours spent per person per year with daily newspapers declined steadily from the late 1990s and was projected to continue that fall: 185 hours in 1998 to an estimated 169 hours per person per year in 2006 (U.S. Bureau of the Census, 2003).

Slightly more than 84% of American daily newspaper circulation comes from morning editions. Between 1995 and 2002, the number of morning dailies rose by 18.4% to 777,000. Evening dailies declined during the same period by 22.3% to 692,000 (U.S. Bureau of the Census, 2003).

Shifts from analog production techniques to digital methods strengthened the relationships between newspaper publishers and printers. Flexibility from digital methods increased newspapers' ability to deliver zoned editions that reduced unprofitable fringe readership in areas far from the central newspaper production facilities. By the end of the 20th century, more than two-thirds of U.S. daily newspapers maintained Web sites that offered classified advertising. The availability of classified advertising on the Web is significant, as classified ads accounted for more than 40% of newspaper advertising revenues. "Despite growth in readership of newspapers on the Internet, an expanding U.S. population continues to prefer to purchase the printed edition rather than viewing the electronic edition" (U.S. Department of Commerce/International Trade Association, 2000, p. 25-6).

Periodicals

By the end of 1999, the U.S. periodicals industry included 4,700 firms that employed 120,000 people. Consumer magazines accounted for 63% of industry revenues, and business magazines generated 28%. Consumer magazines reached an estimated 365 million readers in 1999, nearly 75% of the more than 500 million in annual U.S. periodical circulation. Consumer magazine circulation through subscriptions enjoyed gradual annual increases after 1980, with exceptions in 1992 and 1996. However, sales of single copies "have been in a long-term decline for decades" (U.S. Department of Commerce/International Trade Association, 2000, p. 25-10).

"The first colonial magazines appeared in Philadelphia in 1741, about 50 years after the first newspapers" (Campbell, 2002, p. 310). Few Americans could read in that era, and periodicals were costly to produce and circulate. Magazines were often subsidized and distributed by special interest groups, such as churches (Huntzicker, 1999). *The Saturday Evening Post*, the longest running magazine in U.S. history, began in 1821 and became the first magazine to both target women as an audience and distribute to a national audience. By 1850, nearly 600 magazines were operating.

By early in the 20th century, national magazines became popular with advertisers who wanted to reach wide audiences. No other medium offered such opportunity. However, by the middle of the century, the many successful national magazines began dying in face of advertisers' preferences for the new medium of television and the increasing costs of periodical distribution. Magazines turned to smaller niche audiences that were more effective at reaching these audiences. Table 2.2 and Figure 2.3 show the number of American periodical titles by year, revealing that the number of new periodical titles nearly doubled from 1958 to 1960.

In the 10 years beginning in 1990, the average annual gain in the number of periodical titles reached only 20, despite the average of 788 new titles published annually in the 1990s. The difference occurred from the high mortality rate, as evidenced by a loss in total titles in six of the 10 years in the decade. The rebound in 2000 and 2001 did not continue in 2002. "Approximately two-thirds of all new titles fail to survive beyond four or five years (U.S. Department of Commerce/ International Trade Association, 2000, p. 25-9).

The issuing of periodical titles has a strong positive correlation with the general economic health of the country. Other important factors include personal income, literacy rates, leisure time, and attractiveness of other media forms to advertisers. With the decline in network television viewership throughout the 1990s, magazines became more popular with advertisers, particularly those seeking consumers under age 24 and over age 45. Both of those groups seem likely to increase in numbers.

Table 2.2
Published Periodical Titles, 1904-2002

Year	Titles	Year	Titles	Year	Titles
1904	1,493	1954	3,427	1989	11,556
1909	1,194	1958	4,455	1990	11,092
1914	1,379	1960	8,422	1991	11,239
1919	4,796	1965	8,990	1992	11,143
1921	3,747	1970	9,573	1993	11,863
1923	3,829	1975	9,657	1994	12,136
1925	4,496	1980	10,236	1995	11,179
1927	4,659	1981	10,873	1996	9,843
1929	5,157	1982	10,688	1997	8,530
1931	4,887	1983	10,952	1998	12,448
1933	3,459	1984	10,809	1999	11,751
1935	4,019	1985	11,090	2000	13,019
1937	4,202	1986	11,328	2001	13,878
1939	4,985	1987	11,593	2002	13,848
1947	4,610	1988	11,229		

Note: The data from 1904 through 1958 are from U.S. Bureau of the Census (1976). The data from 1960 through 1985 are from U S. Bureau of the Census (1986). The data from 1987 through 1988 are from U S. Bureau of the Census (1995). The data for 1988 through 1991 are from U.S. Bureau of the Census (1997). The data from 1991 through 1996 are from U.S. Bureau of the Census (1999). The data from 1996 through 2000 are from U.S. Bureau of the Census (2001). The data after 2000 are from U. S. Bureau of the Census (2003).

Figure 2.3
Published Periodical Titles

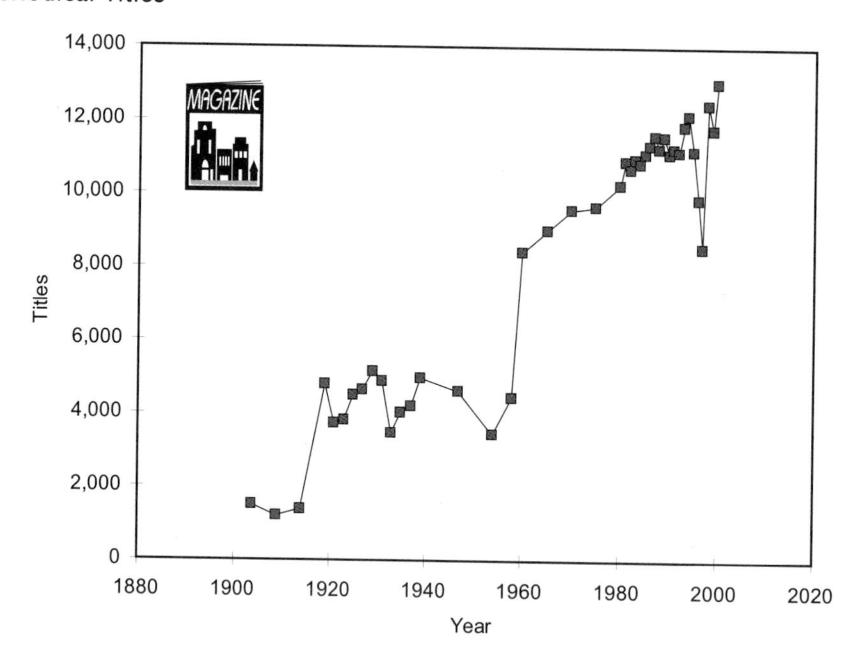

As newspapers flock to the Internet by building Web sites, so do almost all magazine publishers. These sites offer content, solicit audience interaction, and peddle print wares. In an environment that becomes increasingly cluttered with media choices, Web sites provide opportunities to engage readers and win them as subscribers to print media (U.S. Department of Commerce/International Trade Association, 2000).

Books

Books obviously enjoy a history spanning many centuries. Stephen Daye printed the first book in colonial America, *The Bay Psalm Book*, in 1640 (Campbell, 2002). Books remained relatively expensive and rare until after the printing process benefited from the industrial revolution. Linotype machines developed in the 1880s allowed for mechanical typesetting. After World War II, the popularity of paperback books helped the industry expand.

Table 2.3 shows new book titles published by year from the late 1800s through 2000. These data show a remarkable, but potentially deceptive, increase in the number of new book titles published annually, beginning in 1997. The U.S. Bureau of the Census reports that provided the data were based on material from R. R. Bowker, which changed its reporting methods beginning with the 1998 report. Ink and Grabois (2000) explained the increase as resulting from the change in the method of counting titles "that results in a more accurate portrayal of the current state of American book publishing" (p. 508). Data for previous years came from databases compiled, in part by hand, by the R. R. Bowker Company. The older counting process included only books included by the Library of Congress Cataloging in Publication program. This program included publishing by the largest American publishing companies but omitted such books as "inexpensive editions, annuals, and much of the output of small presses and self publishers" (Ink & Grabois, 2000, p. 509). Ink and Grabois observed that the U.S. ISBN (International Standard Book Number) Agency assigns more than 10,000 new publisher ISBN prefixes annually.

Figure 2.4 shows trends in publishing of book titles through 1996. Figure 2.5 shows similar trends from 1997 through 2000, after the change in the methods of counting book titles published annually. Both figures show increasing numbers.

More than 40% of U.S. adults (79,218,000) reported reading books at least once as leisure activity during the previous 12 months in 1999, and more than 20% (43,919,000) reported such participation at least twice each week. Annual expenditures for consumer books per American consumer between 1998 and 2001 held mostly steady, increasing from about $84 per person in 1998 to about $86 per person in 2001 (U.S. Bureau of the Census, 2003).

The outlook for growth of printed matter reading includes both positive and negative indicators. The number of Americans older than 45 years is expected to expand to 10.1 million by 2005. Growth is also occurring in the number of households with Internet access, as is documented later in this chapter. Relatively new book forms, such as Internet sales and print-on-demand, promise to aid publishers' profitability by making books more accessible to consumers (Publishing 2002, 2002).

Table 2.3
Published New Book Titles, 1880-2000

Year	Total Titles	Year	Total Titles	Year	Total Titles	Year	Total Titles
1880	2,076	1911	11,123	1942	9,525	1980	42,377
1881	2,991	1912	10,903	1943	8,325	1981	48,793
1882	3,472	1913	12,230	1944	6,970	1982	46,935
1883	3,481	1914	12,010	1945	6,548	1983	53,380
1884	4,088	1915	9,734	1946	7,735	1984	51,058
1885	4,030	1916	10,445	1947	9,182	1985	50,070
1886	4,676	1917	10,060	1948	9,897	1986	52,637
1887	4,437	1918	9,237	1949	10,892	1987	56,057
1888	4,631	1919	8,594	1950	11,022	1988	55,483
1889	4,014	1920	8,422	1951	11,255	1989	53,446
1890	4,559	1921	8,329	1952	11,840	1990	46,738
1891	4,665	1922	8,638	1953	12,050	1991	48,146
1892	4,862	1923	8,863	1954	11,901	1992	49,276
1893	5,134	1924	9,012	1955	12,589	1993	49,756
1894	4,484	1925	9,574	1956	12,538	1994	51,663
1895	5,469	1926	9,925	1957	13,142	1995	62,039
1896	5,703	1927	10,153	1958	13,462	1996	68,175
1897	4,928	1928	10,354	1959	14,876	1997	119,262
1898	4,886	1929	10,187	1960	15,012	1998	120,244
1899	5,321	1930	10,027	1961	18,060	1999	119,357
1900	6,356	1931	10,307	1962	21,904	2000	96,080
1901	8,141	1932	9,035	1963	25,784		
1902	7,833	1933	8,092	1964	28,451		
1903	7,865	1934	8,198	1965	28,595		
1904	8,291	1935	8,766	1966	30,050		
1905	8,112	1936	10,436	1967	28,762		
1906	7,139	1937	10,912	1968	30,387		
1907	9,620	1938	11,067	1969	29,579		
1908	9,254	1939	10,640	1970	36,071		
1909	10,901	1940	11,328	1975	39,372		
1910	13,470	1941	11,112	1979	45,182		

Note: The data for 1880-1919 include pamphlets; 1920-1928, pamphlets included in total only; thereafter, pamphlets excluded entirely. Beginning 1959, the definition of "book" changed, rendering data on prior years not strictly comparable with subsequent years. Beginning 1967, the counting methods were revised, rendering prior years not strictly comparable with subsequent years. The data from 1904 through 1947 are from U.S. Bureau of the Census (1976). The data from 1975 through 1983 are from U.S. Bureau of the Census (1984). The data from 1984 are from U.S. Bureau of the Census (1985). The data from 1985 and 1989 through 1992 are from U.S. Bureau of the Census (1995). The data from 1986 and 1987 are from U.S. Bureau of the Census (1990). The data from 1988 are from U.S. Bureau of the Census (1992). The data from 1989 through 1993 are from U.S. Bureau of the Census (1997). The data from 1993 through 1996 are from U.S. Bureau of the Census (1999). The data from 1997 through 1999 are from R. R. Bowker (2001).

Figure 2.4
Published Book Titles

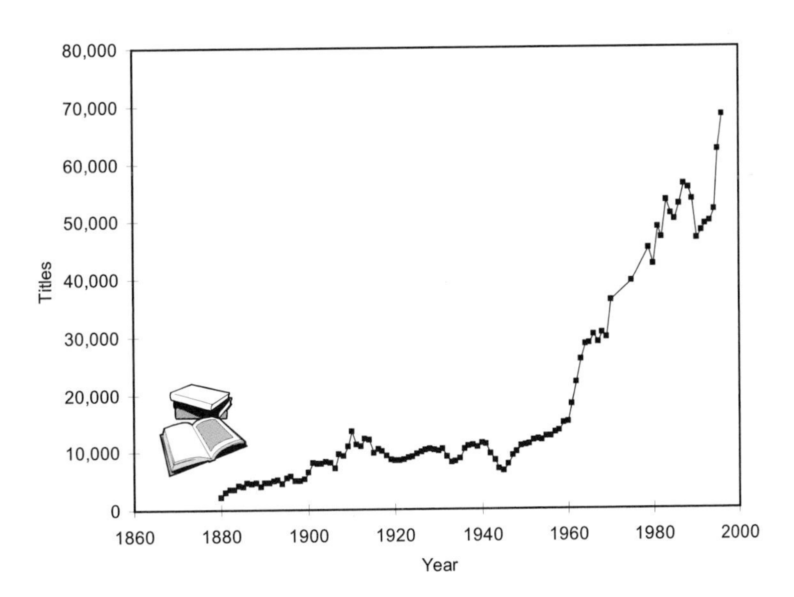

Figure 2.5
Published Book Titles, 1997-2000

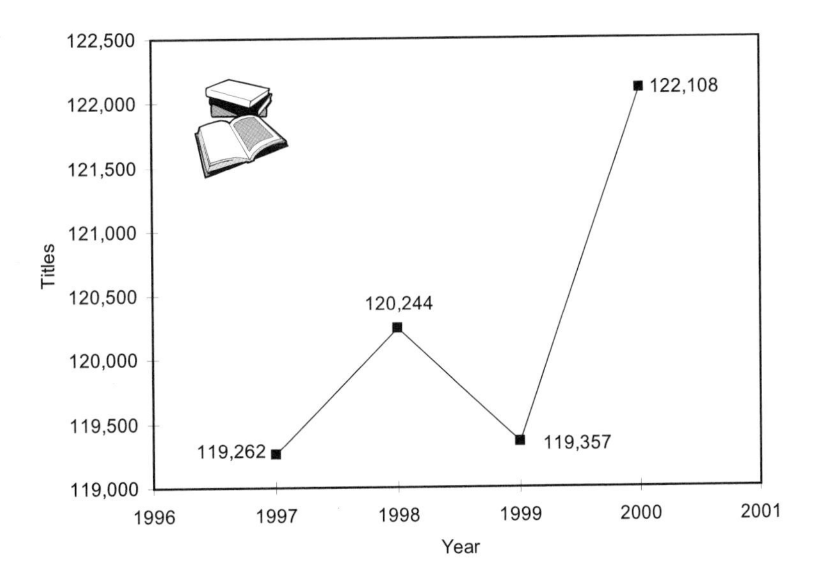

Telephone

Alexander Graham Bell became the first to transmit speech electronically, that is, to use the telephone, in 1876. By June 30, 1877, 230 telephones were in use, and the number rose to 1,300 by the end of August, mostly to avoid the need for a skilled interpreter of telegraph messages. The first exchange connected three company offices in Boston beginning on May 17, 1877, reflecting a focus on business rather than residential use during the telephone's early decades. Hotels became early adopters of telephones as they sought to reduce the costs of employing human messengers, and New York's 100 largest hotels had 21,000 telephones by 1909. After 1894, non-business telephone use became ordinary, in part, because business use lowered the cost of telephone service. By 1902, 2,315,000 telephones were in service in the United States (Aronson, 1977). Table 2.4 and Figure 2.6 document the growth to near ubiquity of telephones in U.S. households and the expanding presence of wireless telephones.

Guglielmo Marconi sent the first wireless data messages in 1895. The growing popularity of telephony led many to experiment with Marconi's radio technology as another means for interpersonal communication. By the 1920s, Detroit police cars had mobile radiophones for voice communication (International Telecommunications Union, 1999). The Bell system offered radio telephone service in 1946 in St. Louis, the first of 25 cities to receive the service. Bell engineers divided reception areas into cells in 1947, but cellular telephones that switched effectively among cells as callers moved did not arrive until the 1970s. The first call on a portable, hand-held cell phone occurred in 1973. However, by 1981, only 24 people in New York City could use their mobile phones at the same time, and only 700 customers could have active contracts. The Federal Communications Commission (FCC) began offering cellular telephone system licenses by lottery in June 1982 (Murray, 2001). Other countries, such as Japan in 1979 and Saudi Arabia in 1982, operated cellular systems earlier than the United States (ITU, 1999). Table 2.4 and Figure 2.6 show the growth in American cellular systems and subscribers from 1983 through 2002.

The number of cellular telephone systems grew from 751 employing 21,382 people in 1990 to 2,481 systems with 192,410 employees in 2002. During the same period, the average monthly bill declined from $80.90 to $48.40, although average monthly bills increased over the previous yearly average every year after 1998 (U.S. Bureau of the Census, 2003). By the end of 2001, 130 million wireless subscribers represented nearly double the number from 1998, and they talked an average of 422 minutes each month (Selingo, 2002). That increase represented a 75% increase from the average of 242 minutes per month two years earlier.

However, some sources reported that cellular phones were approaching the saturation point (Stellin, 2002), noting a slowing of the growth rate. One factor could be the high rate of problems with using cell phones relative to using other communications media. J.D. Power & Associates reported that cellular telephone companies conducted more than three times as many service calls as local telephone service providers over a year-long period that ended in spring 2001, and the FCC received more than 3,000 complaints about cellular telephone service providers in the third quarter of 2001. About one-third of the complaints addressed poor service or false advertising. During the same period, only 48 complaints reached the FCC regarding cable television operators (Selingo, 2002). Census data in Table 2.4 show reductions in the growth rate during 1994-2002, except for a small increase in the rate from 24% to 27% from 1999 to 2000.

Table 2.4
Telephone Penetration and Cellular Systems and Subscribers, 1920-2003

Year	HHs with Telephones (%)	Year	HHs with Telephones (%)	Cellular Systems	Cellular Subscribers (000s)
1920	35.0	1959	78.0		
1921	35.3	1960	78.3		
1922	35.6	1961	78.9		
1923	37.3	1962	80.2		
1924	37.8	1963	81.4		
1925	38.7	1964	82.8		
1926	39.2	1965	84.6		
1927	39.7	1966	86.3		
1928	40.8	1967	87.1		
1929	41.6	1968	88.5		
1930	40.9	1969	89.8		
1931	39.2	1970	90.5		
1932	33.5	1975			
1933	31.3	1979			
1934	31.4	1980	93.0		
1935	31.8	1981			
1936	33.1	1982			
1937	34.3	1983			1
1938	34.6	1984	91.8		100
1939	35.6	1985	92.2		350
1940	36.9	1986	92.2		682
1941	39.3	1987	92.5		1,231
1942	42.2	1988	92.9	517	2,069
1943	45.0	1989	93.0	584	3,509
1944	45.1	1990	93.3	751	5,283
1945	46.2	1991	93.6	1,252	7,557
1946	51.4	1992	93.9	1,506	11,033
1947	54.9	1993	94.2	1,529	16,009
1948	58.2	1994	93.9	1,581	24,134
1949	60.2	1995	93.9	1,627	33,786
1950	61.8	1996	93.8	1,740	44,043
1951	64.0	1997	93.9	2,228	55,312
1952	66.0	1998	94.1	3,073	69,209
1953	68.0	1999	94.2	3,518	86,047
1954	69.6	2000	94.1	2,440	109,478
1955	71.5	2001	94.6	2,587	128,375
1956	73.8	2002	95.5	2,481	140,766
1957	75.5	2003	95.5		
1958	76.4				

Note: 1950-1982 data applies to principal earners filing reports with FCC; earlier data applies to Bell and independent companies. Beginning in 1959, data includes figures from Alaska and Hawaii. The data for 1986 and 1987 are estimates. The data to 1970 are from U.S. Bureau of the Census (1976). The data from 1970 through 1982 are from U.S. Bureau of the Census (1986). The data after 1982 are from U.S. Department of Commerce (1987). The data from 1986 and 1987 are from U.S. Bureau of the Census (1992, 1993). The data for 1987 through 1989 are from U.S. Bureau of the Census (1997). The telephone households data from 1989 through 1998 are from U.S. Bureau of the Census (1999), except that households with telephones for 1998 are from FCC (2000). Cellular telephone data from 1990 through 1994 are from U.S. Bureau of the Census (1999). Households with telephones for 1999 and cellular subscribers for 1994 through 1999 are from U.S. Bureau of the Census (2001). Households with telephones after 1999 are from U.S. Bureau of the Census (2001). Cellular systems from 1990 and 1994 through 2000 are from U.S. Bureau of the Census (2001). Cellular systems and subscribers after 2000 are from U.S. Bureau of the Census (2003). Cellular systems from 1991 and 1992 are from U.S. Bureau of the Census (1998).

Figure 2.6
Telephone Penetration and Cellular Telephone Systems and Subscribers

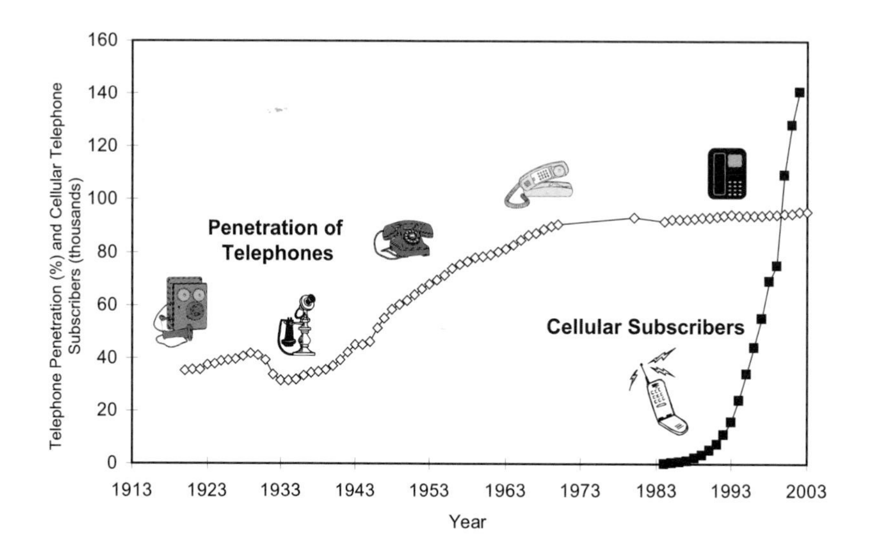

Other competitors exist among the various telephone technologies. The Telecommunications Act of 1996 encouraged competition among communications providers. For example, the act allowed cable television operators to provide telephone services as part of efforts to promote competition with telephone companies. Hafner (2002) pointed out that major wireless service providers announced high-speed wireless data services by early 2002. However, these wireless data services involve multiple incompatible technologies, often entail extra fees, sometimes offer access speeds no faster than telephone dial-up modems, and may require purchasing different phones than the ones used for standard wireless voice communications.

By late 2003, several companies offered telephone plans using Internet technology. In December 2003, AT&T announced a plan to offer local and long distance calls using the Internet as part of a strategy to avoid access fees charged by traditional telephone carriers. The new service would require broadband access, available to only about 23 million households that existed in the 100 largest U.S. markets, although approximately 100,000 households were already using Internet-based phone service (Richtel, 2003a).

Also in December 2003, cable giant Time Warner announced deals with Sprint and MCI to offer telephone service in up to 27 states, offering calling that would take advantage of Internet technology. Time Warner customers in Portland (Maine) began doing so in May 2003. Those customers paid between $39.95 and $49.95 per month for unlimited local and long distance calls (Richtel, 2003b). About eight million households around the world were already using cable services to make telephone calls (Richtel, 2003b), and market analysts predicted that more than 10 million U.S. households would get telephone capability from cable companies by 2007, compared with three million such homes at the end of 2003 (Schiesel, 2004). By early 2004, hotels, airports, and approximately 4.5 million households were using a type of wireless networking called Wi-Fi. This service eliminated the requirement for phone connectors and enabled wireless Internet communication (Dougherty, 2004).

Motion Pictures

In the 1890s, George Eastman improved on work by and patents purchased from Hannibal Goodwin in 1889 to produce workable motion picture film. The Lumière brothers projected moving pictures in a Paris café in 1895, hosting 2,500 people nightly at their movies. William Dickson, an assistant to Thomas Edison, developed the kinetograph, an early motion picture camera, and the kinetoscope, a motion picture viewing system. A New York movie house opened in 1894, offering moviegoers several coin-fed kinetoscopes. Edison's Vitascope, which expanded the length of films over those shown via kinetoscopes and allowed larger audiences to simultaneously see the moving images, appeared in public for the first time in 1896. In France in that same year, Georges Méliès started the first motion picture theater. Short movies became part of public entertainment in a variety of American venues by 1900 (Campbell, 2002), and average weekly movie attendance reached 40 million people by 1922.

Average weekly motion picture theater attendance, as shown in Table 2.5 and Figure 2.7, rose annually from the earliest available census reports on the subject in 1922 until 1930. After falling dramatically during the Great Depression, attendance regained growth in 1934 and continued until 1937. Slight declines in the prewar years were followed by a period of strength and stability throughout the World War II years. After the end of the war, average weekly attendance reached its greatest heights, averaging 90 million attendees weekly from 1946 through 1949. After the beginning of television, weekly attendance would never again reach these levels.

Although a brief period of leveling off occurred in the late 1950s and early 1960s, average weekly attendance continued to plummet until a small recovery began in 1972. This recovery signaled a period of relative stability that lasted into the 1990s. Through the last decade of the century, average weekly attendance enjoyed small but steady gains.

Box office revenues, which declined generally for 20 years after the beginning of television, began a recovery in the late 1960s. Box office revenues then began to skyrocket in the 1970s, and the explosion continued until after the turn of the new century. However, much of the increase in revenues came from increases in ticket prices and inflation, rather than from increased popularity of films with audiences, [illegible] total motion picture revenue from box office receipts declined during recent years, as studi[illegible]zed revenues from television and videocassettes (U.S. Department of Commerce/Internat[illegible]de Association, 2000).

Nielsen EDI reported that motion picture theater attendance in the United States declined in 2003 for the first year since 1991, slipping to 1.53 billion tickets from 1.6 billion in 2002. Box office revenue also declin[illegible]9.27 billion, a drop of about 1% from the previous year (Holson, 2004). Motion picture studios [illegible] 54.8% of their total revenue in 2000 from home video (FCC, 2002), and consumer spending [illegible]ome video was expected by the FCC (2004) to exceed $23 billion in 2003, more than double motion picture box office revenue.

Table 2.5
Motion Picture Attendance and Box Office Receipts, 1922-2002

Year	Average Weekly Attendance (millions)	Receipts ($ million)	Year	Average Weekly Attendance (millions)	Receipts ($ million)	Hours Per Person Per Year
1922	40		1964	44.0	913	
1923	43		1965	44.0	927	
1924	46	1.7	1966		964	
1925	46		1967		989	
1926	50		1968		1,045	
1927	57		1969		1,099	
1928	65		1970	18.0	1,162	
1929	80	720	1971	14.0	1,214	
1930	90	732	1972	15.0	1,583	
1931	75	719	1973	16.0	1,524	
1932	60	527	1974	18.0	1,909	
1933	60	482	1975	20.0	2,115	
1934	70	518	1976	20.0	2,036	
1935	80	556	1977	20.0	2,372	
1936	88	626	1978	22.0	2,643	
1937	88	676	1979	22.0	2,821	
1938	85	663	1980	20.0	2,749	
1939	85	659	1981	21.0	2,966	
1940	80	735	1982	23.0	3,453	
1941	85	809	1983	23.0	3,766	
1942	85	1,022	1984	23.0	4,030	
1943	85	1,275	1985	20.3	3,749	
1944	85	1,341	1986	19.6	3,780	
1945	85	1,450	1987	20.9	4,250	
1946	90	1,692	1988	20.9	4,460	
1947	90	1,594	1989	21.8	5,030	
1948	90	1,506	1990	22.8	5,020	
1949	70	1,451	1991	21.9	4,800	
1950	60	1,376	1992	22.6	4,870	11
1951	54	1,310	1993	23.9	5,200	12
1952	51	1,246	1994	24.8	5,400	12
1953	46	1,187	1995	24.3	5,500	12
1954	49	1,228	1996	26.3	5,900	12
1955	46	1,326	1997	26.7	6,366	12
1956	47	1,394	1998	28.5	6,949	13
1957	45	1,126	1999	28.2	7,448	13
1958	40	992	2000	27.3	7,670	12
1959	42.0	958	2001	28.7	8,410	13
1960	40.0	951	2002	31.3	9,500	13

Note: The data to 1970 are from U.S. Bureau of the Census (1976). The data from 1970, 1975, and 1979 through 1985 are from U.S. Bureau of the Census (1986). The box office data from 1971 are from U.S. Bureau of the Census (1975). The box office receipts data from 1972 through 1988 are from U.S. Department of Commerce (1988), and the data for 1988 through 1992 are from U.S. Department of Commerce (1994). The 1991 attendance came from U.S. Bureau of the Census (1996). The data for 1993 through 1996 are from U.S. Department of Commerce (1998). The data for 1997 through 1999 are from U.S. Bureau of the Census (2001). Data for hours per person per year from 1992 through 1997 are from U.S. Department of Commerce/International Trade Association (1999). Hours per person per year after 1998 are from U.S. Bureau of the Census (2003), and the figure for 2002 represents projected estimate. Attendance and receipt data for 2000-2002 are from National Association of Theater Owners (2004a and b).

Figure 2.7
Motion Picture Attendance and Box Office Receipts

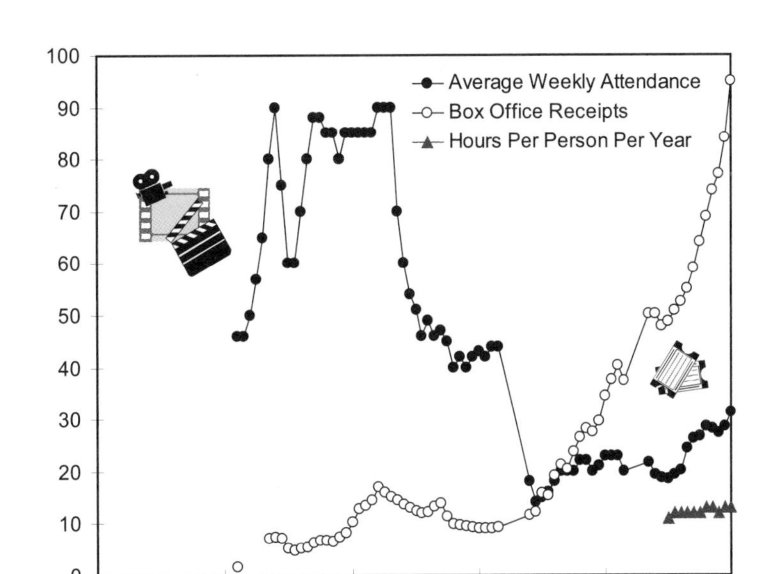

Recording

Thomas Edison expanded on experiments from the 1850s by Leon Scott de Martinville to produce a talking machine or phonograph in 1877 that played back sound recordings from etchings in tin foil. Edison later replaced the foil with wax. In the 1880s, Emile Berliner created the first flat records from metal and shellac designed to play on his gramophone, providing mass production of recordings. The early standard recordings played at 78 revolutions per minute (rpm). After shellac became a scarce commodity because of World War II, records were manufactured from polyvinyl plastic. In 1948, CBS Records produced the long-playing record that turned at 33-1/3 rpm, extending the playing time from three to four minutes to 10 minutes. RCA countered in 1949 with 45 rpm records that were incompatible with machines that played other formats. After a five-year war of formats, record players were manufactured that would play recordings at all of the speeds (Campbell, 2002).

The Germans used plastic magnetic tape for sound recording during World War II. After the Americans confiscated some of the tapes, the technology became a boon for Western audio editing and multiple track recordings that played on bulky reel-to-reel machines. By the 1960s, the reels were encased in cassettes, which would prove to be deadly competition in the 1970s for single song records playing at 45 rpm and long-playing albums playing at 33-1/3 rpm. At first the tape cassettes were popular in 8-track players. As technology improved, high sound quality was obtainable on tape of smaller width, and 8-tracks gave way to smaller audiocassettes. Thomas Stockholm began recording sound digitally in the 1970s, and the introduction of compact disc (CD) recordings in 1983 decimated the sales performance of earlier analog media types (Campbell, 2002). Table 2.6 and Figure 2.8 trace the rise and fall of the sales of audio recordings of these respective types.

Table 2.6
Recorded Music Unit Shipments, 1943-2002 (Millions)

Year	Singles	LPs/EPs	Cassettes	8-Tracks	CDs	Total
1973	228.0	280.0	15.0	91.0		614
1974	204.0	276.0	15.3	96.7		592
1975	164.0	257.0	16.2	94.6		531.8
1976	190.0	273.0	21.8	106.1		590.9
1977	190.0	344.0	36.9	127.3		698.2
1978	190.0	341.3	61.3	133.6		726.2
1979	195.5	318.3	82.8	104.7		701.3
1980	164.3	322.8	110.2	86.4		683.7
1981	154.7	295.2	137.0	48.5		635.4
1982	137.2	243.9	182.3	14.3		577.7
1983	125.0	210.0	237.0	6.0	1	579
1984	132.0	205.0	332.0	6.0	6	681
1985	121.0	167.0	339.0	4.0	23	654
1986	93.9	125.2	344.5		53	616.6
1987	82.0	107.0	410.0		102.1	701.1
1988	65.6	72.4	450.1		149.7	737.8
1989	36.6	34.6	446.2		207.2	724.6
1990	27.6	11.7	442.2		286.5	768
1991	22.0	4.8	360.1		333.3	720.2
1992	19.8	2.3	336.4		407.5	766
1993	15.1	1.2	339.5		495.4	851.2
1994	11.7	1.9	345.4		662.1	1,021.1
1995	10.2	2.2	272.6		722.9	1,007.9
1996	10.1	2.9	225.3		778.9	1,017.2
1997	7.5	2.7	172.6		753.1	935.9
1998	5.4	3.4	158.5		847.0	1,014.3
1999	5.3	2.9	123.6		938.9	1,070.7
2000	4.8	2.2	76.0		942.5	1,025.5
2001	5.5	2.3	45.0		881.9	934.7
2002	4.4	1.7	31.1		803.3	840.5

Note: "Singles" refers to vinyl singles. The data for all years prior to 1983 are from U.S. Department of Commerce (1986). The data from 1983 through 1985 and 1986 through 1989 are from U.S. Bureau of the Census (1986). The data for 1986 through 1989 are from U.S. Bureau of the Census (1995). The data after 1989 are from U.S. Bureau of the Census (2001). Data after 2000 are from U.S. Bureau of the Census (2003).

Figure 2.8
Recorded Music Unit Shipments

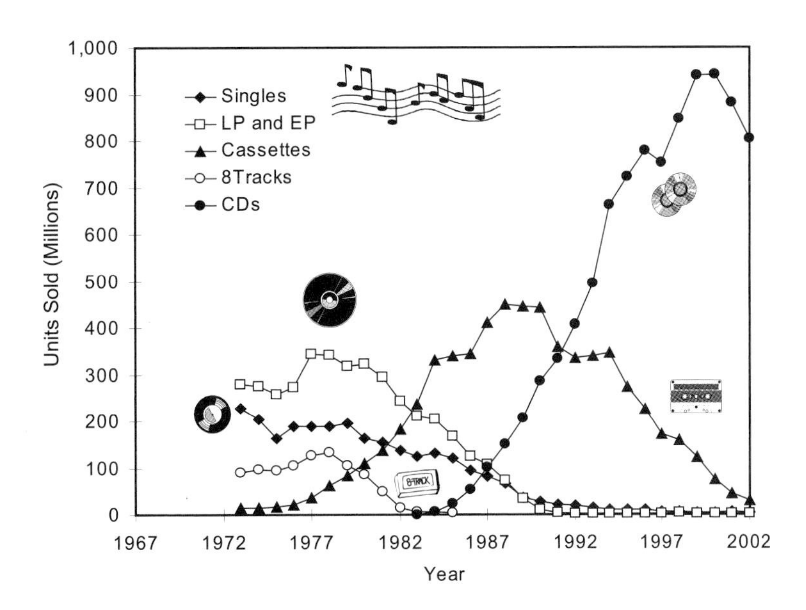

Figure 2.9
Recorded Music Total Sales

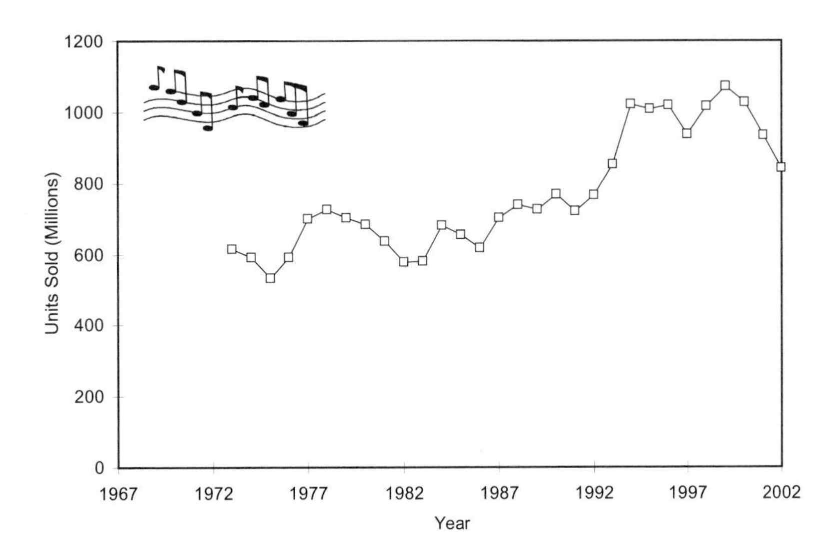

Figure 2.9 shows that total sales of recorded music rose during the early 1990s before fluctuating toward the end of the decade and generally declining every year after 1999. The Recording Industry Association of America (RIAA) (2004) announced that unit recorded music shipments to retailers declined by 2.7% in 2003 over the previous year, a somewhat smaller decline than the 7.8% drop in

2002. However, signs of a rebound occurred in early 2004, as album sales for the first 13 weeks of the year exceeded those of the comparable period in 2003 by 13% (Signs of, 2004). Retail sales were influenced by the advantages of digital recording that allowed distribution and piracy of music on the Internet. Despite aggressive RIAA strategies such as filing lawsuits against unauthorized music file trading services on the Internet, the leading sources of such activities still enjoyed 5.6 million traders, up from 3.9 million at the end of 2002. Users of these services were trading about 250 million tunes weekly (Nelson, 2004). The future of traditional recording companies remains in question.

Radio

Guglielmo Marconi's wireless messages in 1895 on his father's estate led to his establishing a British company to profit from ship-to-ship and ship-to-shore messaging. He formed a U.S. subsidiary in 1899 that would become the American Marconi Company. Reginald A. Fessenden and Lee De Forest independently transmitted voice by means of wireless radio in 1906, and a radio broadcast from the stage of a performance by Enrico Caruso occurred in 1910. Various U.S. companies and Marconi's British company owned important patents that were necessary to the development of the infant industry, so the U.S. firms formed the Radio Corporation of America (RCA) to buy out the patent rights from Marconi.

The debate still rages over the question of who became the first broadcaster among KDKA in Pittsburgh, WHA in Madison (Wisconsin), WWJ in Detroit, and KQW in San Jose (California). In 1919, Dr. Frank Conrad of Westinghouse broadcast music from his phonograph in his garage in East Pittsburgh. Westinghouse's KDKA in Pittsburgh announced the presidential election returns over the airways on November 2, 1920. By January 1, 1922, the Secretary of Commerce had issued 30 broadcast licenses, and the number of licensees swelled to 556 by early 1923. By 1924, RCA owned a station in New York, and Westinghouse expanded to Chicago, Philadelphia, and Boston. In 1922, AT&T withdrew from RCA and started WEAF in New York, the first radio station supported by commercials. In 1923, AT&T linked WEAF with WNAC in Boston by the company's telephone lines for a simultaneous program. This began the first network, which grew to 26 stations by 1925. RCA linked its stations with telegraph lines, which failed to match the voice quality of the transmissions of AT&T. However, AT&T wanted out of the new business and sold WEAF in 1926 to the National Broadcasting Company, a subsidiary of RCA (White, 1971).

The 1930 penetration of radio sets in American households reached 40% and approximately doubled over the next 10 years, passing 90% by 1947. Table 2.7 and Figure 2.10 show the rapid rate of increase in the number of radio households from 1922 through the early 1980s, when the rate of increase declined. However, the increases continued until 1993, when they began to level off.

Although thousands of radio stations were transmitting via the Internet by 2000, Channel1031.com became the first station to cease using FM and move exclusively to the Internet in September 2000 (Raphael, 2000). Many other stations were operating only on the Internet when questions about fees for commercial performers and royalties for music played on the Web arose. In 2002, the Librarian of Congress ruled by setting rates for such transmissions of sound recordings, a decision whose appeals by several organizations remained pending at the end of 2003 (U.S. Copyright Office, 2003). A federal court upheld the right of the Copyright Office to levy fees on streaming music over the Internet (*Bonneville v. Peters*, 2001).

In March 2001, the first two American digital audio satellites were launched, offering the promise of hundreds of satellite radio channels (Associated Press, 2001). Consumers were expected to pay about $9.95 per month for access to commercial-free programming that would be targeted to automobile receivers. The system included amplification from about 1,300 ground antennas. By the end of 2003, 1.621 million satellite radio subscribers tuned to the two top providers, XM and Sirius, up 51% from the previous quarter (Schaeffler, 2004).

The estimated average number of hours of radio listening per person per year increased by 5% from 936 (1998) to 983 (2001). Time spent with radio is projected to increase annually, reaching 1,062 hours per person per year in 2006, a 13.5% increase (U.S. Bureau of the Census, 2003).

Table 2.7
Radio Households & Penetration, 1922-2001

Year	HHs with Sets (000s)	%	Year	HHs with Sets (000s)	%	Year	HHs with Sets (000s)	%
1922	60		1949	39,300	93.4	1976	74,000	
1923	400		1950	40,700		1977	75,800	
1924	1,250		1951	41,900		1978	77,800	
1925	2,750		1952	42,800		1979	79,300	
1926	4,500		1953	44,800		1980	79,968	99.0
1927	6,750		1954	45,100		1981	81,600	99.0
1928	8,000		1955	45,900	95.9	1982	82,691	99.0
1929	10,250		1956	46,800	95.7	1983	83,078	99.0
1930	13,750	40.3	1957	47,600	95.8	1984	84,553	99.0
1931	16,700		1958	48,500	96.1	1985	85,921	99.0
1932	18,450		1959	49,450	96.1	1986		99.0
1933	19,250		1960	50,193	95.1	1987		99.0
1934	20,400		1961	50,695	94.7	1988	91,100	99.0
1935	21,456		1962	51,305	93.7	1989	92,800	99.0
1936	22,869		1963	52,300	94.6	1990	94,400	99.0
1937	24,500		1964	54,000	96.2	1991	95,500	99.0
1938	26,667		1965	55,200	96.1	1992	96,600	99.0
1939	27,500		1966	57,200	97.6	1993	97,300	99.0
1940	28,500	80.3	1967	57,500	97.1	1994	98,000	99.0
1941	29,300		1968	58,500	96.2	1995	98,000	99.0
1942	30,600		1969	60,600	97.4	1996	98,000	99.0
1943	30,800		1970	62,000	97.8	1997	98,000	99.0
1944	32,500		1971	65,400		1998		99.0
1945	33,100		1972	67,200		1999	103,000	99.0
1946	33,998		1973	69,400		2000	104,000	99.0
1947	35,900	91.8	1974	70,800		2001	107,000	99.0
1948	37,623		1975	72,600	98.6			

Note: Authorization of new radio stations and production of radio sets for commercial use was stopped from April 1942 until October 1945. 1959 is the first year for which Alaska and Hawaii are included in the figures. The data prior to 1970 are from U.S. Bureau of the Census (1976). The households with sets data from 1970-1972 are from U.S. Bureau of the Census (1972). The households with sets data from 1973 and 1974 are from U.S. Bureau of the Census (1975). The households with sets data from 1975 through 1977 are from U.S. Bureau of the Census (1978). The households with sets data from 1978 and 1979 are from U.S. Bureau of the Census (1981). Households for 1994 through 1999 are from U.S. Bureau of the Census (2003). Penetration for 1999 through 2001 are from U.S. Bureau of the Census (2003). Households with sets for 1998 through 2001 were estimated from U.S. Bureau of the Census (2003).

Figure 2.10
Radio and Television Households

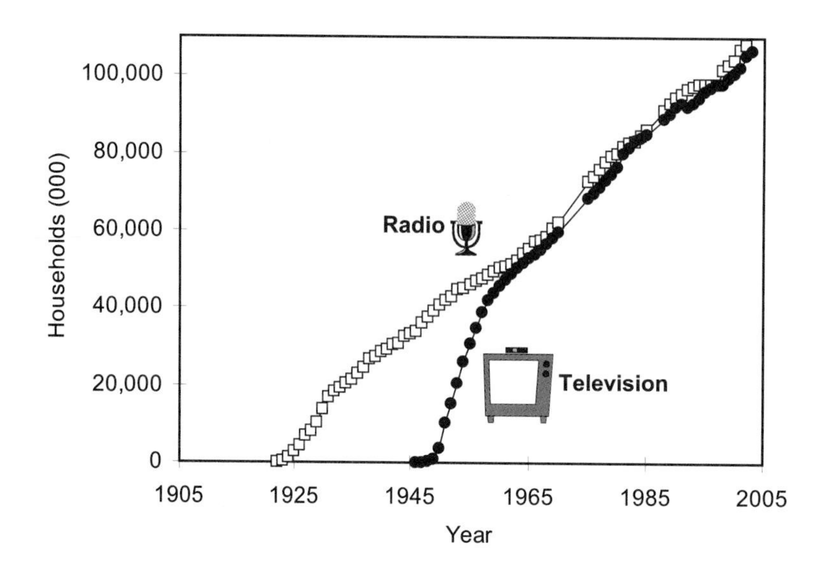

Television

Paul Nipkow invented a scanning disk device in the 1880s that provided the basis from which other inventions would develop into television. In 1927, Philo Farnsworth became the first to electronically transmit a picture over the air. Fittingly, he transmitted the image of a dollar sign. In 1930, he received a patent for the first electronic television, one of many patents for which RCA would be forced, after court challenges, to negotiate. By 1932, Vladimir Zworykin discovered a means of converting light rays into electronic signals that could be transmitted and reconstructed at a receiving device. RCA offered the first public demonstration of television at the 1939 World's Fair.

The FCC designated 13 channels in 1941 for use in transmitting black-and-white television, and the commission issued almost 100 television station broadcasting licenses before placing a freeze on new licenses in 1948. The freeze offered time to settle technical issues, and it ran longer because of U.S. involvement in the Korean War (Campbell, 2002). As shown in Table 2.8, nearly 4,000 households had television sets by 1950, a 9% penetration rate that would escalate to 87% a decade later. Penetration has remained steady at about 98% since 1980. Figure 2.10 illustrates the meteoric rise in the number of households with television by year from 1946 through the turn of the century.

American television standards set in the 1940s provided for 525 lines of data composing the picture. By the 1980s, Japanese high-definition television (HDTV) increased that resolution to more than 1,100 lines of data in a television picture. This increase enabled a much higher quality image to be transmitted with less electromagnetic spectrum space per signal. In 1996, the FCC approved a digital television transmission standard and authorized broadcast television stations a second channel for a

10-year period to allow the transition to HDTV. That transition will eventually make all older analog television sets obsolete because they cannot process HDTV signals (Campbell, 2002).

Table 2.8
Television Households and Penetration, 1946-2003

Year	HHs with Sets (000s)	%	Year	HHs with Sets (000s)	%
1946	8		1976	69,600	
1947	14		1977	71,200	
1948	172		1978	72,900	
1949	940		1979	74,500	
1950	3,875	9	1980	76,300	98
1951	10,320		1981	79,900	98
1952	15,300		1982	81,500	98
1953	20,400		1983	83,300	98
1954	26,000		1984	83,800	98
1955	30,700		1985	84,900	98
1956	34,900		1986	85,900	98
1957	38,900		1987	87,400	98
1958	41,924		1988	89,000	98
1959	43,950		1989	90,000	98
1960	45,750	87	1990	92,000	98
1961	47,200		1991	93,000	98
1962	48,855		1992	92,000	98
1963	50,300		1993	93,000	98
1964	51,600		1994	94,000	98
1965	52,700		1995	95,000	98
1966	53,850		1996	96,000	98
1967	55,130		1997	97,000	98
1968	56,670		1998	98,000	98
1969	58,250		1999	99,400	98
1970	59,550	95	2000	100,802	98
1972			2001	102,185	98
1973			2002	105,444	98
1974			2003	106,642	
1975	68,500	97			

Note: 1959 is the first year for which Alaska and Hawaii are included in the figures. The data dealing with households with television to 1971 are from U.S. Bureau of the Census (1976). The data dealing with households with television from 1980 through 1984 are from U.S. Bureau of the Census (1985). The data about penetration for all other pre-1987 years and all data for 1985 and 1986 are from U.S. Bureau of the Census (1986), and data from 1987 through 1991 are from U.S. Bureau of the Census (1995). The data from 1991 through 1996 are from FCC (1999). The penetration data from 1996 through 1999 are from U.S. Bureau of the Census (2001). The households with sets data from 1996 through 2001 are from FCC (2001). The penetration data from 2000 through 2001 are from U.S. Bureau of the Census (2003). The data after 2001 are from FCC (2004b).

The FCC set May 2002 as the deadline by which all U.S. commercial television broadcasters were required to be broadcasting digital television signals. Progress toward digital television broadcasting fell short of FCC requirements that all affiliates of the top four networks in the top 10 markets transmit digital signals by May 1, 1999. Within the 10 largest television markets, all except one network affiliate had begun HDTV broadcasts by August 1, 2001. By that date, 83% of American television stations had received construction permits for HDTV facilities or a license to broadcast HDTV signals, and 229 stations were already broadcasting digital television signals (FCC, 2002).

By September 2003, 38 of 40 stations in the largest 10 television markets in the United States began broadcasting digital television signals. Among the licensed television stations, 80% were broadcasting digitally, and virtually all of the others had obtained digital broadcasting construction permits or licenses (FCC, 2004a). Despite broadcast HDTV signals and having cable systems carrying HDTV programming available to more than 60 million households (FCC, 2004a), only about 1.5 million households were watching high-definition television by early 2004 (In-Stat/MDR, 2004).

From 1993 through 2003, the total number of U.S. commercial and noncommercial broadcast television stations grew by 13.7% to 1,733 (FCC, 2004b). However, the audience for broadcast network programming declined to an average share of 49% of prime-time viewers, compared with 74% a decade earlier. The full-day network averages also declined to 45% from 71% during the period (FCC, 2004a).

Cable Television

Cable television began as a means to overcome poor reception for broadcast television signals. John Watson claimed to have developed a master antenna system in 1948, but his records were lost in a fire. Robert J. Tarlton of Lansford (Pennsylvania) and Ed Parsons of Astoria (Oregon) set up working systems in 1949 that used a single antenna to receive programming over the air and distribute it via coaxial cable to multiple users (Baldwin & McVoy, 1983). At first, the FCC chose not to regulate cable, but after the new medium appeared to offer a threat to broadcasters, cable became the focus of heavy government regulation. Under the Reagan administration, attitudes swung toward deregulation, and cable began to flourish. Table 2.9 and Figure 2.11 show the growth of cable systems and subscribers, with penetration remaining below 25% until 1981 but passing the 50% mark before the 1980s ended.

The rate of growth in the number of cable subscribers slowed over the last half of the 1990s. Penetration, after consistently rising every year from cable's outset, declined every year after 1997. The FCC reports annually to Congress regarding the status of competition in the marketplace for video programming. The 2001 report (FCC, 2002) revealed that cable television remained the primary mode of consumer access to video programming, but franchise cable operator share of multichannel video programming distributors (MVPDs) slipped from 80% in 2000 to 78% in 2001.

During the decade beginning in 1994, the number of U.S. households passed by cable increased from 91.6 million to about 103.5 million in 2003. Put differently, about 96.3% of American households had access to cable television by the end of 2002 (FCC, 2004a). The National Cable Television Association (2004) Industry Overview cites data from A.C. Nielsen Media Research showing 73,365,880 basic cable customers in November 2003, a somewhat higher figure than that reported by the FCC (2004). See Table 2.9 and Figure 2.11 for data showing cable television growth.

Table 2.9
Cable Television Households and Penetration, 1946-2003

Year	Systems	Subscribers (000s)	Penetration (%)	Year	Systems	Subscribers (000s)	Penetration (%)
1952	70	14		1984	6,200	30,000	41.2
1955	400	150		1985	6,600	31,275	44.6
1960	640	650		1986	7,600	36,933	46.8
1965	1,325	1,275		1987	7,900	41,100	48.7
1967	1,770	2,100		1988	8,500	41,100	49.4
1968	2,000	2,800		1989	9,050	48,600	52.8
1969	2,260	3,600		1990	9,575	52,600	56.4
1970	2,490	4,500		1991	10,704	54,900	58.9
1971	2,639	5,300		1992	11,075	55,800	60.2
1972	2,841	6,000		1993	11,217	56,400	61.4
1973	2,991	7,300		1994	11,230	57,200	62.4
1974	3,158	8,700		1995	11,126	58,000	63.4
1975	3,506	9,800		1996	11,119	65,300	67.8
1976	3,681	10,800		1997	10,950	66,500	68.2
1977	3,832	11,900		1998	10,845	66,000	67.2
1978	3,875	13,000		1999	10,700	67,000	67.5
1979	4,150	14,100		2000	10,400	68,500	66.4
1980	4,225	16,000		2001	10,300	69,000	66.3
1981	4,375	18,300	25.3	2002	9,947	68,800	68.4
1982	4,825	21,000	29.0	2003	9,339	70,490	71.3
1983	5,600	25,000	37.2				

Note: Cable penetration refers to the proportion of TV households with cable. The systems and subscribers data are from U.S. Bureau of the Census (1986). The penetration data through 1986 are from U.S. Bureau of the Census (1986), except for 1970, 1980, and 1987 through 1995. Of the latter, 1987 data are from U.S. Bureau of the Census (1988), and data from 1970, 1980, and 1988 through 1994 are from U.S. Bureau of the Census (1995). Systems and penetration data through 2001 are from U.S. Bureau of the Census (2003). Subscriber data from 2001 through 2003 are from FCC (2004). Cable systems and penetration for 2003 are from National Cable & Telecommunications Association (2004). Penetration for 2002 is estimated from NCTA (2004).

Although the most popular broadcast networks continued to draw more viewers than the most popular cable television networks, audience share of cable television networks in prime-time increased to 51% in the 2002-2003 season. Prime time shares were even during the previous season. A decade earlier, cable's combined share of total day viewing was 29%, but that total reached 55% in the 2002-2003 season, up from 53% the season before (FCC, 2004).

Figure 2.11
Cable Television Households and Penetration

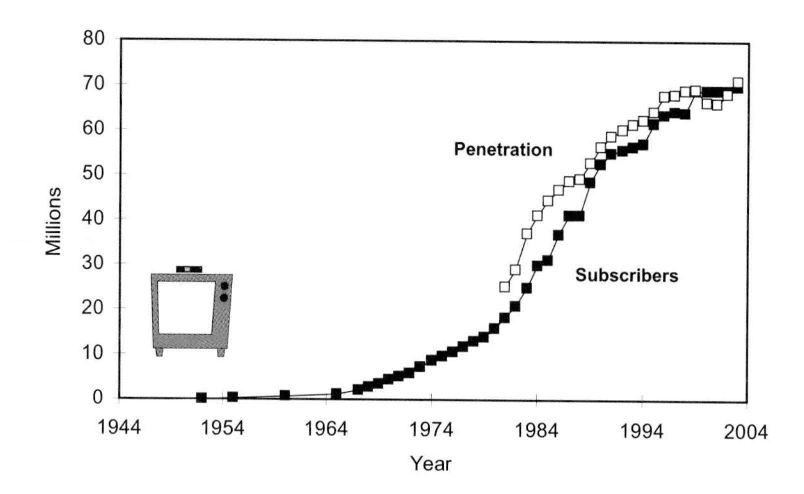

During the decade beginning in 1993, the cost of cable television subscriptions grew by 53.1%, more than double the 25.5% increase in the consumer price index. The National Cable Television Association (NCTA) claimed that, although prices for cable were up, cost per viewing hour declined, as the public spent more time watching cable programming.

The U.S. General Accounting Office blamed rising cable fees on increased programming costs for cable operators, increased spending for infrastructure, and efforts by cable operators to provide improved customer service and more kinds of services. For example, cable operators' move away from analog transmission toward digital service brought a variety of offerings, including video on demand, two-way communication, telephone service, and high-speed Internet access (FCC, 2004).

Almost all MSOs (multiple system operators) offer Internet access via cable modem. In addition, about half of smaller cable operations offer the service, and more than 13.8 million households subscribed by June 2003 (FCC, 2004).

Digital cable subscriptions enjoyed rapid growth from the 8.7 million subscribers at the end of 2000. Within another six months, the estimated count reached 12 million, with projections for 15.1 million digital cable subscribers by the end of 2001 (FCC, 2002), and 21.5 million by late 2003 (NCTA, 2004).

The FCC (2002) noted that video on demand (VOD) services showed signs of growing popularity by the end of 2001, when the services were estimated to have generated revenues exceeding $65 million. The commission cited projections of $420 million in 2002 VOD revenue. Market researchers reported that 12.5 million American households were using VOD via cable by the end of 2003 and forecast that the total would rise to 31.9 million by the end of 2006 (Schiesel, 2004).

HDTV distribution via cable television began in October 2001, serving more than 1.3 million cable subscribers (FCC, 2002). Broadcast network and premium cable channels offered HDTV programming. Because few consumers owned HDTV television sets capable of receiving the high-definition content over the air, cable operators began providing HDTV through set-top boxes.

Direct Broadcast Satellite and Other Cable TV Competitors

Competitors for the cable industry include a variety of technologies. FCC reports (FCC, 2002) distinguish between home satellite dish (HSD) systems and direct broadcast satellite (DBS) systems. Both are included as MVPDs, which include cable television, wireless cable systems called multichannel multipoint distribution services (MMDS), and private cable systems called satellite master antenna systems (SMATV). Along with these video sources, the FCC includes the Internet, but the agency does not yet find that Internet video has become a competitor to traditional video sources.

The FCC (2004) reported a 56% growth of households with MVPD services, including both cable and non-cable sources between 1993 (60.3 million homes) and 2003 (94.1 million homes). The proportion of these services accounted for by cable television dropped from 94.9% in 1993 to 74.9% in 2003. The growth in MVPD households was due to an increase in subscribers to DBS, which accounted for 21.6% of the 2003 MVPD subscribers. The number of DBS subscribers grew by 12.1% in the year following June 2002 to 20.4 million households. This increase was the smallest proportional increase in a decade of steadily rising absolute numbers of DBS subscribers, as shown in Table 2.10. The FCC (2004) cited DirecTV, the largest DBS service, as reporting that about 70% of its subscribers came to the service as cable TV subscribers. The company claimed 12.2 million subscribers overall at the end of 2003 (Amdur, 2004).

The FCC (2004a) attributed DBS growth partly to the Satellite Home Viewer Improvement Act of 1999, which granted permission to DBS providers to carry local broadcast stations in their local markets. The agency also cited the increased number of markets in which DBS service became available. Notably, only four firms were licensed by the end of 2003 to offer DBS programming to the United States.

Home satellite dishes in the early 1980s spanned six feet or more in diameter. Sales of dishes topped 500,000 only twice (1984 and 1985) between 1980 and 1995 (Brown, 2002), when home satellite system owners apparently peaked before a steady decline in numbers. The number of 2003 larger dish owners was less than 25% of the 1995 total. Conversely, smaller dish DBS subscribers have increased explosively every year since their numbers were first reported by census data in 1993. By the end of 2003, programming on C-band transmissions that was available to these large dishes consisted of about 500 channels, 350 of which were freely accessible, with the remaining 150 channels requiring subscription fees (FCC, 2004).

SMATV subscriptions generally grew after 1993, although declines occurred in 1998 and again in 2003 after a brief resurgence. MMDS subscriptions peaked in 1996, and declined in 2003 to the lowest levels since 1991. The FCC (2000) report forecast that the approval by the commission of two-way and digital MMDS services would eventually result in this part of the spectrum's use primarily for data transmission services. Table 2.10 and Figure 2.12 track trends of these non-cable video delivery types.

Table 2.10
Multichannel Video Program Distribution (Thousands), 1991-2003

Year	Home Satellite Dish Subscribers	DBS Subscribers	MMDS Subscribers	SMATV Subscribers
1991	764		180	965
1992	1,023		323	984
1993	1,612	<70	397	1,004
1994	2,178	602	600	850
1995	2,341	1,675	800	950
1996	2,300	3,500	1,200	1,100
1997	2,184	5,047	1,100	1,163
1998	2,028	7,200	1,000	940
1999	1,783	10,078	821	1,450
2000	1,477	12,987	700	1,500
2001	1,000	16,070	700	1,500
2002	701	18,200	490	1,600
2003	502	20,400	200	1,200

Note: Home satellite dishes represent the four- to eight-foot dishes. DBS uses smaller dishes. The data through 1995 are from FCC (1995). Subscriber data for 1996 and 1997 are from FCC (1998). Data from 1997 through 2001 are from FCC (2002). Data after 2001 are from FCC (2004a).

Figure 2.12
Non-Cable MVPD TV Households

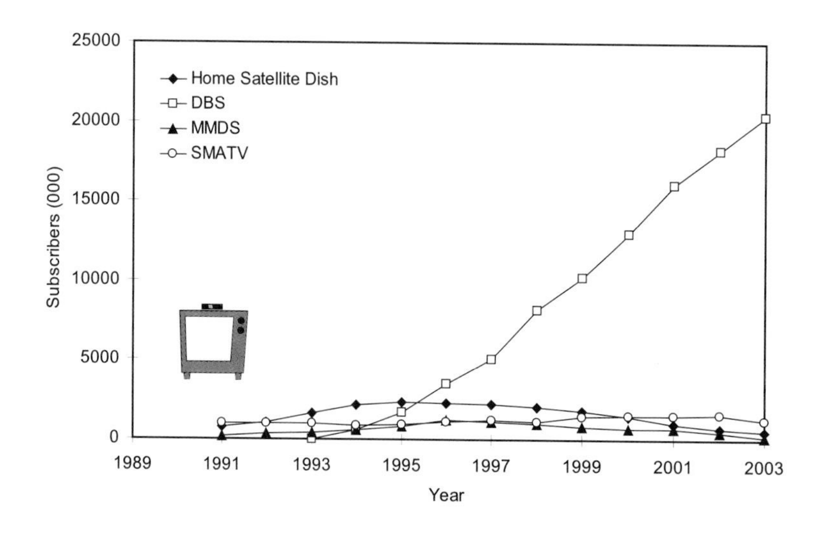

The FCC also considers several types of services as potential MVPD operators. These services include home video sales and rentals, the Internet, electric and gas utilities, and local exchange carriers (LECs). The latter category includes telephone service providers that were allowed by the Telecommunications Act of 1996 to provide video services to homes, at the same time that the act allowed cable operators to provide telephone services. The expected competition from LECs for other video services did not occur, and the FCC noted (2002) that existing LECs had left the arena of providing video, with the exceptions of reselling DBS services and efforts by new LECs to offer video services via telephone lines.

In 2001, the FCC began reporting on a new category of video program distributor that began in 1998: broadband service providers. These providers offer video, voice, and telephone capability to their customers. In that first year, such services passed 7.2 million households, but, by 2003, they were authorized by franchises to serve 17.7 million homes with 1.4 million of those households subscribing, despite the obstacle of availability in only a few areas of the country (FCC, 2004a).

Both cable television and other MVPDs deliver a variety of communications services to households (FCC, 2000a). Beyond basic video channels, DBS operations provide Internet access, and MMDS and SMATV operators offer Internet access along with local and long distance telephone communications. DirecTV offers both one- and two-way satellite Internet access, and it teamed with BellSouth to explore operations involving satellite and digital subscriber line (DSL) Internet access (FCC, 2004a).

Market structures and regulations have inhibited growth by non-cable MVPD companies. Vertically integrated cable program providers often refuse to sell to these companies, which also have difficulty in obtaining programming from other providers that have exclusive contracts with vertically-integrated suppliers. Local governments often refuse to grant franchise agreements to these non-cable entities, whose delivery options even with franchises are often hindered by lack of access to utility structures that support program transmission (FCC, 2004a).

The FCC attempted to reduce barriers to new services (FCC, 2000b) that might serve residential and multiple dwelling units. The commission required utilities to provide reasonable, nondiscriminatory access to telecommunications and cable providers that would facilitate delivery of services to potential customers. For example, utility poles and conduits in rights-of-way became available to content providers other than the utilities themselves.

Whether consumers live in multiple- or single-family dwellings, they usually have few choices of providers of MVPD or video services. Most cable operators enjoy monopolies through franchise agreements that restrict competitors. The FCC reported effective cable competition in just 878 (2.6%) of 33,485 communities that offer cable television (FCC, 2004a). DBS offers a form of competition for cable for most viewers other than people in multiple-family buildings, and advertisements for the two types of services tout the differences between the services.

Video Recorders

Although VCRs became available to the public in the late 1970s, competing technical standards slowed the adoption of the new devices. After the longer taping capacity of the VHS format won greater public acceptance over the higher-quality images of the Betamax, the popularity of home recording and playback rapidly accelerated, as shown in Table 2.11 and Figure 2.13.

Table 2.11
Households with VCRs, VCR Penetration, and Households with DVD Players, 1978-2002

Year	VCRs (000s)	VCR Penetration (%)	Year	VCRs (000s)	VCR Penetration (%)	DVD Players (000s)
1978	402		1991	67,000	71.9	
1979	478		1992	69,000	75.0	
1980	804	1.1	1993	72,000	77.1	
1981	1,330	1.8	1994	74,000	79.0	
1982	2,030	3.1	1995	77,000	81.0	
1983	4,020	5.5	1996	79,000	82.2	
1984	7,143	10.6	1997	82,000	84.2	
1985	18,000	20.8	1998	83,000	84.6	500
1986		36.0	1999	84,000	84.6	2,500
1987	43,000	48.7	2000	86000	85.1	3,300
1988	51,000	58.0	2001	88,000	86.2	7,900
1989	58,000	64.6	2002			10,700
1990	63,000	68.6				

Note: VCR penetration refers to the proportion of TV households with VCRs. The data from 1978 are from U.S. Bureau of the Census (1982). The data from 1979 through 1984 are from U.S. Bureau of the Census (1984). The data from 1985 are from U.S. Bureau of Census (1986). The penetration data are from U.S. Bureau of the Census (1986). The VCR sales data for 1985 and 1988 through 1994 are from the U.S. Bureau of the Census (1995). The VCR sales data from 1987 are from U.S. Bureau of the Census (1990). The VCR penetration data from 1987 are from U.S. Bureau of the Census (1988). The VCR penetration data from 1988 through 1992 are from U.S. Bureau of the Census (1995). The data from 1993 through 1997 are from U.S. Bureau of the Census (1999). The data from 1998 through 2000 are from U.S. Bureau of the Census (2001). VCR, VCR penetration, and DVD sales for 2000 through 2002 are from U.S. Bureau of the Census (2003).

The FCC (2002) report on competition in the video marketplace listed VCR penetration at about 90%, with multiple VCRs in nearly 46 million households. Spending for rented or purchased recorded home video exceeded $19 billion in 2000, an increase of nearly 10% over spending in the previous year. The popularity of home video prompted the commission to view home video as a viable competitor to the MVPD options described earlier. Other types of home video devices threaten the long-term growth of VCRs, but strengthen options for consumers.

About 25 million digital videodisc (DVD) players existed in American homes by the end of 2001 (FCC, 2002), and penetration reached nearly 50% by 2003 (FCC, 2004a). Approximately 10.3 million players sold in the first six months of 2003, 57% more than in the same period a year earlier. The Bureau of the Census reports of DVD player units sold fell far short of popular media reports of 31.1 million units sold in 2003 (Walker, 2004).

Americans now spend more on home video than on tickets to motion pictures in theaters. During 2002, total consumer rentals and purchases of videos jumped 11.5% over the 2001 total to $20.3 billion. Video purchases alone increased by 25% over the previous year to $12.1 billion, of which $8.7 billion was spent to buy DVDs. In June 2003, DVD rentals outnumbered VHS rentals for the first time (Bossong-Martinez, 2003a). The subscriber base of Netflix, the leading DVD rental firm, grew to 1,147,000 in 2003 from 670,000 in 2002, an increase of 71% (FCC, 2004a). Table 2.11 and Figure 2.14 trace the rapid growth in DVD sales.

Figure 2.13
Households with VCRs and VCR Penetration

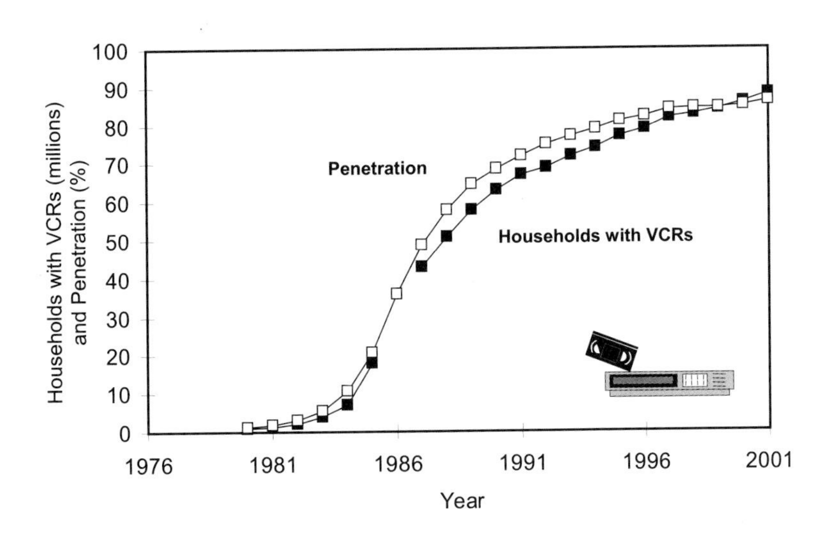

Personal video recorders (PVRs, also called digital video recorders, DVRs) debuted during 2000, and about 500,000 units were sold by the end of 2001 (FCC, 2002). The devices save video content on computer hard drives, and their penetration reached 2%, or more than 2.1 million, of U.S. TV households by 2003 (FCC, 2004a). Both DVDs and DVRs offer digital audio and video with superior quality and functionality to analog VCRs. By early 2002, major satellite content providers, such as EchoStar and DirecTV, began building DVRs into satellite television receivers as a means of enhancing their competitiveness with cable television (Lee, 2002). The FCC (2004a) cited EchoStar as reporting the sale of about one million DVR-equipped receivers by the end of 2003, and DirecTV reported about 500,000 subscribers using that service. The arrangement generates additional revenue for the providers by charging households $10 to $20 per month, and households using the service tend to be less likely to drop their satellite subscriptions.

One of the last advantages offered by the VCR over other forms of video playback is the ability to cheaply record both audio and video. However, by early 2004, the cost of DVD recorders in set-top form dropped below $300 (Pogue, 2004). Such prices could doom the VCR to the fate of the 8-track audiocassette. Perhaps the most fundamental importance of personal video recording reflects the ability of consumers to make their own programming decisions about when and what they watch. This flexibility threatens the revenue base of network television (Bossong-Martinez, 2003a).

Figure 2.14
Households with DVDs

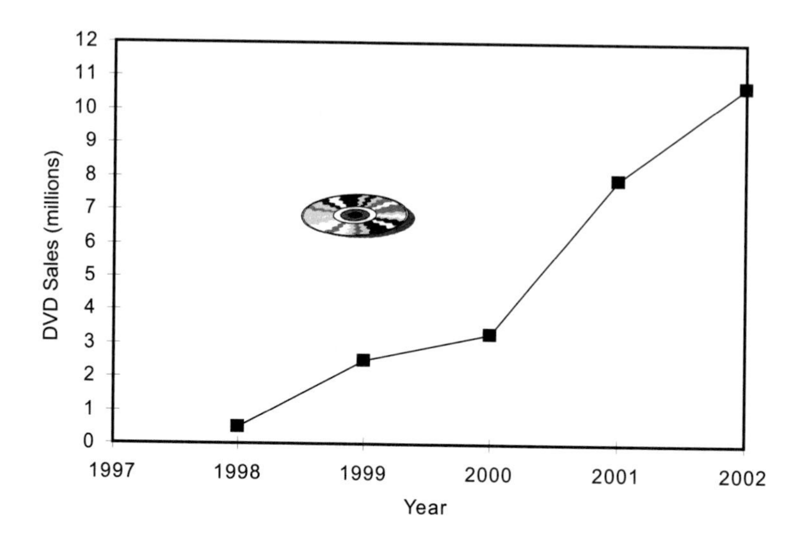

Personal Computers

The history of computing traces its origins back thousands of years to such practices as using bones as counters (Hofstra University, 2000). Intel introduced the first microprocessor in 1971. The MITS Altair, with an 8080 processor and 256 bytes of RAM, sold for $498 in 1975, introducing the desktop computer to individuals. In 1977, Radio Shack offered the TRS80 home computer, and the Apple II set a new standard for personal computing, selling for $1,298. Other companies began introducing personal computers, and by 1978, 212,000 personal computers were shipped for sale.

Table 2.12 and Figure 2.15 trace the rapid and steady rise in computer shipments and home penetration. By 2001, 56.5% of American households owned personal computers (U.S. Bureau of the Census, 2003). About 61% of white households, 37.1% of black households, 40% of Hispanic households, and 89% of households with incomes greater than $75,000 owned computers.

Table 2.12
Personal Computer Shipments and Home Use, 1978-2002

Year	PCs Shipped (Millions)	Homes with PCs (Millions)	% of Homes with PCs*
1978	0.2		
1979	0.2		
1980	0.4		
1981	1.1	0.8	1%
1982	3.5	3.0	4%
1983	6.9	7.6	9%
1984	7.6	12.0	14%
1985	6.8	15.0	18%
1986	7.0	17.4	20%
1987	8.3	20.0	23%
1988	9.5	22.4	25%
1989	9.3		
1990	9.8		
1991	10.9	25.0	25%
1992	12.5	25.0	27%
1993	14.8	31.0	30%
1994	18.6	32.0	33%
1995	22.6		
1996	25.0		
1997	30.0	35.0	37%
1998	35.4	44.2	42%
1999	44.3		
2000	47.8	53.7	51%
2001	53.0	60.2	57%
2002	55.4	63.4	59%

* The percentages of homes with PCs before 1988 are calculated by dividing the number of homes with computers by the number of homes with TVs from Table 2.9. The percentage of homes with PCs for 1997 is Newburger (1999). The homes with PCs for 1997 are calculated by multiplying the percentage of homes with PCs times the number of homes with TVs from Table 2.9. The percentage of homes with computers for 1998 is from FCC (2000a). The percentage of homes with computers for 2000 is from U.S. Bureau of the Census (2001). Homes with computers for 2000 are from the National Telecommunications & Information Administration (2000). Homes with computers and percentage of homes with computers for 2001 are from NTIA (2002). PC shipments for 1999 through 2002 are estimated from reports of the annual rates of growth in Bossong-Martinez (2003b). Penetration for 2002 is from Bossong-Martinez (2003c), and homes with PCs for 2002 are estimated from the penetration rate.

Note: Shipments from 1978 are from U.S. Bureau of the Census (1983); from 1979 and 1980 are from U.S. Bureau of the Census (1984).The data from 1981 through 1988 are from U.S. Bureau of the Census (1992). Shipments from 1989 and 1990 are from U.S. Bureau of the Census (1993), for 1991 through 1993 are from U.S. Bureau of the Census (1995), and for 1994 through 1995 are from U.S. Bureau of the Census (1997). The computer shipments for 1996 are from U.S. Department of Commerce (1998), and shipments for 1997 are from U.S. Department of Commerce/International Trade Association (1999). Shipments for 1998 are from U.S. Department of Commerce/International Trade Association (2000).

Figure 2.15
Homes with Personal Computers

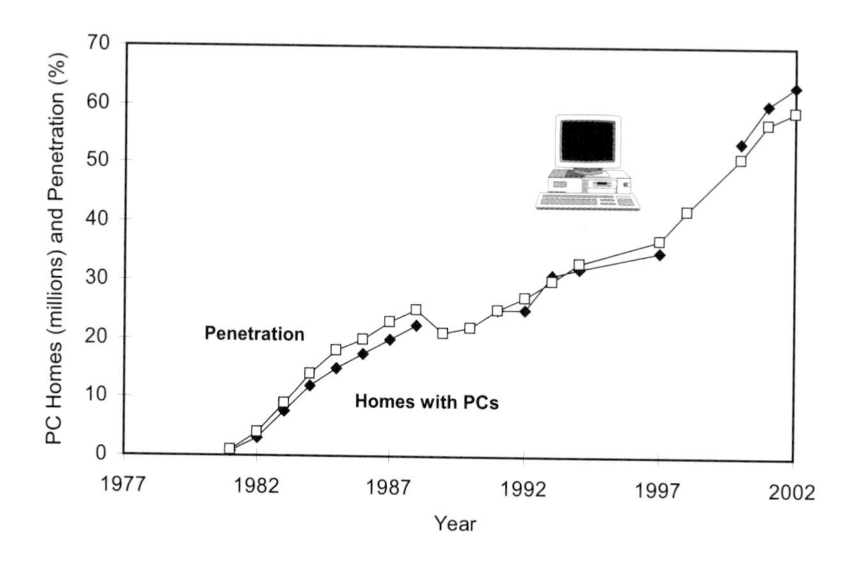

Internet

The Internet began in the 1960s with ARPANET, or the Advanced Research Projects Agency (ARPA) network project under the auspices of the U.S. Defense Department. The project intended to serve the military and researchers with multiple paths of linking computers together for sharing data in a system that would remain operational even when traditional communications might become unavailable. Early users, mostly university and research lab personnel, took advantage of electronic mail and posting information on computer bulletin boards. Usage increased dramatically in 1982 after the National Science Foundation supported high-speed linkage of multiple locations around the United States. After the collapse of the Soviet Union in the late 1980s, military users abandoned ARPANET, but private users continued to use it, and multimedia transmissions of audio and video became available. More than 150,000 regional computer networks and 95 million computer servers hosted data for Internet users (Campbell, 2002).

Pastore (2002) cited *Nielsen/NetRatings Reports* in stating that 498 million people around the world had home access to the Internet during the last quarter of 2001, the rate of Internet growth reached 15 million new users in the third quarter, and the rate nearly doubled that in the fourth quarter. Pastore cited *eMarketer* in reporting that, in 2001, 27% of the world's Internet users or 119 million people resided in the United States. U.S. Census data fail to report such spectacular numbers, in part, because they are not as recent. In keeping with the philosophy of this chapter, however, only U.S. Bureau of the Census reports contribute to the data shown in Table 2.13 and Figures 2.16 and 2.17 that report Internet usage.

Table 2.13
Internet Use by Persons 18 and Over, 1990-2003

Year	Home Access (%)	Home or Work Access (000s)	Used Internet Last 30 Days (000s)	Accessed Internet (%)	Hours/ Person/ Year	Spending for Internet Access/ Person/Year
1990					1	
1991					1	
1992					2	$4.39
1993					2	$5.35
1994					3	$6.20
1995					7	$11.33
1996				9.4	16	$17.13
1997					28	$25.52
1998	26.2	37,047	34,227	32.5	35	$27.63
1999		83,677	53,052		39	$41.77
2000	41.5	112,949	75,409	45.4	43	$50.63
2001	58.0				44	$62.08
2002		150,852	114,230	57.8	157	$75.10
2003		162,408	123,734		174	$88.96

Note: All data are from U.S. Bureau of the Census (1995, 1998, 1999, 2001, and 2002). Hours/person/year after 1997 and spending/person/year after 1997 are from the U.S. Bureau of the Census (2003), and figures after 2001 are projections.

Figure 2.16
Internet Use by Persons 18 and Over

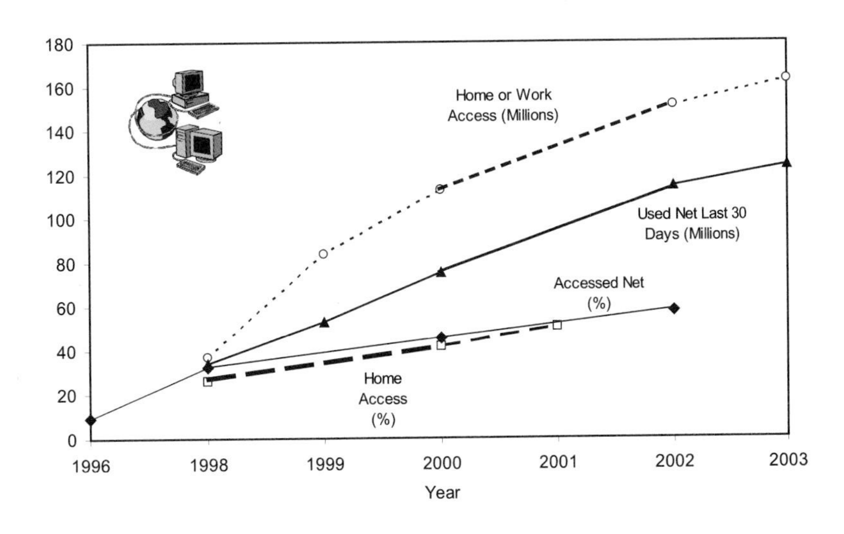

Figure 2.17
Internet Use and Spending

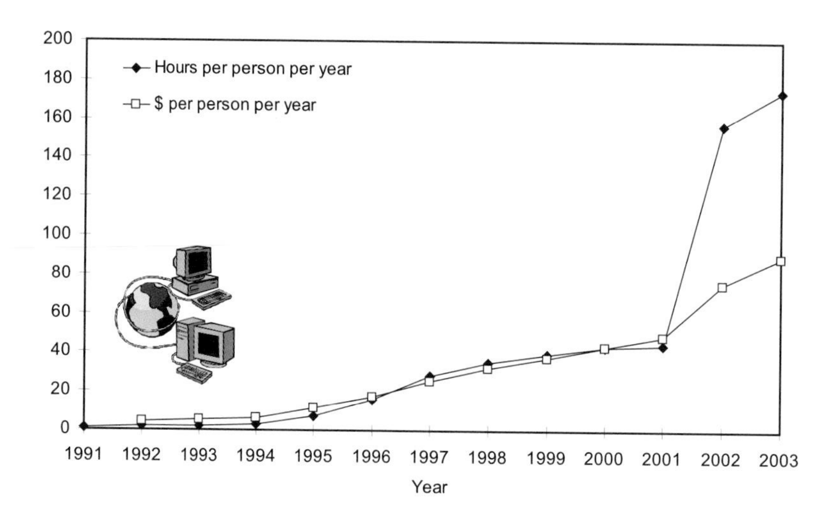

Early visions of the Internet (see Chapter 13) did not include the emphasis on entertainment and information to the general public that has emerged. The combination of this new medium with older media belongs to a phenomenon called "convergence," referring to the merging of functions of old and new media. The FCC (2002) reported that the most important type of convergence related to video content is the joining of Internet services. The report also noted that companies from many business areas were providing a variety of video, data, and other communications services.

As of 2003, 59.8% of adults connected to the Internet either at home or at work within 30 days of being surveyed, and more than 50% of American households had Internet access (U.S. Bureau of the Census, 2003). As of 2002, about 74% of Americans accomplished their Internet contact with dial-up telephone modems (FCC, 2004a). The FCC predicted that dial-up would remain the most frequently used connection means until 2004, when an estimated 55.7% of Internet households will connect through broadband equipment.

Nearly all major MSOs offer Internet access via cable modems, and the National Cable Television Association (2004) reported that 15 million households were using them by late 2003. Cable modems offer the most commonly used means of broadband access, as shown in Table 2.13 and Figure 2.18, but their share of broadband access is declining, primarily from competition of DSL telephone services.

Figure 2.18
Households with DSL or Cable Modem

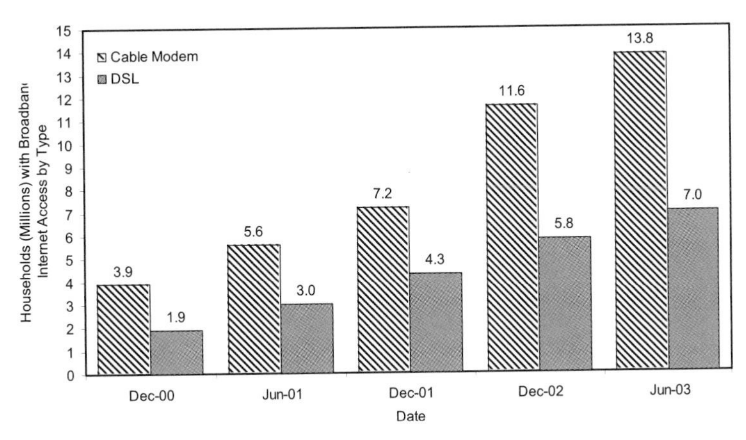

Video Games

Competition for the attention of home entertainment audiences has included commercial video games since the 1970s, and research into the form began in the 1950s, as details elsewhere in this volume show. Early versions were quite sedate, compared with modern versions that incorporate complex story lines, computer animation, and/or motion picture footage. Table 2.14 and Figure 2.19 trace trends in video game popularity, as measured by units purchased, spending per person, and time spent per person. Classifying games by media category will become increasingly difficult because they are available for dedicated gaming consoles that qualify as specialized computers, and both game software and hardware allow players to compete via the Internet. Since 1996, the number of home devices that play video games has more than doubled, and the amount of spending per person per year has more than tripled.

Table 2.14
Video Games, 1996-2003

Year	Hours/Person Per Year	Spending Person/Year	Console & PC Units Sold (Millions)
1996	25	$11.47	105
1997	36	$16.45	133
1998	43	$18.49	181
1999	61	$24.45	215
2000	75	$25.93	219
2001	78	$27.96	225
2002	84	$30.46	
2003	90	$33.60	

Note: Usage and spending per year data for 1996 and 1997 are from U.S. Bureau of the Census (2002). The figures for 1998 through 2003 are from U.S. Bureau of the Census (2003), and figures after 2001 are projections. Units sold for all years are from Levy, Ford-Livene & Levine (2002).

Figure 2.19
Video Game Usage, 1996-2003

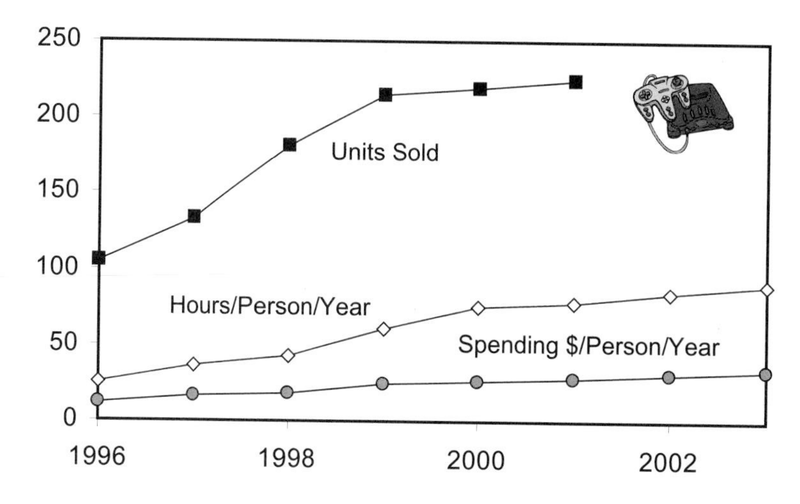

Synthesis

Just as media began converging nearly a century ago when radios and record players merged in the same appliances, media in recent years have been converging at a much more rapid pace. Wireless telephones enjoyed explosive growth, and cable and telephone companies toyed with offering competing personal communications opportunities. More than 98% of U.S. households have television, and more than 80% have either cable or satellite television, suggesting that Americans have accepted the idea of paying for television entertainment (Lee, 2002). In fact, 85% (91.4 million) of U.S. households paid for television by the end of 2003, compared with 63% (60.3 million) in 1993 (FCC, 2004a).

Popularity of print media forms generally declined throughout the 1990s, perhaps, in part, after newspapers, magazines, and books began appearing in electronic and audio forms. Recorded music sales, including CDs, declined, although the controversies over digital music indicated no loss of public popularity of listening to music. Although copyright questions slowed the expansion of radio to the Internet, satellite radio gathered nearly two million subscribers in two years after starting in 2001. Motion picture box office receipts continued to climb through 2002, although the number of people going to theaters seems to be somewhat giving way to home viewing. New ways of recording and playing back video material encouraged home viewing of motion pictures, and wealthier patrons began reproducing the theater viewing experience at home with digital systems equipped with large screens and surround-sound.

These trends illustrate that, although some ebb-and-flow occurs across various media forms, American usage of media for entertainment, information, and communication continues to expand. A thorough understanding of American life requires studying how Americans use media.

Bibliography

Amdur, M. (2004, March 9). DirecTV finishes 2003 with 12.2 million subs. *Videobusiness.com*. Retrieved March 9, 2004 from http://www.videobusiness.com/article.asp?articleID=6799&catID=5.

Aronson, S. (1977). Bell's electrical toy: What's the use? The sociology of early telephone usage. In I. Pool (Ed.). *The social impact of the telephone*. Cambridge, MA: The MIT Press, pp. 15-39.

Associated Press. (2001, March 20). Audio satellite launched into orbit. *The New York Times*. Retrieved March 20, 2001 from http://www.nytimes.com/aponline/national/AP-Satellite-Radio.html?ex=986113045&ei=1&en=7af33c7805ed8853.

Baldwin, T., & McVoy, D. (1983). *Cable communication*. Englewood Cliffs, NJ: Prentice-Hall.

Bonneville International Corp., et al. v. Marybeth Peters, as Register of Copyrights, et al. Civ. No. 01-0408, 153 F. Supp.2d 763 (E.D. Pa., August 1, 2001).

Bossong-Martinez, E. (2003a). Movies and home entertainment. *Standard & Poor's Industry Surveys, 171* (35), 1-26.

Bossong-Martinez, E. (2003b). PC outlook brightening. *Standard & Poor's Industry Surveys, 171* (50), 1-33.

Bossong-Martinez, E. (2003c). The Internet: An important communications and content medium. *Standard & Poor's Industry Surveys, 171* (36), 1-48.

Brown, D. (2002). Communication technology timeline. In A. E. Grant & J. H. Meadows (Eds.). *Communication technology update* (8th ed.). Boston: Focal Press, pp. 7-45.

Brown, D., & Bryant, J. (1989). An annotated statistical abstract of communications media in the United States. In J. Salvaggio & J. Bryant (Eds.), *Media use in the information age: Emerging patterns of adoption and consumer use*. Hillsdale, NJ: Lawrence Erlbaum Associates, pp. 259-302.

Campbell, R. (2002). *Media & culture*. Boston, MA: Bedford/St. Martins.

Dougherty, J. (2004, February 18). Wi-Fi could make cell phones obsolete. *NewsMax.com*. Retrieved February 18, 2004 from http://newsmax.com/archives/articles/2004/2/14/160956.shtml.

Federal Communications Commission. (1995, December 11). *Annual assessment of the status of competition in markets for the delivery of video programming*. CS Docket No. 95-61, Washington, DC: Author.

Federal Communications Commission. (1998, January 13). *Annual assessment of the status of competition in markets for the delivery of video programming*. CS Docket No. 97-141, Washington, DC: Author.

Federal Communications Commission. (1999, July 31). *Broadcast station totals as of July 31, 1999*. Available online at http://www.fcc.gov/Bureaus/Mass_Media/News_Releases/1999/ nrmm9022.wp.

Federal Communications Commission. (2000a, January 14). *Annual assessment of the status of competition in markets for the delivery of video programming*. CS Docket No. 99-230, Washington, DC. Author.

Federal Communications Commission. (2000b). *Promotion of competitive networks in local telecommunications markets*. 15 FCC Rcd 17521 (2000).

Federal Communications Commission. (2001, August). *Trends in telephone service*. Washington, DC: Industry Analysis Division, Common Carrier Bureau. Retrieved February 27, 2002 from http://www.fcc.gov/Bureaus/Common_Carrier/Reports/index.html.

Federal Communications Commission. (2002, January 14). *In the matter of annual assessment of the status of competition in the market for the delivery of video programming* (eighth annual report). CS Docket No. 01-129. Washington, DC 20554. Retrieved February 25, 2002 from http://www.fcc.gov/csb/.

Federal Communications Commission. (2004a). *In the matter of annual assessment of the status of competition in the market for the delivery of video programming* (tenth annual report). CS Docket No. 03-172. Retrieved February 26, 2004 from http://www.fcc.gov/mb/.

Federal Communications Commission. (2004b, February 24). *Broadcast station totals as of December 31, 1999*. Retrieved March 31, 2004 from http://www.fcc.gov/mb/audio/totals/bt031231.html.

Glaskowsky, P. (2004, January). The TV of tomorrow: Will 2004 be the year of HDTV? *Electronic Business, 30* (1), 12.

Hafner, K. (2002, February 14). The future of cell phones is here. Sort of. *The New York Times*. Retrieved February 14, 2002 from http://www.nytimes.com/2002/02/14/technology/circuits/14FUTU.html?ex=1014701434&ei=1&en=699ed96f5b641493.

Hofstra University. (2000). *Chronology of computing history*. Retrieved March 13, 2002 from http://www.hofstra.edu/pdf/CompHist_9812tla1.pdf.

Holson, L. (2004, January 6). Middle earth eclipses a sailor and a samurai. *The New York Times*. Retrieved January 6, 2004 from http://www.nytimes.com/2004/01/06/movies/06BOX.html?ex=1074400506&ei=1&en=6603e16c561db49a.

Huntzicker, W. (1999). *The popular press, 1833-1865*. Westport, CT: Greenwood Press.

Ink, G., & Grabois, A. (2000). *Book title output and average prices: 1998 final and 1999 preliminary figures*. (45th edition). D. Bogart (Ed.). New Providence, NJ: R. R. Bowker, pp. 508-513.

In-Stat/MDR. (2004, April 5). *High-definition TV services finally establish a foothold*. Retrieved April 8, 2004 from http://www.instat.com/press.asp?ID=925&sku=IN0401241MB.

International Telecommunications Union. (1999). *World telecommunications report 1999*. Geneva, Switzerland: Author.

Lee, A. (1973). *The daily newspaper in America*. New York: Octagon Books.

Lee, J. (1917). *History of American journalism*. Boston: Houghton Mifflin.

Lee, J. (2002). Interactive TV arrives. Sort of. *The New York Times*. Retrieved April 4, 2002 from http://www.nytimes.com/2002/04/04/technology/circuits/04INTE.html?ex=1019032996&ei=1&en=6eb6bb3127ddcfd2.

Levy, J., Ford-Livene, M., & Levine, A. (2002, September). *Broadcast television: Survivor in a sea of competition*. OPP Working Paper Series. Washington, DC: Federal Communications Commission.

Murray, J. (2001). *Wireless nation: The frenzied launch of the cellular revolution in America*. Cambridge, MA: Perseus Publishing.

National Association of Theater Owners. (2004a). *Total U.S. admissions*. Retrieved March 9, 2004 from http://www.natoonline.org/statisticsadmissions.htm.

National Association of Theater Owners. (2004b). *Total U.S. box office grosses*. Retrieved March 9, 2004 from http://www.natoonline.org/statisticsboxoffice.htm.

National Cable and Telecommunications Association. (2004). *Industry overview*. Retrieved April 8, 2004 from http://www.ncta.com/Docs/PageContent.cfm?pageID=86.

National Telecommunications & Information Administration. (2000, October 16). *Falling through the net, toward digital inclusion*. Washington, DC: U.S. Department of Commerce. Retrieved February 22, 2002 from http://www.ntia.doc.gov/ntiahome/digitaldivide/.

National Telecommunications & Information Administration. (2002, February). *A nation online: How Americans are expanding their use of the Internet*. Washington, DC: U.S. Department of Commerce. Retrieved February 28, 2004 from http://www.ntia.doc.gov/ntiahome/dn/index.html.

Nelson, C. (2004, February 23). CD sales rise, but industry is still wary. *The New York Times*. Retrieved on February 23, 2004 from http://www.nytimes.com/2004/02/23/business/media/23music.html?ex=1078557757&ei=1&en=9a3dd7a953ac3508.

Newburger, E. (1999, September). *Computer use in the United States*. Washington, DC: U.S. Government Printing Office. Available online: http://www.census.gov/prod/99pubs/p20-522.pdf.

Pastore, M. (2002, March 6). At home Internet users approaching half-billion. *The Big Picture Geographics*. Retrieved March 12, 2002 from http://cyberatlas.internet.com/big_picture/geographics/article/0,,5911_986431,00.html.

Pogue, D. (2004, February 5). Recording the VCR's swan song. *The New York Times*. Retrieved February 5, 2004 from http://www.nytimes.com/2004/02/05/technology/circuits/05stat.html?ex=1076991859&ei=1&en=150cebe2887ba048.

Publishing 2002: Where the buck stops; Eight top executives talk about where the business is now and where it is going. (2002, January 7). *Publisher's Weekly, 2* (49), 24-36.

R. R. Bowker. (2001). *The Bowker annual library and book trade almanac, 2001*. Medford, NJ: Information Today, Inc.

Raphael, J. (2000, September 4). Radio station leaves earth and enters cyberspace. Trading the FM dial for a digital stream. *The New York Times*. Retrieved September 4, 2000 from http://www.nytimes.com/library/tech/00/09/biztech/articles/04radio.html.

Recording Industry Association of America. (2004, March 4). RIAA announces 2003 year-end shipment numbers. *RIAA.com*. Retrieved March 9, 2004 from http://www.riaa.com/news/newsletter/030404.asp.

Richtel, M. (2003a, December 11). AT&T joins fray for cheaper calls through the Web. *The New York Times*. Retrieved December 11, 2003 from http://www.nytimes.com/2003/12/11/technology/11PHON.html?ex=1072142040&ei=1&en=8d9cc0f5be6166a4.

Richtel, M. (2003b, December 9). Time Warner to use cable lines to add phone to Internet service. *The New York Times*. Retrieved December 9, 2003 from http://www.nytimes.com/2003/12/09/technology/09PHON.html?ex=1072074798&ei=1&en=46a03fbead3c2af1.

Schaeffler, J. (2004, February 2). The real satellite radio boom begins. *Satellite News*, *27* (5). Retrieved April 7, 2004 from Lexis-Nexis.

Schiesel, S. (2004, February 16). For Comcast, it's about bundling services. *The New York Times*. Retrieved on February 16, 2004 from http://www.nytimes.com/2004/02/16/business/media/16pipe.html?ex=1077944089&ei=1&en=9f9d47d0e6069ce9.

Selingo, J. (2002, February 14), Talking more but enjoying it less. *The New York Times*. Retrieved February 14, 2002 from http://www.nytimes.com/2002/02/14/technology/circuits/14CELL.html?ex=1014692874&ei=1&en=0b7def632252279c.

Signs of a comeback. (2004, February 22). *The New York Times*. Retrieved February 22, 2004 from http://www.nytimes.com/pages/business/index.htm.

Stellin, S. (2002, February 14). Cell phone saturation. *The New York Times*. Retrieved February 14, 2002 from http://www.nytimes.com/pages/business/media/index.html.

Sullivan, S. (2002). Prices of U.S. and foreign published materials. In D. Bogart (Ed.), *The Bowker annual library and book trade almanac* (47th Ed.). Medford, NJ: Information Today, Inc., pp. 525-564.

U.S. Bureau of the Census. (1972). *Statistical abstract of the United States: 1972* (93rd Ed.). Washington, DC: U.S. Government Printing Office.

U.S. Bureau of the Census. (1975). *Statistical abstract of the United States: 1975* (96th Ed.). Washington, DC: U.S. Government Printing Office.

U.S. Bureau of the Census. (1976). *Statistical history of the United States: From colonial times to the present.* New York: Basic Books.

U.S. Bureau of the Census. (1978). *Statistical abstract of the United States: 1978* (99th Ed.). Washington, DC: U.S. Government Printing Office.

U.S. Bureau of the Census. (1981). *Statistical abstract of the United States: 1981* (102nd Ed.). Washington, DC: U.S. Government Printing Office.

U.S. Bureau of the Census. (1982). *Statistical abstract of the United States: 1982-1983* (103rd Ed.). Washington, DC: U.S. Government Printing Office.

U.S. Bureau of the Census. (1983). *Statistical abstract of the United States: 1984* (104th Ed.). Washington, DC: U.S. Government Printing Office.

U.S. Bureau of the Census. (1984). *Statistical abstract of the United States: 1985* (105th Ed.). Washington, DC: U.S. Government Printing Office.

U.S. Bureau of the Census. (1985). *Statistical abstract of the United States: 1986* (106th Ed.). Washington, DC: U.S. Government Printing Office.

U.S. Bureau of the Census. (1986). *Statistical abstract of the United States: 1987* (107th Ed.). Washington, DC: U.S. Government Printing Office.

U.S. Bureau of the Census. (1988). *Statistical abstract of the United States: 1989* (109th Ed.). Washington, DC: U.S. Government Printing Office.

U.S. Bureau of the Census. (1990). *Statistical abstract of the United States: 1991* (111th Ed.). Washington, DC: U.S. Government Printing Office.

U.S. Bureau of the Census. (1992). *Statistical abstract of the United States: 1993* (113th Ed.). Washington, DC: U.S. Government Printing Office.

U.S. Bureau of the Census. (1993). *Statistical abstract of the United States: 1994* (114th Ed.). Washington, DC: U.S. Government Printing Office.

U.S. Bureau of the Census. (1995). *Statistical abstract of the United States: 1996* (116th Ed.). Washington, DC: U.S. Government Printing Office.

U.S. Bureau of the Census. (1996). *Statistical abstract of the United States: 1997* (117th Ed.). Washington, DC: U.S. Government Printing Office.

U.S. Bureau of the Census. (1997). *Statistical abstract of the United States: 1998* (118th Ed.). Washington, DC: U.S. Government Printing Office.

U.S. Bureau of the Census. (1998). *Statistical abstract of the United States: 1999* (119th Ed.). Washington, DC: U.S. Government Printing Office.

U.S. Bureau of the Census. (1999). *Statistical abstract of the United States: 1999* (119th Ed.). Washington, DC: U.S. Government Printing Office.

U.S. Bureau of the Census. (2001). *Statistical abstract of the United States: 2001* (121st Ed.). Washington, DC: U.S. Government Printing Office. Retrieved February 7, 2002 from http://www.census.gov/prod/2002pubs/01statab/.

U.S. Bureau of the Census. (2003). *Statistical abstract of the United States: 2003* (123rd Ed.). Washington, DC: U.S. Government Printing Office. Retrieved February 13, 2004 from http://www.census.gov/prod/www/statistical-abstract-03.html.

U.S. Copyright Office. (2003). *106th Annual report of the Register of Copyrights for the fiscal year ending September 30, 2003*. Washington, DC: Library of Congress.

U.S. Department of Commerce. (1986). *U. S. industrial outlook 1986.* Washington, DC: U.S. Department of Commerce, U.S. Bureau of Economic Analysis and U.S. Bureau of Labor Statistics.

U.S. Department of Commerce. (1987). *U. S. industrial outlook 1987.* Washington, DC: U.S. Department of Commerce, U.S. Bureau of Economic Analysis and U.S. Bureau of Labor Statistics.

U.S. Department of Commerce. (1988). *U. S. industrial outlook 1988.* Washington, DC: U.S. Department of Commerce, U.S. Bureau of Economic Analysis and U.S. Bureau of Labor Statistics.

U.S. Department of Commerce. (1994). *U. S. industrial outlook 1994.* Washington, DC: U.S. Department of Commerce, U.S. Bureau of Economic Analysis and U.S. Bureau of Labor Statistics.

U.S. Department of Commerce. (1998). *U. S. industry and trade outlook 1998.* New York: McGraw-Hill.

U.S. Department of Commerce/International Trade Association. (1999). *U. S. industry and trade outlook 1999.* New York: McGraw-Hill.

U.S. Department of Commerce/International Trade Association. (2000). *U. S. industry and trade outlook 2000.* New York: McGraw-Hill.

Walker, R. (2004, March 7). The Apex DVD player. *The New York Times*. Retrieved on March 7, 2004 from http://www.nytimes.com/2004/03/07/magazine/07CONSUMED.html?ex=1079754378&ei=1&en=9a808c0450312d14.

White, L. (1971). *The American radio.* New York: Arno Press.

II Electronic Mass Media

Digital technologies are revolutionizing virtually all aspects of mass media. Digital audio and video transmission, interactivity, and new business opportunities are fueling an explosion in the number of mass media, and the programming they provide.

The changes are most evident in multichannel video distribution services. As the following chapter indicates, cable television and direct broadcast satellite services continue to reinvent themselves, incorporating digital technology to provide new services and increase channel capacity. Chapter 4 then explains how direct broadcast satellite (DBS) services have emerged as the most aggressive competitor to cable television.

The factor shared by all of these multichannel distribution services is programming. Most of these services depend, in large part, on revenues from the pay television services explored in Chapter 5, including premium cable channels and various types of pay-per-view television, including numerous forms of video on demand. Not all technologies have been as successful as pay-TV. As discussed in Chapter 6, interactive television efforts that have been discussed as the "future of television" for more than 20 years continue to disappoint inventors and investors.

Chapter 7 explores how digital technology is forcing the biggest change ever in broadcast television, as the TV industry and audience cope with the difficult reality of converting from analog to digital transmission. This chapter explores how a variety of factors, from regulation to economics, is affecting the diffusion of the eye-catching innovation that spawned digital television—high-definition television. Chapter 8 explores broadcasting's "low-definition" challenger: streaming media.

Finally, Chapter 9 explains how radio is undergoing its own digital revolution. That revolution may take longer than the television revolution, but digital technology promises the same degree of change in radio as it has offered to all areas of television broadcasting.

In reading these chapters, you should consider two basic communication technology theories. Diffusion theory helps us understand that the introduction of innovations is a process that occurs over time among members of a social system (Rogers, 1983). Different types of people adopt a technology at different times, and for different reasons. The smallest group of adopters is the innovators: They are first to adopt, but it is usually for reasons that are quite different from later adopters. Hence, it is dangerous to predict the ultimate success, failure, diffusion pattern, gratifications, etc. of a new technology by studying the first adopters.

Diffusion theory also suggests five attributes of an innovation that are important to its success: compatibility, complexity, trialability, observability, and relative advantage (Rogers, 1983). In studying or predicting diffusion of a technology, use of these factors suggests analysis of competing technologies is as important as attributes of the new technology.

A second theory to consider is the "principle of relative constancy" (McCombs, 1972; McCombs & Nolan, 1992). This theoretical perspective suggests that, over time, the aggregate disposable income devoted to the mass media, as a proportion of gross national product, is constant. In simple terms, people spend a limited proportion of their disposable income on the media discussed in this section, and that amount rarely increases when new media are introduced. In applying this theory to the electronic mass media discussed in the following chapters, consider which media will win a share of audience income, and what will happen to the losers.

Bibliography

McCombs, M. (1972). Mass media in the marketplace. *Journalism monographs*, 24.

McCombs, M., & Nolan, J. (1992). The relative constancy approach to consumer spending for media. *Journal of Media Economics*, *5* (2), 43-52.

Rogers, E. M. (1983). *Diffusion of innovations,* 3rd edition. New York: Free Press.

3

Cable Telecommunications

Larry Collette, Ph.D.*

What began as a fairly complex enterprise built around delivering television signals to underserved households has become much more than simply a video service provider. Although in the minds of many cable subscribers that simple coaxial cable represents an indispensable link to a host of popular video entertainment, today's cable telecommunications companies are much more. Many cable companies participate in services that reach well beyond their original industry boundaries into areas such as broadband data services and telephony. These companies, as part of a nearly $50 billion a year business, are continually upgrading their infrastructures, providing advanced services, and competing head-to-head with a variety of firms in the telecom marketplace. The pace of change within the industry is perhaps only matched by rapid changes in the technology itself.

This chapter examines the development of cable telecommunications, explores its current status, and highlights its potential future directions.

Background

The Technology

The five primary components of a traditional cable system are:

1) The *headend*, where television signals are brought in from a variety of satellite and over-the-air sources for distribution across the cable plant.

* Larry Collette is a telecommunications consultant in Denver, Colorado.

2) The *trunk cable* brings these signals to the neighborhood, with broadband amplifiers arrayed about every 2,000 feet or so to boost signal strength.

3) The *distribution or feeder cable* extends the signal from the trunk into the neighborhood, going past subscriber homes and providing service to feeder sections.

4) The *subscriber drop* taps a signal off the feeder cable and routes it to the individual subscriber residence.

5) The *terminal equipment* located in the subscriber's home. This can be a cable modem, a television set, or other equipment (Cicora, et al., 1999).

Figure 3.1
Traditional Cable TV Network Tree and Branch Architecture

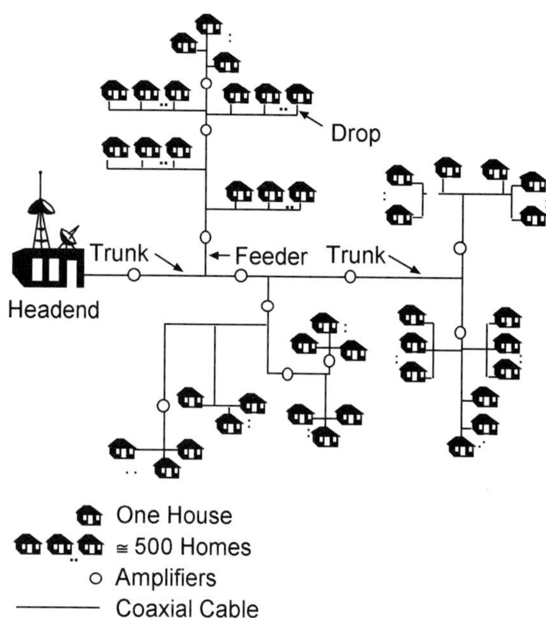

Source: Technology Futures, Inc.

Most cable systems in the United States have upgraded or are now upgrading their systems to accommodate expanded and future services. Fiber optic cable, with its large, seemingly limitless bandwidth capacity, has played an important role in these upgrades. At this point, deploying a full fiber network is not regarded as economically sound because of the high capital costs and low expected return from the deployment. In the interim, most cable systems are upgrading to hybrid fiber/coax (HFC) networks. HFC is based on a three-level hierarchy and has some limited features in common with the traditional cable architecture. In the case of HFC, a headend, a distribution hub (typically 20,000 or so homes passed), and a fiber node (500 or so homes) define the network. Rather than using the tree-and-branch architecture of the traditional cable system, fiber nodes may be connected to a distribution hub in a star network configuration (Adams, 2000). A fiber node converts the optical signal carried over the network into a radio frequency closer to the subscriber home. HFC makes it possible to provide a variety of advanced services, such as high-speed (broadband) Internet access, to cable subscribers.

Figure 3.2
Hybrid Fiber/Coax Cable TV System

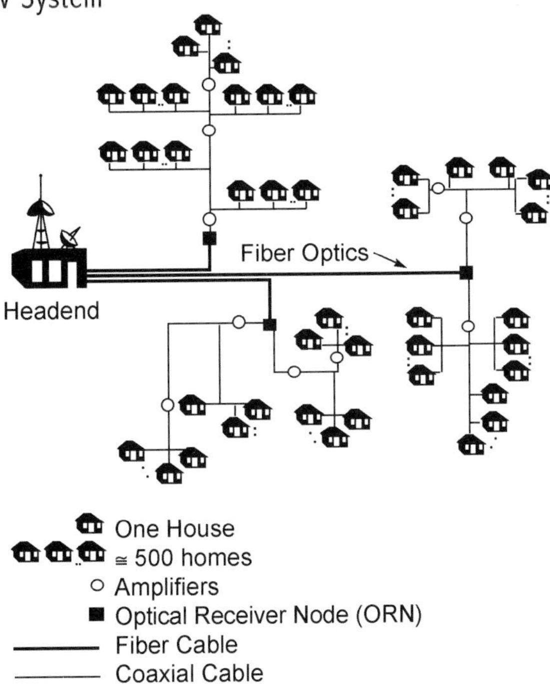

Source: Technology Futures, Inc.

Cable infrastructure expenditures leveled off to around $11 billion in 2003, after averaging roughly $16 billion over the previous three years (NCTA, 2003). This slowdown in spending is due primarily to the completion of many large-scale projects by cable operators and a weakened economy that has encouraged the delay of some other ambitious projects.

Historical Overview

The initial patent for sending pictures over a wire was granted in 1937 to Louis H. Cook, a professor of aeronautical engineering at Catholic University in Washington, DC (Taylor, 2000). Ironically, this patent would be granted a full three years before the Federal Communications Commission (FCC) even adopted technical standards for over-the-air transmission of television signals. The patent described a relatively primitive method for delivering "installed" television transmissions to houses through use of "an enclosed metal tube" to be received on "a conventional television receiving instrument" (Taylor, 2000).

By most accounts, cable television first came on the scene in the United States in 1949. That very same year saw the introduction of Silly Putty, the Berlin Airlift, the formation of NATO, and Americans rushing out to buy 100,000 television sets a week. For some, getting the television home merely ended up in frustration, for no matter how they adjusted the rabbit ears or rooftop antennas, their new electronic toy did nothing but glow. These expensive nightlights were out of reach of the over-the-air signals needed to bring the likes of *Kukla, Fran, & Ollie* and *Texaco Star Theater* directly into the living room. The solution lay in a wire—coaxial cable—developed by AT&T in 1938. While there was no apparent shortage of people desiring televised entertainment in those post WWII years, there definitely was a shortage of television stations. Even where those stations might have existed, the signal strength needed to bring in an acceptable signal could be lacking.

The idea for cable television as a business enterprise and its subsequent deployment can be credited to entrepreneurs wishing to bring television services to remote, rural areas of the United States. These regions had been deprived of over-the-air television due to a shortage of licensed TV stations and because these towns were simply too far from existing stations, making television signals out of reach. Beginning in 1949, Robert Tarlton, along with other appliance dealers, first delivered retransmissions of Philadelphia television stations to the town of Lansford (Pennsylvania) (Baldwin & McVoy, 1988). This early attempt was aimed at generating local demand for television sets that had been sitting in appliance dealers' showrooms. CATV (community antenna television), as it was first known, thus came into existence. Subscribers paid a few dollars a month to access a coaxial cable that delivered signals from a large antenna perched on a summit in the nearby Allegheny Mountains.

At around the same time, a similar story was being played out in the Pacific Northwest. Ed Parsons created a similar television service there to import a Seattle television signal, from 125 miles away, into Astoria, Oregon (Cicora, et al., 1999). These early events, and others like them, demonstrated that people would willingly pay directly for the television services they wished to receive. This important premise remains essential to the multichannel video business today.

Early systems were often a scattered patchwork of coaxial cables strung across trees, utility poles, and roofs throughout the local community. In contrast to current cable telecommunications technology, where large corporations manufacture a vast array of equipment for very specific applications, most of the early equipment consisted of ad hoc adaptations of existing equipment designed primarily for other purposes. The content on those early cable systems was limited to retransmissions of distant television stations. Agreements between the city and cable service providers were often informal "handshake deals" rather than the formalized agreements that would evolve much later.

According to the FCC, 70 communities had cable systems in 1950, serving 14,000 homes (FCC, 1996). The period from 1950 to 1955 is characterized as the "mom-and-pop" era of the cable industry, and these small enterprises faced a variety of challenges (Parsons & Frieden, 1998). Many of the early systems faced problems in finding necessary capital, locating reliable equipment suppliers, securing public rights-of-way, maintaining systems, and attracting customers. By the end of 1955, there were 400 cable systems in operation in the United States.

Most of the growth in cable during the 1960s occurred in smaller communities that were denied full access to broadcast television programming. Cable continued to have difficulties penetrating the larger metropolitan markets where up to seven VHF stations and a sprinkling of UHF stations provided "all the television anyone would ever need." There was a clear, inverse relationship between the number of channels already in a market and the demand for cable service. Improved reception, though, could motivate some to subscribe even where an adequate number of over-the-air channel options existed. The 1960s also brought the first wave of consolidation in cable systems, and with that, the appearance of the first MSOs (multiple system operators) such as Westinghouse Broadcasting, Teleprompter, and Tele-Communications, Inc. (TCI).

The development of a unique programming identity in the 1970s was, in many ways, responsible for cable first meeting some of the "blue sky" projections associated with the medium. In contrast to broadcasting, a medium where scarcity of spectrum would forever be a problem, cable and its non-radiating, sealed spectrum method of local signal delivery promised a medium of abundance.

The first major programming innovation that took advantage of the surplus channel space available over cable systems occurred in the 1970s, when Home Box Office (HBO) began using a communications satellite to make its novel service available to cable companies across the United States. (For more on HBO's history, see Chapter 4.)

Ted Turner was one of the first to realize the potential for delivery of television signals via satellite for distribution on cable television systems. He uplinked the signal of his small, independent Atlanta television station (then known as WTCG, later WTBS) to the same satellite carrying the HBO signal, allowing any cable system that picked up and distributed HBO to distribute his station's signal as well. This prescient move gave birth to the first "superstation," which inspired the creation of countless new cable channels and a number of competing superstations.

These services (and others that followed) would be instrumental in creating a unique programming identity for cable, differentiating it from broadcast competitors and driving increased consumer demand. For example, from 1976 to 1987, the number of satellite-delivered programming services grew from four to 70. At the close of the 1970s, 19% of American households subscribed to cable television.

The so-called cable revolution in the 1980s resulted from the creation of a unique programming identity, improved technology and capacity, and a somewhat more favorable regulatory environment. Major municipalities such as Cleveland, Detroit, Los Angeles, and others were wired for cable television. Even greater numbers of programming services came on the scene. Networks such as the USA Network (a general interest, broadly-targeted program service), CNN (a 24-hour news channel), and Cinemax (a specialized premium movie service) were created to take advantage of the new opportunity. The emergence of MTV, a partnership of Warner Bros. and American Express, gave a new definition to "niche" service, as this broadcast service targeted a young demographic with certain lifestyle characteristics.

The 1980s also saw a big change in the business dynamic within the programming business. Programmers who had once paid cable operators to carry their service now charged the operators fees for those same programming services. The Cable Communications Policy Act of 1984, a set of amendments to the Communications Act of 1934, dealt with a broad range of cable-related issues. To the cable industry's advantage, franchise fees that cable companies paid to local governments were lowered, cable rates were largely deregulated, and general franchise and re-franchising procedures were fixed in statute. By 1987, for the first time, the majority of television households in the United States would be receiving their television through a cable.

The 1990s saw cable companies move aggressively into new geographic and content markets. Importantly, this decade saw two substantial pieces of legislation that had profound impacts on the cable telecommunications industry. Congress passed The Cable Communications Consumer Protection and Competition Act of 1992, which re-regulated cable rates for a period of time in response to accusations of price gouging by cable companies. The act also mandated signal carriage requirements, permitting local broadcasters to opt for "must-carry" or "retransmission consent." Under must-carry, the station was assured a position on the local cable system lineup. If the broadcaster chose retransmission consent, it could negotiate a fee for carriage of their signals, but the cable system was not obligated to carry the station. A "range war" flared, with many cable and broadcast interests refusing to meet each other's demands.

One cable system in Corpus Christi (Texas) failed to come to terms with the Big Three network affiliates in the market in negotiating for retransmission consent and, as a result, its cable subscribers could not view the local television stations affiliated with CBS, ABC, and NBC. For the most part, however, cable operators and broadcasters recognized that mutual interests were served for the broadcaster to be carried on local cable systems. Though little money changed hands, cable operators in some cases provided "compensation" in the form of local advertisements for the broadcast stations and other forms of promotional tradeoffs. The large MSOs most often refused to pay money for retransmission consent. However, deals were struck to open channel capacity for some services on some MSO cable systems. For example, TCI provided Fox with space for its new cable channel, FX, in exchange for consent to carry the Fox broadcast network.

The Telecommunications Act of 1996 was aimed largely at increasing the prospects of competition in the telecommunications marketplace (Shaw, 2002). For the first time, telephone companies were permitted four entry paths to provide multichannel video service within their service regions. Telcos could set up a separate multichannel video subsidiary, operate an open video system (over which they could provide a set percentage of programming themselves), operate as a common carrier simply providing video carriage to others, or operate a wireless or RF multichannel service. Despite a great deal of initial enthusiasm for the idea, most telephone companies have been reluctant to enter the video services market, and competition there has been slow to emerge (FCC, 2002).

Over time, several large mergers and acquisitions have changed the face of the cable telecommunications industry. This industry consolidation has been driven, in large part, by the desire to expand the size of companies, thus giving individual companies more clout in various aspects of the business and allowing synergies that result from vertical and horizontal integration. In 1995, Time Warner acquired Turner Broadcasting System in an $8 billion stock swap. This acquisition gave Time Warner a vast array of cable networks, production units, extensive film libraries, and other assets. Several years later, that company proved to be an attractive target for America Online (AOL) in a merger of new media and old media that was the hoped-for model for the new century. Serious economic hard times befell the company despite its strategic assets in production, distribution, exhibition, and operations in a variety of markets. Billions of dollars in losses in the aftermath of this merger have signaled heightened caution among others seeking to merge new with old media.

A series of cooperative "franchise swaps," in which larger MSOs attempted to cluster cable systems within a region in hopes of expanding services such as telephony, data services, and Internet access, also occurred in the late 1990s. Programmers continued international expansion, with MTV and HBO, among others, offering regional and international variations of their services. At the close of the 20th century, 68% of American households took cable service from one of 10,700 cable systems (NCTA, 2000). There were also 214 national cable video networks providing service to cable operators. In November 2002, Comcast, with a $72 billion bid, acquired AT&T Broadband. After undergoing regulatory scrutiny, the company emerged with 22 million subscribers in 41 states, by far the largest MSO, with 25% of all multichannel homes in the United States (Higgins, 2001). The newly combined company had over five million digital subscribers, 2.2 million high-speed broadband subscribers, and around one million cable telephony customers from AT&T. The company also has serious negotiating clout in its dealings with programmers: witness its recent go around with Starz!/Encore in 2003 (Romano, 2003).

In other MSO-related activity, Microsoft, through its subsidiaries, purchased a $500 million stake in Cox Communications in October 2001, making the computer giant a bigger player in the cable business (Higgins, 2002). Comcast Cable Communications emerged as the largest MSO, with over 21 million subscribers, double that of its next closer competitor Time Warner Cable (NCTA, 2003).

As cable faced competition from DBS, system upgrades accelerated. First, the cable plant (city-level cable network) had to be upgraded. Prior to the 1990s, a decent cable network could deliver 300 to 350 MHz of capacity to the home. At 6 MHz per station, this translates to about 50 channels. In the 1990s, capacity jumped to 750 MHz on more than 60% of cable systems. In addition to doubling existing capacity, digital technology could compress more signals in the same bandwidth. Compression eventually allowed 12 standard-definition digital signals to be transmitted in the space of one analog channel. To receive these digital channels, the consumer needed new cable boxes, but digital converters provided a dramatic increase in capacity.

An analysis of the FCC data revealed that 105 cable networks were created prior to 1992 and 344 from 1992 to 2002 (McDowell & Dick, 2004). According to the National Cable & Telecommunications Association (NCTA), the greatest growth came between 1997 and 1999 when 111 national program services were launched (see Figure 3.1). According to Nielsen Media Research, channel availability for the typical American home (cable and non cable combined) surged from 33.2 channels in 1990 to 102 channels in 2002 (Nielsen Report 2003).

Enthusiasm for interactive television via cable started to waver in mid-2001, as many MSOs scaled back deployment plans. Industry caution was prompted by a downturn in the economy and the billions of dollars lost in Internet-related businesses, suggesting that careful reviews of business models were in order (Grotticelli & Kerschbaumer, 2001). Thinking had changed, and expenditures where revenues could be clearly identified were now the desirable pattern. Still, five of the eight largest MSOs made aggressive deployment of video on demand (VOD) a priority for the future.

Figure 3.3
Growth in Video Channels to the Home 1994-2002

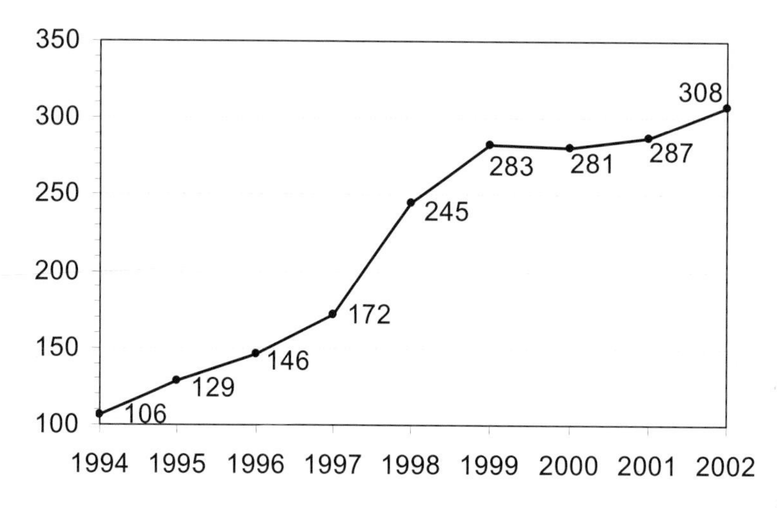

Source: NCTA, 2004

Recent Developments

Several major trends have spurred important changes within the cable telecommunications industry and have provided the basis for expansion into new varieties of service. These technological trends include digital cable upgrades, transitions to high-definition television, and development of ITV (interactive television) services. Clearly, upgrades into digital cable have expanded programming capacity. Equally as important, however, they have given cable telecommunications companies greater flexibility in provision of new, advanced services such as Internet service. High-speed Internet access via cable modem, which competes with digital subscriber line (DSL) and dial-up for business, was available to 80% of cable households by 2003 (Morgan Stanley, 2003). The cable telecommunications industry has ramped up efforts in this competitive sphere, and, by 2003, there were approximately 12 million cable modem subscribers in the United States. (For more on cable modem service, see Chapter 20.)

Digital Cable Upgrades

The continued rollout of digital cable is perhaps the single-most important technological development for the cable industry. On the back of this technology, programming tiers have expanded. The basic digital tier on most systems usually includes around 40 or so channels of audio and video. Within the first quarter of 2003, the number of subscribers with digital cable crossed the 20 million threshold for the first time (NCTA, 2003). Large MSOs such as Time Warner, Adelphia, Comcast, and Charter have increased their digital subscriber bases substantially over the past several years. Comcast planned to spend $3 billion on digital upgrades in 2003 and offer video on demand to nearly half of its subscriber base (Deloitte & Touche, 2004). Morgan Stanley Research projects that, by 2005, there will be over 29 million digital cable subscribers in the United States (Special report, 2002). Digital tier expansion continues to play a vital role in cable's strategy in competing with direct broadcast satellite in the multichannel video program distributor (MVPD) marketplace. Cable still dominates with a market share of 76% of the business.

Transition to High Defintion

High-definition television (HDTV) over cable TV has occupied a good deal of the attention of cable operators, industry engineers, and policy makers since 2002. This technology promises up to a ten-fold improvement over standard analog television resolution. The first HDTV programs were delivered by some cable systems to their subscribers in 2001, and the number of systems delivering HDTV has increased over that period as larger MSOs have begun providing such channels.

Time Warner Cable has been among the most aggressive in rolling out HDTV, with most of its 50 markets now carrying HD signals. Cablevision, in its New York service region, offered HD set-top boxes to its digital customers with HD-ready sets in 2003. As of mid-2004, the largest MSO, Comcast, offered five or more HDTV channels to approximately half of its subscribers. Large Comcast markets such as Atlanta, Washington DC, Detroit, and Denver began HD offerings in the summer of 2003. As of mid-2004, Cox Communications provided HDTV to its Oklahoma City, Cleveland, Phoenix, San Diego, and Fairfax (Virginia) cable franchises. The number of cable operators providing HDTV services passed over 55 million TV household by mid-2003. These cable-enabled HD services reach into 78 of the top 100 television markets in the United States (NCTA, 2003). In late 2003, the FCC, as part of the "plug-and-play" standards, ordered that cable-ready TVs be equipped with technology allowing digital television to be received without set-top boxes (McConnell, 2003).

Technology offers only a partial picture of cable's effort in the arena of HDTV. Programmers themselves have been active in producing programming that takes advantage of the HD delivery capabilities of cable operators. 2003 was an important year for sports programmers who were among the most active in HD programming. For example, ESPN HD delivered live broadcasts of Major League Baseball games beginning in the 2003 season. The same network delivered HD versions of the National Football League and the National Hockey League during their seasons. HBO, one of the early HDTV programming services on cable, currently offers nearly 70% of its programming content in high-definition form. That trend is expected to continue on that network, and other premium movie services have followed suit. Meanwhile, in the basic cable sphere, Discovery committed $65 million to the production of a series of programs earmarked for Discovery HD, which commenced regular service in the summer of 2002.

On the consumer side of the technology equation, cable has responded to various calls for a more consumer friendly environment for subscribers. In a more proprietary vein, Comcast began co-branding an HDTV kit as a companion to HD-ready sets being sold in a few Best Buy stores in the United States. A far-reaching agreement between major consumer electronics and large MSOs was struck in December 2002, with the end result being a push toward digital cable connection without the need for a set-top box. With the FCC's blessing, this long anticipated "plug-and-play" standard is expected to have profound impacts on the spread of HDTV technology in cable telecommunications.

Interactive Services

The exact definition and the very nature of interactive television has been cussed and discussed for over 20 years now. Differing levels of interactivity are now possible through upgraded digital cable systems as never before. Current interactive offerings through two-way cable plants and digital offer a glimpse of more exciting prospects for the future. Interactive capabilities have the potential to change the way people utilize television or the Internet, access and experience feature films, play games, and communicate with one another. Less clear is the viability of many of these services in a yet-to-be-defined marketplace. Interactive services as they currently exist fall into the following general categories:

- *Video on demand*—A service that enables subscribers to both order and watch motion pictures or other content on demand. Features allow subscribers to pause, rewind, and advance content as they please.
- *Subscription video on demand*—Similar to VOD, except that a flat rate is charged to subscribers for unlimited access to a set number of programs from the video library of the cable network.
- *Interactive program guides*—A complement to the digital cable package. These electronic guides enable the subscriber to choose, set reminders, and sort programming according to time, category, channel, or other criteria. Information about program content, rating, and year of production is often provided as well.
- *Enhanced TV services*—A range of offerings that provide supplemental information about television programs. These may also include the ability to play along with game shows, order a product as information about it appears on screen, participate in surveys or polls, and obtain the latest in news or information services.
- *Digital video recorders*—Set-top devices that allow cable subscribers to record, store, and playback content via a hard drive that provides users with VCR-like capabilities. DVRs provide the subscriber with greater personal control over viewing activity and help to leverage temporal programming power away from programming networks.

The host of new services includes video on demand, interactive program guides and access functions, enhanced television (ETV) features, and TV-enabled Web access. Interactive program guides are regarded by some as indispensable in the crowded information universe of cable. Nearly 20 million cable customers now have them, and they are part-and-parcel of all digital video services. They have been popular with subscribers, which suggests a willingness to embrace increasingly complex interactive technologies that enhance cable's potential. (Interactive television is discussed in more detail in Chapter 6.)

The reality of VOD is beginning catch up with some of its earliest projections. Often cloaked in unrealistic hype or industry exuberance, VOD is no longer viewed as the panacea for cable's revenue future and the cash cow that will alone justify substantial upgrade costs. Nevertheless, the appeal of "anytime, as you want it" video seems undeniable. VOD is now available in 32 of Time Warner's franchise areas; New York City is the largest with over one million subscribers (NCTA, 2003). All the large MSOs are either offering or testing VOD services. (For more on VOD, see Chapter 5.)

Charter Communications has been aggressive in deploying its Charter i-Channels that include interactive games, news, weather, and other offerings to more than 650,000 subscribers. Cablevision Systems introduced ESPN Today, an interactive channel amid a suite of other interactive offerings that provide on-demand access to sports scores, video clips, and news (Broadband databook, 2002).

Current Status

In 2003, cable penetration in the United States stood at 68%, reflecting relative stability in cable's customer base. Yet, the number of cable systems providing service in the United States actually shrank to 9,947 due to both industry and technological consolidation (NCTA, 2003). Advanced services such as high-speed cable modems and cable telephony have been rolled out to customers on an increasing scale. The number of customers taking high-speed cable modem service surpassed 12 million in the first quarter of 2003 (NCTA, 2003). By comparison, cable telephony, which has been slower to advance in its competition with incumbent telcos, found its way into 2.5 million cable homes over the same period.

As stated previously, there are 9,947 cable systems in operation in the United States and its territories (NCTA, 2003). This represents a decline of 295 cable systems since early 2002 A good deal of this attrition is due to new combinations of cable systems that now share facilities, such as headends, which adds to technological efficiencies in operation. There are over 73 million cable subscribers in the United States (NCTA, 2003). Among states, California has the most cable subscribers, with slightly less than 8 million, followed by New York with 5,345,000, Florida with 5,081,000, and Texas with 4,354,000 (NCTA, 2003). Four states represent a cable operator's nirvana with cable penetration rates (as a percentage of all TV households) of at least 85%: Hawaii at 91%, Connecticut 90%, and Massachusetts and New Jersey at 85%. The largest single cable system serves Houston, with 721,000 basic cable subscribers; a system operated by Time Warner Cable provides service to 492,487 basic cable subscribers (NCTA, 2004).

In total, there are more than 1,307,138 miles of cable plant in the United States and its territories, and 68% of U.S. television households subscribe. The average monthly rate paid by cable subscribers is $34.53 for expanded basic cable (NCTA, 2003). The largest MSOs are listed in Table 3.1 (NCTA, 2003). The top cable networks as of 2003 in terms of numbers of subscribers are listed in Table 3.2 (NCTA, 2003).

The cable industry operates with a dual revenue stream, with revenues coming from both subscribers and advertisers. In 2002, subscriber revenues within the U.S. cable industry were estimated to be $49.4 billion, with advertising yielding another $14.7 billion (NCTA, 2003). Although steadily increasing through the years in its importance, advertising continues to remain less lucrative than direct subscriber payments: 60% of cable's subscriber revenues are derived from basic cable subscriptions, with 12% from pay or premium services. For cable operators, "churn" (the number of people who disconnect from cable service) is a constant concern.

Table 3.1
Largest U.S. MSOs as of 2003

MSO	Basic Subscribers
Comcast Cable Communications	21,305,100
Time Warner Cable	10,914,000
Charter Communications	6,578,800
Cox Communications	6,280,800
Adelphia Communications	5,775,400
Cablevision Systems	2,963,000
Advance/Newhouse	2,099,000
Mediacom Communications	1,592,000

Source: NCTA

Table 3.2
Leading U.S. Cable Networks in 2003

Network	U.S. Subscribers
TBS Superstation	87,700,000
ESPN	86,700,000
CSPAN	86,600,000
Discovery	86,500,000
USA Network	86,300,000
CNN	86,200,000
TNT	86,200,000
Lifetime	86,000000
Nickleodeon	86,000,000
A&E	85,900,000
Spike	85,800,000
The Weather Channel	85,800,000
MTV	84,900,000
QVC	84,900,000

Source: NCTA

Cable television service is usually divided into bundled programming tiers for marketing purposes. These configurations vary by cable system, but usually include basic, extended basic, and premium or pay service tiers. The basic tier generally includes over-the-air signals, public access channels, and whatever additional channels the operator may choose to include. Expanded basic includes the basic tier, plus cable networks such as ESPN, USA, TNN, and the like. Pay or premium services may be sold per-channel or bundled in multi-pay packages, and include services such as Showtime, HBO, and their kind. Pay-per-view is offered separately from pay services and is a one-time, one-program purchase. Some cable operators taking advantage of digital technology are also offering expanded digital programming tiers at an additional fee for subscribers. Included here may be a variety of program offerings ranging from Discovery Science to BBC America.

Cable operators pay the cable programmers directly for their program services. The price is calculated on a cost-per-subscriber basis, with larger cable systems (e.g., MSOs) enjoying the best price breaks. Cable operators spent nearly $11 billion in the acquisition and production of programming in 2002 (NCTA, 2003). Discounts are usually given to cable operators who take multiple programming services from the same programmer, for example, ESPN, ESPN 2, ESPN News, and ESPN Classic. As of 2003, there were 308 different national programming services offering programming to multichannel services in the United States (NCTA, 2003)

Factors to Watch

- *The reality of converged distribution and content*—More large cable MSOs, those not linked with production capabilities of their own, will attempt to converge with content providers through acquisition or merger.

- *The full-service mantra* —Even though the term "full service network" seems to have fallen away, the concept of a network that provides cross-platform flexibility in providing a range of services retains its fundamental appeal and can justify upgrade costs. High-speed Internet service and cable telephony taking advantage of voice over Internet protocol (VoIP) technology are just two examples of how bundled services in the telecommunications world have become the order of the day. Positioning a cable company as simply a video service is no longer a wise business strategy amid the present-day communications environment.

- *Arrival of true à la carte*—Expansion of program channels has carried the dual effect of expanding both *desired* and *undesired* channels for the consumer/subscriber. Rising cable rates make subscribers increasingly aware of the actual utility (or lack of utility) of a cable subscription. Increased churn amid the present structure of tiered cable programming offerings seems likely to make cable executives consider the possibility of personalized bundles of channels tailored to individual household tastes. Some have even suggested that this may be the stimulus for now-stagnant cable penetration rates.

- *Home networks focused on broadband*—This will gain momentum as the connectivity possible with home networking technology coupled with broadband provides better access and greater flexibility in both video and information services. In the coming years, storage and appliances will increasingly rule the home entertainment environment. By 2007, we may very well have terabyte storage within our homes, making it possible to store thousands of hours of video, a few thousand hours of digital music, and an enormous amount of other information (Saracco, 2002). The Multimedia over Coax Alliance represents an effort to build home networking efforts around coaxial cable (Kershbaumer, 2004). Companies such as Time Warner, Toshiba, Comcast, Panasonic, Intel, Cisco, and others are working together to network television, DVRs, and other in-home technologies over coaxial cable. MSOs see advantages in utilizing already-existing infrastructure combined with chip sets that will turn coaxial cable into a 270 megabit per second (Mb/s) network that will support Ethernet, MPEG transport streams, and FireWire.

- *Greater flexibility in consumer choice*—Interactivity and DVRs will place more control in the subscriber's hands, thus permitting time shifting and more flexibility in tailoring programming schedules. Expanded channel capacity combined with this flexibility is expected to improve the subscriber viewing experience.

- *Increased internationalization for programmers*—Expansion into international markets by programmers will continue at an even more rapid pace. Overseas appetites for channels represent added revenue streams for programmers as they continue to bicker with domestic multichannel program distributors for fees. The History Channel, for example, is available now in over 70 countries and in 20 languages. Clones of A&E are available in Latin America and Canada and have increased that channel's presence and revenues in a positive way.

- *Impacts of a tightened credit market*—Cable operators now face a tightened credit market due to downward trends in Standard & Poor's ratings of the cable industry (Higgins, 2003). Concerns over the inroads made by direct broadcast satellite service precipitated this action, and this suggests reduced access to capital markets needed for growth. Highly-leveraged cable companies are likely to feel the sting the most. The long-term implications of this for an industry dependent on capital spending to upgrade systems and for future acquisitions are uncertain.

- *Dual must-carry*—In the transition phase from broadcast analog to broadcast digital, dual must-carry continues to be a vexing problem for policy makers and the cable industry. The FCC ruled in 2001 that requirements that cable operators carry "the primary video" should be interpreted to mean a single video stream, not the multiple streams made possible in the digital environment. The FCC also tentatively concluded that cable operators were not required to carry both the broadcaster's analog and digital signals. Some suggested that possible extension of must-carry laws into this new area may pose a threat to the first amendment rights of the cable operator. More recently, the FCC's Media Bureau announced a plan that would require cable systems, after 2008, to carry the digital signal of all must-carry eligible broadcasters and to "down-convert" such signals to analog at the cable headend. Meanwhile, the broadcast industry has asked that must-carry be extended to both digital signals during the transition and open the possibility of multiple must-carry afterward. Cable interests see this as a threat that could result in a reduction of available channels for other services. The FCC is set to rule on this matter in 2004.

- *À la carte pricing*—In March 2004, Senator John McCain, the chairman of the Senate Commerce Committee, expressed his view that it was time for a government-mandated experiment allowing cable subscribers to pick and choose their own channels. À la carte pricing has become a hot topic amid Congressional concerns over the rise in cable rates that has occurred since 1996. Cable industry executives are largely opposed to such a move, saying that channel diversity, consumer choice, and industry economics would ultimately be seriously threatened.

Cable telecommunications companies remain vital to a society that seems to enjoy the entertainment and information mix they can deliver to the American home. We might, however, expect that, in the future, consumers are likely to become increasingly indifferent to the manner in which these services are delivered. As part of an increasingly competitive marketplace, these firms must respond not only to changes in technology, but changes in subscriber needs. The cable industry has been responsible for developing programming that, in many ways, is driving the programming models in directions that never would have been possible without it. Looking beyond these video services into new products and services, today's cable telecommunications companies are continually reevaluating the market and attempting to respond to consumer demands. Flexibility in the technology itself now means that capabilities that were only on the cable industry's "wish list" five years ago are being rapidly rolled out. The implications of these capabilities for both companies and consumer/end users will be increasingly understood through time, and are best viewed as part of a dynamic information environment where change has become the norm.

Bibliography

Adams, M. (2000). *Open cable architecture*. Indianapolis: Cisco Press.

Baldwin, T., & McVoy, D. (1988). *Cable communication*. 2nd edition. Englewood Cliffs, NJ: Prentice-Hall.

Broadband databook. (2002, December). *Multichannel News*.

Cicora, W., Framer, J., & Large, D. (1999). *Modern cable television technology: Video, voice, and data communication*. San Fransisco: Morgan Kaufmann.

Deloitte & Touche. (2004). *Convergence and choice: Deloitte & Touche survey on the future of cable television*. Deloitte and Touche USA LLP.

Federal Communications Commission. (1996). *Cable television information bulletin*. Washington, DC: FCC.

Federal Communications Commission. (2001). *FCC adopts rules for cable carriage of TV signals*. Retrieved from http://www.fcc.gov/Bureaus/Cable/News_Releases/2001/ncrb0103.htm.

Federal Communications Commission. (2002). *Eighth annual report on competition in video markets*. Washington: FCC.

Grotticelli, M., & Kerschbaumer, K. (2001, July 9). Slow and steady. *Broadcasting & Cable*, 32.

Higgins, J. (2001, December 31). No holiday for consolidation. *Broadcasting & Cable*, 6.

Higgins, J. (2002, January 28). Gates takes stake in Cox. *Broadcasting & Cable*, 13.

Higgins, J. (2003, Octoner 13). S&P toughens up on cable's credit rating. *Broadcasting &Cable, 30.*

Kershbaumer, K. (2004, January 12). A home network based on coaxial. *Broadcasting & Cable, 48.*

McConnell, B. (2003, September 22). Court tunes in digital debate. *Broadcasting & Cable,* 19.

Morgan Stanley. (2003, April 7) *Fruit salad*. 42.

National Cable & Telecommunications Association. (2000). *Cable television developments 2000*. Washington: NCTA.

National Cable & Telecommunications Association. (2003). *Cable developments 2003*. Washington: NCTA.

National Cable & Telecommunications Association. (2004). *Cable developments 2004*. Washington: NCTA.

Parsons, P., & Frieden, R. (1998). *The cable and satellite television industries*. Boston: Allyn & Bacon.

Romano, A. (2003, September 29). Comcast gets what it wants. *Broadcasting & Cable*, 3.

Saracco, R. (2002). *Is there a future for telecommunications?* Rome: Future Center Telecom Italia Lab, 70.

Shaw, J. (2002). *Telecommunication deregulation and the information economy*, Second edition. Boston: Artech House.

Special report: Digital cable. (2002, March 4). *Broadcasting & Cable,* 18.

Taylor, A. (2000). *History between their ears*. Denver: The Cable Center.

4

Direct Broadcast Satellites

Ted Carlin, Ph.D.*

What is the best way to get multichannel video programming today? Like long-distance phone service of the past few years, the answer might just depend on the last marketing campaign you have seen. As discussed in Chapter 3, cable television continues to evolve, upgrading in many communities to a digital service that offers more channels, better picture quality, and extras such as high-speed Internet service and digital video recorders (DVRs). However, continual cable price hikes have accompanied this evolution, primarily due to increased programming and digital upgrade costs. From 1993 to 2003, cable rates rose 53%, while the consumer price index (CPI) increased just 25% (FCC, 2004a). This has given consumers reason to consider other options for their television service. Already frustrated by past cable problems of fluctuating signal quality, limited channel capacity, and lackluster customer service, higher cable prices have pushed many consumers to search for alternatives.

Waiting for these consumers are the multichannel video technologies of direct broadcast satellite (DBS) systems, wireless cable systems (MMDS), private cable systems (SMATV), home satellite dishes (HSDs), local telephone exchange carriers (LECs), and broadband service providers. With the increasing development of digital production, transmission, and reception equipment, the multichannel video program distribution (MVPD) field is bustling with activity as companies race to develop and market digital systems that will entice consumers away from the more familiar local cable television operators. The digital revolution in television technology is now giving consumers what the Federal Communications Commission (FCC) has always promoted: a level playing field of multichannel video program distribution services from which consumers can pick and choose.

---

* Associate Professor and Chair, Department of Communications/Journalism, Shippensburg University (Shippensburg, Pennsylvania).

This chapter will focus on the second largest multichannel video program distribution service: DBS systems. Providers of DBS systems are actively competing with cable television for subscribers, and currently offer the most "cable-like" programming alternatives for consumers.

Background

When it was originally conceived in 1962, satellite television was never intended to be transmitted directly to individual households. After the FCC implemented an "open skies policy" to encourage private industry to enter the satellite industry in 1972, satellite operators were content to distribute programming between television networks and stations, cable programmers and operators, and business and educational facilities (Frederick, 1993). The FCC assigned two portions of the Fixed Satellite Service (FSS) frequency band to be used for these satellite relay services: the low-power C-band (3.7 GHz to 4.2 GHz) and the medium-power Ku-band (11.7 GHz to 12.2 GHz).

In late 1975, Stanford University engineering professor Taylor Howard was able to intercept a low-power C-band transmission of Home Box Office (HBO) on a makeshift satellite system he designed (Parone, 1994). In 1978, Howard published a "low-cost satellite-TV receiving system" how-to manual. Word spread rapidly among video enthusiasts and ham radio operators, and, by 1979, there were about 5,000 of these television receive-only (TVRO) HSD systems in use.

Today, these 6- to 12-foot HSDs are still commonplace throughout the United States, especially in rural areas not served by cable television; there are fewer than 500,000 in use today. This number is significantly less than the peak of 2.4 million HSD users in 1995 (FCC, 2004a). The large receiving dish is required to allow proper reception of the low-power C-band transmission signal. A number of factors, including the high cost of the HSD system (around $2,000), the large size of the dish, city and county zoning laws, and the scrambling of C-band transmissions by program providers, prevented HSD systems from becoming a realistic, national alternative to cable television for MVPD service.

In the 1980s, a few entrepreneurs turned to the medium-power Ku satellite band to distribute television directly to consumers. By utilizing unused transmission space on existing Ku-band relay satellites, these companies aimed to be the first to create a direct-to-home (DTH) satellite television service that would use a much smaller receive dish than existing HSDs (Whitehouse, 1986). The initial advantages of these DTH systems included the higher frequencies and power of the Ku-band, which resulted in less interference from other frequency transmissions and stronger signals to be received on smaller, three-foot dishes.

However, many factors contributed to the eventual failure of these medium-power Ku-band DTH services in the 1980s, including:

- High consumer entry costs for the DTH equipment ($1,000 to $1,500).
- Potential signal interference from heavy rain and/or snow.
- Limited channel capacity (compared with existing cable systems).

- Restricted access to familiar cable programming (Johnson & Castleman, 1991).

DTH ventures by Comsat, United Satellite Communications, Inc., Skyband, Inc., and Crimson Satellite Associates failed to get off the ground during this period.

During this time, between 1979 and 1989, the World Administrative Radio Conference (WARC) of the International Telecommunications Union (ITU) authorized and promoted the use of a different section of the FSS frequency band. The ITU, as the world's ultimate authority over the allocation and allotment of all radio transmission frequencies (including radio, TV, microwave, and satellite frequencies), allocated the high-power Ku-band (12.2 GHz to 12.7 GHz) for "multichannel, nationwide satellite-to-home video programming services in the Western hemisphere" (Setzer, et al., 1980, p. 1). These high-power Ku-band services were to be called direct broadcast satellite services. Specific DBS frequency assignments for each country, as well as satellite orbital positions to transmit these frequencies, were allocated at an ITU regional conference in 1983 (RARC, 1983).

A basic description of a DBS service, based on ITU's specifications, was established by the FCC's Office of Plans and Policy in 1980:

> A direct broadcast satellite would be located in the geostationary orbit, 22,300 miles above the equator. It would receive signals from earth and retransmit them for reception by small, inexpensive receiving antennas installed at individual residences. The receiver package for a DBS system will probably consist of a parabolic dish antenna, a down-converter, and any auxiliary equipment necessary for encoding, channel selection, and the like (Setzer, et al., 1980, p. 7).

The FCC established eight satellite orbital positions between 61.5°W and 175°W for the new DBS satellites. Only eight orbital positions are available for DBS transmission to the United States, because a minimum of 9° of spacing between each satellite is necessary to prevent interference among signal transmissions. The FCC also assigned a total of 256 analog TV channels (transponders) for DBS to use in this high-power Ku-band, with a maximum of 32 DBS channels per orbital position. Only three of these eight orbital positions (101°W, 110°W, and 119°W) can provide DBS service to the entire continental United States. Four orbital positions (148°W, 157°W, 166°W, and 175°W) can provide DBS service only to the western half of the country, while the orbital position at 61.5°W can provide service only to the eastern half.

The FCC received 15 applications for these DBS orbital positions and channels in 1983, accepted eight applications, and issued conditional construction permits to the eight applicants. The FCC granted the construction permits "conditioned upon the permitee's due diligence in the construction of its system" (see 27 C.F.R. Sect. 100.19b). DBS applicants had to do two things to satisfy this FCC "due diligence" requirement:

1) Begin construction or complete contracting for the construction of a satellite within one year of the permit being granted.

2) Begin operation of the satellite within six years of the construction contract.

The original eight DBS applicants were CBS, Direct Broadcast Satellite Corporation (DBSC), Graphic Scanning Corporation, RCA, Satellite Television Corporation, United States Satellite Broadcasting Company (USSB), Video Satellite Systems, and Western Union. During the 1980s, some of these applicants failed to meet the FCC due diligence requirements and forfeited their construction permits. Other applicants pulled out, citing the failures of the medium-power Ku-band DTH systems, as well as the economic recession of the late 1980s (Johnson & Castleman, 1991).

In August 1989, citing the failures of the eight DBS applicants to launch successful services, the FCC revisited the DBS situation to establish a new group of DBS applicants (FCC, 1989). This new group of applicants included two of the original applicants, DBSC and USSB. These were joined by Advanced Communications, Continental Satellite Corporation, Direcsat Corporation, Dominion Satellite Video, EchoStar Communications Corporation, Hughes Communications, Inc., and Tempo Satellite Services.

From 1989 to 1992, none of these DBS services was able to launch successfully. In addition to raising capital, most were awaiting the availability of programming and the development of a reliable digital video compression standard. Investors were unwilling to invest money in these new DBS services unless these two obstacles were overcome (Wold, 1996).

Cable operators, fearing the loss of their own subscribers and revenue, were placing enormous pressure on cable program networks to keep their programming off the new DBS services. Cable operators threatened to drop these program networks if they chose to license their programming to any DBS service (Hogan, 1995). These program networks were seen as the essential programming needed by DBS to launch their services because the DBS companies had little money or expertise for program production of their own (Manasco, 1992).

In late 1992, DBS companies had this programming problem solved for them through the passage of the Cable Television Consumer Protection and Competition Act. The act guaranteed DBS companies access to cable program networks, and it "[forbade] cable television programmers from discriminating against DBS by refusing to sell services at terms comparable to those received by cable operators" (Lambert, 1992, p. 55). This provision, which was upheld in the Telecommunications Act of 1996, finally provided DBS companies with the program sources they needed to attract investors and future subscribers.

The other obstacle—the establishment of a digital video compression standard—was solved by the engineering community in 1993 when MPEG-1 was chosen as the international video compression standard. By using MPEG-1, DBS companies could digitally compress eight program channels into the space of one analog transmission channel, thus greatly increasing the total number of program channels available on the DBS service to consumers. (For example, the FCC had assigned DirecTV 27 analog channels. Using MPEG-1, DirecTV could actually provide their subscribers with 216 channels of programming.) In 1995, DBS companies upgraded their systems to MPEG-2, an improved, broadcast-quality compression standard.

With these obstacles behind them, two DBS applicants, Hughes Communications and USSB, were the first to launch their DBS services in June 1994. Utilizing the leadership and direction of Eddie Hartenstein (DirecTV) and Stanley Hubbard (USSB), Hughes established a subsidiary, DirecTV, to operate its DBS system, and then agreed to work with USSB to finance, build, deploy, and market their DBS systems together (Hogan, 1995). Hughes launched three satellites to the 101°W orbital position from 1993 to 1995. Both companies then signed a contractual agreement with Thomson Consumer Electronics to use Thomson's proprietary Digital Satellite System (DSS) to transmit and receive DirecTV and USSB programming (Howes, 1995).

To ensure its availability in rural areas, DirecTV signed an exclusive agreement with the National Rural Telecommunications Cooperative (NRTC) that allowed NRTC affiliates the right to market and distribute DirecTV in rural markets. A number of affiliates implemented the agreement in 1996, offering sales, installation, billing, collection, and customer service. Consolidation among affiliates became commonplace, and two affiliates emerged as the major players: Pegasus Satellite Television and Golden Sky Systems, Inc. In May 2000, Pegasus purchased Golden Sky for $1 billion in stock to create the third largest provider of DBS and the 10th largest MVPD in the United States. Pegasus is now the only MVPD focused exclusively on rural and underserved areas of the country and delivers DirecTV to 1.4 million households in 41 states (Pegasus Communications, 2004; Golden Sky, 2000).

After the successful launch of DirecTV and USSB in 1994, the FCC tried to force the other DBS applicants to bring their DBS services to the marketplace. In late 1995 and early 1996, the FCC re-evaluated the DBS applicants for adherence to its due diligence requirements. After several hearings, the FCC revoked the application of Advanced Communication Corporation and stripped Dominion Satellite Video of some of its assigned channels for failing to meet the requirements.

The FCC denied appeals by both companies and auctioned the channels in January 1996. MCI and News Corporation, working together in a joint venture, obtained the Advanced Communication DBS channels, while EchoStar obtained the Dominion channels to add to its previously assigned channels. EchoStar also acquired the channels from two other applicants, DirecSat and DBSC, through FCC-approved mergers in 1995 and 1996 (FCC, 1996).

In another merger, R/L DBS (now called Rainbow DBS), a subsidiary of Loral Aerospace Holdings and Cablevision's Rainbow Media Holdings, acquired the DBS channels of Continental Satellite Corporation. Continental was forced to turn over the channels to Rainbow DBS after failing to meet previous contractual obligations with Loral Aerospace for the launch of Continental's proposed DBS satellite (FCC, 1995). In December 2000, the FCC granted an extension to the Rainbow DBS construction permit. The extension required the launch of a satellite and commencement of DBS service by not later than December 29, 2003. Rainbow DBS was able to successfully launch its satellite, Rainbow 1, to 61.5°W in July 2003, and begin a new DBS service, VOOM, in October 2003 (Rainbow DBS Company, 2004). VOOM, primarily an HDTV service, is discussed in more detail in the Current Status section of this chapter.

On March 4, 1996, EchoStar launched its high-power DBS service, the Digital Sky Highway (DISH) Network, using the EchoStar-1 satellite at 119°W. Similar to DirecTV, EchoStar launched a second satellite in September 1996, at the 119°W orbital position, to increase its number of available program channels on the DISH Network to 170.

The DISH Network does not use the same DSS transmission format used by DirecTV. Instead, it uses the international satellite video transmission standard, Digital Video Broadcasting (DVB), which was created after DSS. Like DSS equipment, the DISH Network's DVB equipment uses MPEG-2 for digital video compression. What this means is that DISH Network subscribers can receive only DISH

Network transmissions, and DirecTV subscribers can receive only DirecTV transmissions. Despite differences in transmission standards, both the DSS and DVB systems employ an 18-inch receiving dish, a VCR-sized integrated receiver/decoder, and a multi-function remote control.

Similar to DirecTV, the DISH Network requires subscribers to acquire the DVB system, and then pay a separate amount for monthly or yearly programming packages. Also, like DirecTV, the DISH Network offers a choice of professional installation or a do-it-yourself installation kit.

The cable television industry did not ignore the implementation and growth of these DBS companies. In 1994, Continental Cablevision, intent on establishing a cable "headend in the sky" for consumers living in non-cabled areas of the United States, was able to enlist the support of five other cable operators (Comcast, Cox, Newhouse, Tele-Communications, Inc. [TCI], and Time Warner) and one satellite manufacturer (GE Americom) to launch a successful medium-power Ku-band DTH service (Wold, 1996). The service, named Primestar, transmitted 12 basic cable channels from GE Americom's medium-power K-1 satellite to larger three-foot receive dishes. Primestar offered far fewer channels than any of the cable operators' own local cable systems, so they believed that their cable subscribers would not be interested in Primestar as a replacement for cable service. (Primestar was not a true DBS service because it did not use FCC-assigned high-power DBS channels, although most consumers were unaware of this difference.)

In late 1994, as DirecTV and USSB began to prove that DBS was a viable service, Primestar decided to change its focus and expand and enhance its offerings to compete directly with DBS. Primestar converted its 12-channel analog system to a proprietary DigiCipher-1 digitally compressed service capable of delivering about 70 channels. In 1997, Primestar moved its service to GE Americom's medium-power GE-2 satellite at 85°W, and increased its channel capacity to 160. To differentiate itself from the DBS companies, Primestar decided to market its service just like a local cable TV service does by leasing the equipment and the programming packages together in one monthly fee. Subscribers were not required to purchase the Primestar dish, integrated receiver/decoder, and remote, although equipment purchase was an option.

As of 1998, there were four companies operating DTH services in the continental United States. There were three high-power DBS services (DirecTV, USSB, and the DISH Network) and one medium-power DBS service (Primestar). In April 1999, DirecTV strengthened its position in the market with the purchase of both Primestar and USSB, leaving only two players in the U.S. DBS market. These mergers not only strengthened DirecTV's subscriber base, but also provided satellites and orbital slots that enabled DirecTV to expand its services.

The completion of these transactions gave DirecTV:

- More than seven million U.S. subscribers.
- More than 370 entertainment channels delivered through five high-power DBS spacecraft.
- The broadest distribution network in the DBS industry, combining more than 26,000 points of retail sale with Primestar's rural and small urban-based distribution network.

- High-power DBS frequencies at each of the three orbital slots that provide full coverage of the continental United States: 101°W, 110°W, and 119°W.

- The opportunity for DirecTV to begin "local-into-local" broadcast signal carriage.

- For the first time, DBS service to Hawaii, which had not been previously served by any DBS or DTH provider.

As DirecTV was increasing its market dominance, the DISH network was using aggressive pricing strategies for programming and DVB equipment. These efforts enabled DISH in December 1997 to reach one million subscribers, achieving that number faster than any other DBS/DTH service (Hogan, 1998a). Marketing itself as the best value in satellite television, the DISH Network offered its equipment for only $199, plus installation. DirecTV responded by lowering its DSS prices, resulting in a price war that continues today.

In mid-1998, the DISH Network announced plans to transmit local broadcast stations back into their local markets. By early 1999, the DISH Network finalized an agreement with MCI to acquire its 28 DBS channels at 110°W. This provided the DISH Network with enough channel capacity to provide local-into-local service in major U.S. television markets. On May 19, 1999, the FCC granted the application of MCI and EchoStar for transfer of MCI's license to construct, launch, and operate a DBS system at the 110°W location. On June 16, 1999, EchoStar was also granted authority to temporarily relocate one of its satellites to a new orbital slot in order to improve DBS service to Alaska and to initiate service to Hawaii.

On May 17, 1999, the FCC granted Dominion Video Satellite authority to commence operation of a DBS service using the EchoStar III satellite currently in orbit at 61.5°W. This authorization waived Dominion's due diligence requirement to build and launch a satellite of its own because the lease arrangement with EchoStar was viewed as an efficient method of starting service, as long as Dominion maintained control over the programming (FCC, 1999). Dominion launched its Sky Angel religious programming service in fall 1999 with 16 channels of video and 10 channels of audio for $9 per month. Subscribers must use DISH Network equipment to access Sky Angel's programming.

Recent Developments

Table 4.1 summarizes the status of the DBS licensees as of mid-2004.

DBS companies are aggressively pursuing current cable customers for their services. Various new marketing campaigns by DirecTV and DISH are constantly implemented to attack the cable industry's most observable weaknesses: rate hikes (due largely to increased programming costs that continue to rise faster than inflation), picture and sound quality (due to the all-digital transmission of DBS programming), and customer service problems (due to past monopolistic practices still plaguing the cable industry) (FCC, 2004a; FCC, 2002). VOOM is marketing its unique high-definition television (HDTV) channels, while Sky Angel is promoting its service as "the most effective means of assuring that the Gospel will penetrate every nation" (Skyangel.com, Inc., 2004).

Table 4.1
U.S. DBS Licensees

Orbital Position	61.5°W	101°W	110°W	119°W	148°W	157°W	166°W	175°W	Total
Satellite Deployment	E-3 V-1	D-1R D-2 D-3 D-4S D-8 D-9S	D-1 E-5 E-6 E-8	E-4 E-6 E-7 D-5 D-6 D-7S	E-1 E-2				
DirecTV Channels		32	3	11					46
DISH Channels	11		29	21	24	3			88
Sky Angel Channels	8								8
VOOM Channels	11								11
Unassigned Channels	2				8	29	32	32	103
Total	32	32	32	32	32	32	32	32	256

Source: T. Carlin

Notes: (1) E = EchoStar satellites; D = DirecTV satellites; V = VOOM satellites
(2) 101°W, 110°W, 119°W are the only full-CONUS slots
(3) D-7S to be launched in 2004; D-8 and D-9S to be launched in 2005.
(4) In August 2003, EchoStar launched E-9 to 121°W to provide Ka-band spot beam services.

While attacking the cable industry problems, DBS companies are also trying to solve two main DBS problems, as well as several related issues. The two main problems being addressed are multiple television setups in the home and subscriber access to local broadcast television stations and broadcast networks. Both problems have long been considered major impediments to the development and growth of DBS systems as true competitors to cable television (FCC, 2004a; Hogan, 1998b).

Multiple TV Setups

When DBS companies began operations in the mid-1990s, the goal was to get a basic one-TV DBS system into as many rural subscriber homes as possible (Boyer, 1996). Due to declining costs, increased technology, and a new effort to attract cable customers, the focus has shifted to providing more user-friendly DBS service. DirecTV's customer research has found that the cost of a second receiver has often been a barrier to entry for some first-time subscribers (FCC, 2002; Hogan, 1998c).

As a result, DirecTV and the DISH Network have been marketing multiple TV setups for new subscribers. There are now three types of dishes available to consumers. They include:

1) A single-LNB (low noise block) dish that receives programming from one satellite orbital location.

2) A dual-LNB dish that also receives programming from one satellite orbital location.

3) A multi-location dish that receives programming from multiple satellite orbital locations.

The single-LNB model is the most basic and allows only one DBS receiver to be connected. It receives signals only from the DBS provider's primary orbital location. That means it can receive most of the mainstream programming but not some of the less common programming such as foreign language or HDTV programming. Also, depending on a subscriber's location, a single-LNB model may not allow the reception of local network affiliates, since not all local affiliates are broadcast from the primary orbital location. Use of a single-LNB dish requires a single coaxial cable from the dish to the home to connect the single receiver, so it is the easiest to install.

The dual-LNB model receives programming from a single orbital location, but viewers can connect one, two, or more receivers to it. This may be a better choice for those who want only the most common mainstream programming, but who want more than one receiver connected either now or sometime in the future. Like the single-LNB models, dual-LNB also may not allow the reception of local network affiliates. If subscribers want more than one receiver in the household, two coaxial cables must be run from the dish into the home.

Multi-location dishes are required by those who want to receive signals from satellites in different orbital slots. These dishes are now readily available and are quite popular, especially in those areas where the local broadcast affiliates are not carried on the primary orbital location. Like the dual-LNB model, they actually have two antennas, but they are focused at different positions in the sky. Multi-location dishes require up to four coaxial cables to be run from the dish into the home.

Local-into-Local

The second issue, access to local television stations, has been much more difficult to overcome for DBS in urban and most suburban communities. A provision in the Satellite Broadcasting Act of 1988 prohibited DBS subscribers who live *within* the Grade B coverage area of local broadcast television stations from receiving any broadcast television stations or networks via their DBS system (SHVA, 1988). Subscribers in these areas were forced to connect an over-the-air television antenna to their DBS system or subscribe to a local cable TV system to receive any broadcast television stations. (DBS providers are allowed to provide broadcast stations that are available via satellite to those subscribers living *outside* of local station Grade B coverage areas [i.e., rural, non-cabled areas], and each offers various à la carte packages of independent stations and network affiliates.)

This issue was finally resolved when the Satellite Home Viewer Improvement Act of 1999 (SHVIA) was signed into law on November 19, 1999. SHVIA significantly modifies the 1998 SHVA, the Communications Act, and the U.S. Copyright Act (SHVIA, 1999). SHVIA was designed to promote competition among MVPDs such as DBS companies and cable television operators while, at the same time, increasing the programming choices available to consumers.

Most significantly, for the first time, SHVIA permitted DBS companies to provide local broadcast TV signals to *all* subscribers who reside in the local TV station's market (also referred to as a designated market area [DMA]), as defined by Nielsen Media Research. This ability to provide local broadcast channels is commonly referred to as local-into-local service.

The DBS service provider has the option of providing local-into-local service, but is not required to do so. In addition, a DBS company that has chosen to provide local-into-local service is required to carry, upon request, *all* TV stations in markets where the DBS company carries at least one local TV station. This is now referred to as the "carry one, carry all" rule. However, a DBS company is not required to carry a local broadcast TV station that substantially duplicates the programming of a local broadcast TV station already being carried. In addition, a DBS company is not required to carry more than one local broadcast TV station that is affiliated with a particular TV network unless the TV stations are licensed to communities in different states (FCC, 2001).

SHVIA allows satellite companies to provide distant network broadcast stations to eligible satellite subscribers in unserved areas. (A distant signal is one that originates outside of a satellite subscriber's DMA.) The FCC created a computer model for DBS companies and television stations to use to predict whether a given household is served or unserved. Congress incorporated this model into SHVIA, but also required the FCC to improve the accuracy of the model by modifying it to include vegetation and buildings among the factors to be considered. The DBS company, distributor, or retailer from which subscribers obtained their satellite system and programming are to be able to tell subscribers whether the model predicts that they are served or unserved. (The FCC does not provide these predictions.) If unserved, the subscriber would be eligible to receive no more than two distant network affiliated signals per day for each TV network. For example, the household could receive no more than two *ABC* stations, no more than two *NBC* stations, etc.

SHVIA also permitted DBS companies to distribute a national PBS (Public Broadcasting Service) signal to all subscribers—served and unserved—until January 1, 2002. DBS companies may now choose to provide the local PBS affiliate or another noncommercial station within a local market or may provide the national PBS signal to subscribers that are eligible to receive distant signals.

Current Status

United States DBS/DTH

The United States is the world's number one user of DBS/DTH services. As of January 2004, there were 21.55 million DBS/DTH subscribers in the United States. Table 6.2 summarizes the subscription figures for the industry since its inception in July 1994.

Many of these 21.55 million subscribers are located in rural areas that were not previously served by a local cable system. The FCC currently estimates that 97% of all U.S. homes are now passed by cable (FCC, 2004a). This means that DBS and cable TV operators are now vying for the same homes to sustain subscriber growth.

Table 4.2
U.S. DBS/DTH Subscribers

Date	Total DTH	DirecTV	Primestar	DISH	Sky Angel
7/1994	70,000		70,000		
7/1995	1.15 mil.	650,000	500,000		
7/1996	2.95 mil.	1.6 mil.	1.27 mil.	75,000	
7/1997	5.04 mil.	2.64 mil.	1.76 mil.	590,000	
7/1998	6.60 mil.	3.45 mil.	2.01 mil.	1.14 mil.	
7/1999	9.93 mil.	5.42 mil.	1.94 mil.	2.57 mil.	
7/2000	11.74 mil.	8.24 mil.		3.50 mil.	
7/2001	16.30 mil.	10.13 mil.		6.17 mil.	
12/2001	17.61 mil.	10.75 mil.		6.86 mil.	
12/2002	20.93 mil.	11.85 mil.		9.08 mil.	1.0 mil. Est.
12/2003	21.55 mil.	12.150 mil.		9.40 mil.	1.4 mil. Est.

Source: SkyReports

Total multichannel video household penetration in the continental United States, including 66 million cable television subscribers and 428,400 TVRO users, is about 88 million households. Cable television is still the dominant technology for the delivery of video programming to consumers in the MVPD marketplace, although its market share continues to decline. As of January 2004, 75% of all MVPD subscribers received their video programming from a local franchised cable operator, compared with 80% in 2000, 82% in 1999, and 85% in 1998 (FCC, 2002; FCC, 2000a).

In addition to the recent developments in multiple TV setups and local-into-local, six additional DBS issues have attracted attention since 2002:

- Retransmission consent agreements.
- Public interest obligations.
- Transmission of HDTV signals.
- Internet access.
- Available programming.
- Strategic alliances.

Retransmission consent agreements. In order to deliver local-into-local service, DBS companies were mandated by the SHVIA to seek retransmission consent agreements with the owners of local television stations. (Retransmission consent is discussed in more detail in Chapter 3.) SHVIA required the FCC to revise the existing cable television rules surrounding retransmission consent agreements to encompass all MVPDs including DBS providers. SHVIA prohibits a TV station that provides retransmission consent from engaging in exclusive contracts for carriage or failing to negotiate in "good faith" until January 1, 2006, allowing DBS companies time to fully implement their local-into-local services.

In March 2000, the FCC, by First Report and Order (FCC, 2000b), established a two-part test for good faith negotiations. The first part consists of a brief, objective list of procedural standards applicable to television broadcast stations negotiating an agreement. The second part allows an MVPD to present facts to the FCC that constitute a TV station's failure to negotiate in good faith. The order directs the FCC staff to expedite resolution of good faith and exclusivity complaints, and notes that the burden of proof is on the MVPD complainant.

A major test of DBS retransmission agreements occurred in late 2003 and early 2004. The DISH Network's contracts with several of the Turner Broadcasting System networks, including CNN and the Cartoon Network, and all of the Viacom cable networks and owned-and-operated CBS affiliates, were set to expire in late 2003. In both cases, the DISH Network claimed that the license fees to carry the Turner and Viacom channels were being unfairly increased in the renewal contracts (EchoStar, 2004). Temporary restraining orders requested by the DISH Network were granted in federal district court to allow all parties to continue license fee negotiations.

While continuing to negotiate with Turner, the DISH Network allowed the restraining order with Viacom to expire on March 8, 2004, and subsequently removed all of the Viacom channels from its lineup in cities where CBS owned local television stations. Citing an impasse in negotiations with Viacom, DISH Network CEO Charlie Ergen claimed that pulling the channels was the only way to move negotiations forward. "DISH Network will always have a place for CBS, and we're willing to pay for retransmission rights, but Viacom is holding the public airwaves hostage, trying to extract concessions and higher rates on programming unrelated to CBS" (EchoStar, 2004, p.1). This channel-pulling strategy was previously used by Time Warner Cable to black-out several Disney channels in a retransmission battle over ABC television stations (McClellan, 2000). The strategy worked—within 48 hours, an agreement had been reached, and the Viacom channels were back on DISH.

Public interest obligations. Seeking to further level the competitive environment between DBS and cable, the FCC also adopted rules (FCC 98-307) implementing Section 25 of the Cable Television Consumer Protection and Competition Act of 1992, which imposed certain public interest obligations on DBS providers. The statute requires DBS companies to set aside 4% of their channel capacity exclusively for noncommercial programming of an educational or informational nature. DBS companies cannot edit program content, but must simply choose among qualified program suppliers for the reserved capacity. In 2000, DirecTV challenged these public interest obligation rules by asserting that they cause noncommercial station carriage to occupy a much larger percentage of DBS provider channel capacity relative to any cable system operator. The FCC solicited public comments on the impact of the rules and, in 2001, decided to deny DirecTV's complaint. The FCC retained the rules in an Order on Reconsideration (FCC, 2001) and stated that the rules provided the same degree of carriage by satellite carriers as is provided by cable systems. Basically, the public interest obligation rules promote parity between DBS and cable by assuring that consumers receive via satellite essentially the same local channels they would receive if they subscribed to cable.

As of January 2004, DirecTV offered 15 channels of public interest programming to fulfill this obligation: BYU TV, Church Channel, C-SPAN, C-SPAN2, Daystar, EWTN, Inspirational Network, Maria+Vision, NASA TV, PBS YOU, RFD-TV, TBN, World Harvest TV, WorldLink TV, and the WORD Network. The DISH Network offered 21 channels: BYUTV, C-SPAN, C-SPAN 2, Classic Arts Showcase, Colours TV, Educating Everyone, EWTN, FSTV, Good Samaritan Network, Health TV, HITN, NASA TV, Northern Arizona University, Panhandle Area Educational Network, PBS YOU, Research Channel, RFD-TV, TBN, UCTV, UWTV, and WorldLink TV.

DBS companies must also comply with the political broadcasting rules of Section 312 of the Communications Act, granting candidates for federal office reasonable access to broadcasting stations. They must also comply with Section 315's rules granting equal opportunities to federal candidates at the lowest unit charge.

Transmission of HDTV signals. DirecTV and the DISH Network are now using similar approaches to their HDTV equipment plans. DirecTV HDTV reception is provided by multi-location dishes and receivers that are able to receive both the standard compressed NTSC (National Television System Committee)and HDTV signals from the satellite. Both the NTSC and HDTV signals are sent to an HDTV set with a built-in DirecTV receiver or a DirecTV-enabled HDTV receiver and a slightly larger than normal triple-LNB, 18-inch × 24-inch DirecTV multi-satellite dish antenna. The DirecTV HD Package includes four HD channels (ESPN HD, DiscoveryHD, HD Net, and HD Net Movies) for about $11 a month.

The DISH Network is also using integrated NTSC/HDTV receivers that can deliver HDTV programming onto a 16:9-ratio HDTV screen and supports both 720p and 1080i HD formats. The receivers are intended to provide seamless switching between HDTV and standard TV with accompanying Dolby Digital surround-sound. The DISH Network HD Pak includes the same HD channels as DirecTV for about $10 a month.

Internet access. Similar to cable systems and phone companies, DirecTV and the DISH Network have been actively upgrading their systems to provide Internet services to subscribers. Because DBS signals are sent as digital information, the systems can send video, audio, and computer data in any combination to the receivers. Each DBS channel has a large amount of bandwidth, some of which the DBS companies are using for data services such as Internet or interactive TV services.

DirecTV's first Internet offering, DirecPC, provided DirecTV subscribers Internet service through a separate, second dish and receiver. In 2001, DirecTV upgraded the capabilities of DirecPC and renamed the service DirecWay. The DirecWay service is a two-way (satellite return) system that offers users the power of the satellite for both uploads and downloads, eliminating the need to monopolize a second phone line to surf the Internet. Another Internet option is DirecTV's partnership with Microsoft's Ultimate TV, allowing users to receive DirecTV programming, Internet access, and digital video recording in a specially configured Ultimate TV receiver.

As of mid-2004, DISH Network did not provide Internet access to consumers, although it is expected to rollout a service in late 2004. The DISH Network has partnered with Earthlink to offer DISH programming discounts to consumers who also sign up for terrestrial DSL or dial-up service with Earthlink.

Available programming. In terms of programming, DirecTV and the DISH Network have been able to acquire all of the top cable program networks, sports channels and events, and pay-per-view (PPV) events as envisioned by the 1992 Cable Act. What differentiates one service from the other is how the program services are priced, packaged, and promoted. Each service has on-screen program guides, parental control features, preset PPV spending limits, instant PPV ordering using the remote control and a phone line hookup, favorite channel lists, equipment warranties, and 800 phone numbers for customer service.

DIRECTV. Programming on DirecTV consists of packages costing between $30.99 and $90.99 per month that consist of different combinations of basic cable channels, local broadcast channels, and packages of premium movie channels such as HBO, Cinemax, Showtime, and Starz! It also offers individual PPV movies, concerts, and sporting events as available through DirecTicket (i.e., movies for $3.99; boxing for $19.95). Using the remote control, subscribers can search the interactive program guide to access desired channels or to request PPV events. DirecTV also offers unique packages of college and professional sports (MLB Extra Innings, MLS Shootout, NBA League Pass, NFL Sunday Ticket, NHL Center Ice, and ESPN College Basketball & Football). It also offers 31 CD-quality, commercial-free digital audio channels as part of its Total Choice packages.

DirecTV offers three packages of Spanish-language service, DirecTV Para Todos, to subscribers. DirecTV Para Todos offers more than 22 Spanish language national and international channels including Univision, Discovery en Español, Fox Sports World Español, Galavisión, MTV-S, TVN Chile, and Canal Sur, among others. All packages include seven Music Choice channels of commercial-free Spanish-language music.

To receive any DirecTV programming, subscribers previously had to purchase a DSS equipment package from a variety of retailers and have the DSS system installed. However, the big news in DBS since 2003 has been the ability of consumers to obtain the DSS equipment—up to four receivers, a multi-location dish, and installation—for *free*. Substituting a receiver with TiVo in the order will cost only $99 (TiVo, Inc., 2000). DirecTV has authorized several different companies to manufacture DSS equipment (including GE, HNS, Panasonic, RCA, Sanyo, and Sony), hoping to entice consumers with familiar, reliable brands. Prices vary according to individual retailers, the brand name chosen, and the complexity of the DSS system selected. Equipment with advanced features can also be purchased and range from $99 to over $499 (see http://www.directv.com for the most current information).

THE DISH NETWORK. The DISH Network is to satellite TV what Saturn is to automobiles: The service is highly practical, the packages are the most inexpensive yet comprehensive, and prices range from $24.99 to $82.99 per month. With its large channel capacity and deployed satellites, the DISH Network is marketing itself as the only satellite service to deliver more than 500 video and audio channels to subscribers, using many of these channels to deliver a range of Latino programming as well as pay-per-view movies, sporting events, etc. This capacity allowed the DISH Network to be the first satellite provider to supply local channels ($5.99 per month per city) with a dual-LNB or multi-location dish. An optional second dish can also provide subscribers with international programming in 10 languages or specialty religious or science programming.

Like DirecTV, the DISH Network is offering its equipment for free. Despite the fact that the receiver specifications are developed exclusively by EchoStar, the units offer a wide range of options and advanced equipment options are competitively priced with similar DSS units. Some of the higher-grade receivers offer features such as RF remotes, timed remote control of VCRs, seamless integration with off-air signals and local listings in the channel guide, on-screen caller ID, and an integrated DVR. However, unlike DirecTV's TiVo DVRs, the DISH Network's DVRs do not allow users to record one program while watching another program—a definite disadvantage. All receivers are feature-upgradeable via satellite. Equipment is available directly from EchoStar (via phone or the Web) or through local distributors such as Radio Shack. Like DirecTV, programming sign-up and/or changes are implemented immediately via a 24-hour 800 number. Technical support for installation or hardware issues is also available (for the most current information, see http://www.dishnetwork.com).

SKY ANGEL. Sky Angel is the world's first and only Christian-owned and -operated multichannel DBS service, delivering 36 Christian and "family-friendly" digital channels (20 television and 16 radio) to consumers in the continental United States. As described earlier, a DISH-brand receiver is required to receive Sky Angel programming, but Sky Angel subscribers do not have to be DISH Network subscribers. Sky Angel programming can be seamlessly integrated into DISH, DirecTV, and cable TV systems.

Two affordable subscription plans are available to receive the 36 Sky Angel channels: $11.99 monthly or $119.90 for one year when paid in advance. Programming is organized into the following categories: teaching and ministry, help programs, news from a Christian perspective, home life, kids, movies, special events, and music. Sample channels include INSP, Liberty TV, Spirit, and TBN (for the most current lineup go to http://skyangel.com).

VOOM. Cablevision's subsidiary, Rainbow DBS, has launched VOOM, a DBS service that offers more HD programming than any other satellite or cable service. VOOM features a basic package of 39 commercial-free, 1080i HDTV channels created to meet the demand of today's rapidly growing HDTV audience. VOOM customers can also receive 88 cable favorites such as The Disney Channel, A&E, FX, AMC, and more, as well as over-the-air digital local broadcast channels delivered in standard definition. In addition, Rainbow DBS has teamed with leading programming partners such as NFL Network, Playboy, Discovery HD Theater, and Starz! to offer consumers premium movies, sports, and entertainment channels in high definition.

Sears, one of the nation's leading retailers of HDTV sets, is selling VOOM through its stores across the country. Consumers can also order VOOM by calling 1-800-GET-VOOM or by visiting www.voom.com. Basic VOOM service is $39.90 a month, while a full-featured line-up, called Va Va Voom, is available for $79.90 a month. Somewhat like DirecTV and the DISH Network, VOOM is using a $0 upfront and free installation sign-up campaign to draw new customers.

Strategic alliances. What once was an industry in search of reliable distribution technology and attractive programming is now an industry focused on brand awareness, marketing strategies, and strategic alliances. DirecTV and the DISH Network are consistently using promotions such as "free views" of various channels and sponsorships of major events (i.e., World Wrestling Federation, college basketball tournaments) and concerts (i.e., Toby Keith, Sting).

DirecTV has established strategic alliances with SMATV and MMDS services, such as CS Wireless, Wireless One, and Heartland Communications, to provide DirecTV to multiple dwelling units (MDUs) such as apartments and town homes. In addition, DirecTV has also formed a distribution alliance with Verizon and BellSouth to allow these telecom companies to offer DirecTV program packages bundled with their telecom services.

Other recent DirecTV alliances include several with various business establishments (i.e., bars, restaurants, hotels, hospitals, private offices, malls, and fitness clubs) to provide customized packages of DirecTV services. Clients include Applebee's, Ruby Tuesday's, American Airlines, and Marriott Hotels. Business travelers can even receive DirecTV on selected commercial airlines (Alaska Airlines and JetBlue Airways) and private corporate jets via DirecTV Airborne. Launched in 1999, DirecTV Airborne offers passengers 24 channels of real-time DirecTV sports, news, children's, and general programming. Two of the 24 channels are reserved for specialty programming including concerts and special events. The service is viewed on in-flight entertainment equipment supplied by LiveTV. The low-profile LiveTV antenna, located in the top center of the aircraft's fuselage, maintains constant communication with DirecTV satellites located at the 101°W orbital slot (DirecTV, 2002).

The DISH Network jumped into the interactive services arena quite aggressively in 2002 and developed "DISH Home" with OpenTV (DISH Network, 2002). DISH Home provides interactive TV services such as video replay, interactive TV advertising, and entertainment services including movie information and music news. The DVR also has the ability to record OpenTV-enabled interactive programs and services, allowing the viewer to interact even with recorded TV programs. OpenTV set-top box software is used by more than 40 million viewers in over 50 countries, including subscribers to British Sky Broadcasting (BSkyB) in the United Kingdom and TPS in France (OpenTV, 2002).

International DBS/DTH

Although other countries have used satellites to transmit television signals to stations and cable television systems, Japan was the first country to launch a DBS service in 1984 (Otsuka, 1995). In October 1996, Japan's largest satellite operator, JSAT, launched the country's first digital DBS system, PerfecTV. In March 1998, PerfecTV merged with a competitor, News Corporation's JSkyB. The combined digital DBS service, SkyPerfecTV, delivered about 200 channels to its subscribers (Sky PerfecTV, 2000).

DirecTV Japan, a digital DBS competitor launched by Hughes Electronics Corporation in December 1997, ended service in 1999 by merging its operations into SkyPerfecTV. This added about 400,000 subscribers to SkyPerfecTV's base of 1.7 million. As part of the transaction, Hughes and other shareholders of DirecTV Japan received an equity stake in SkyPerfecTV (JSkyB, 2000). As of February 2004, SkyPerfecTV had 3.61 million subscribers accessing 300 channels of television and audio (Sky Perfect Communications, 2004).

STAR was launched in 1991 in Hong Kong, and is still the driving force for satellite television in Asia. STAR is a wholly-owned subsidiary of News Corporation and is watched by 173 million viewers a week. STAR covers 53 countries, spanning an area from Egypt to Japan and the Commonwealth of Independent States to Indonesia, reaching an estimated audience of 300 million viewers (STAR, 2004). In 2001, the company launched the first 24-hour commercial FM radio network in India. In 2002, the company rebranded from STAR-TV to STAR, reflecting the company's evolution from a television brand to a multi-service, multi-platform brand. STAR offers both subscription and free-to-air television services, using AsiaSat 1 as its primary satellite platform with additional services available on the AsiaSat 2 and Palapa C2 satellites.

In Europe, satellite consortia SES Astra and Eutelsat continue to dominate the European MVPD market, grabbing a 92% share of total MVPD households, including cable TV. SES Astra delivers digital and analog radio and television channels to over 91.8 million cable and DBS households (SES Astra, 2004). In the DBS market alone, four out of every five satellite households in Europe, roughly 34.4 million users, receives its programming from Astra. Eutelsat's Hot Bird Satellite TV service delivers 1,400 channels to over 110 million households (Eutelsat, 2004).

Primary competition to SES Astra and Eutelsat has been from a number of recent national/regional DTH systems including News Corporation's England-based BSkyB, France's Canal Satellite and TPS, Germany's Premiere World, Italy's Sky Italia, Norway's Canal Digital, and Spain's CSD. Digital DBS/DTH operators have done extremely well lately in Germany, France, Italy, and Asia (SkyReport.com, 2004).

Closer to the United States, in Latin America and Canada, DBS systems are also expanding. In Latin America, as deregulation and privatization of the telecommunications markets continue to spread through the region, the result has been fierce competition in satellite services. The leader in DBS in Latin America is DirecTV Latin America, which provides DBS service to over 1.5 million subscribers in 28 markets, including Brazil, Costa Rica, Mexico, and Panama (Hughes, 2004). DirecTV Latin America, which commenced service in mid-1996, is a multinational company owned by Hughes Electronics Corporation, Venezuela's Cisneros Group of Companies, Brasil's Televisao Abril, and Mexico's MVS Multivision. It filed for Chapter 11 in March 2003 and emerged successfully with a new organizational structure in February 2004.

In Canada, Star Choice Television is one of two firms that have satellite systems in operation. Star Choice delivers Canada's largest channel selection—more than 370 channels—to its more than 800,000 subscribers (Star Choice, 2004). Star Choice also launched Canada's first elliptical dish, which facilitates multiple satellite reception. Canadians can purchase Star Choice equipment at more than 4,000 locations across the country, including Radio Shack, Future Shop, Canadian Tire, Sears, Leon's, and The Brick. The other company, Bell ExpressVu, uses EchoStar's DISH Network equipment to operate a 300-channel DBS service. Subscribers must purchase the DVB equipment and a basic tier of programming, and then can add a wide range of specialty programming tiers organized by theme and language (English or French). Now owned by the largest telecom company in Canada (Bell Canada Enterprises), Bell ExpressVu serves over 1.3 million subscribers (BCE, 2004). Both Bell ExpressVu and Star Choice offer programming in English and French.

Factors to Watch

A proposed 2002 merger between DirecTV and the DISH Network fell apart following the December 10, 2002 decision of DISH Network parent company, EchoStar, to end the merger. Rather than fight an uphill battle with the FCC for approval, CEO Charlie Ergen decided to walk away from the negotiations.

This left DirecTV in limbo, but not for very long. Seeking to extend its reach into the U.S. multichannel video distribution market, News Corporation initiated merger talks with Hughes and DirecTV in early 2003. As one of the world's largest DBS providers via its BSkyB, FOXTEL, and

Sky Italia outfits, News Corp.'s partnership with DirecTV is a natural extension of its satellite TV business strategy (The News Corporation Ltd., 2004).

The FCC and the Department of Justice (DOJ) approved News Corp.'s $6.6 billion deal to gain control of DirecTV. The FCC's December 23, 2003 approval, which included several major conditions, gave News Corp. a controlling 34% stake in Hughes, DirecTV's parent company. The conditions, created to ensure that News Corp. does not abuse its new position as both a leading carrier and producer of pay TV programming, included:

- By the end of 2004, DirecTV must offer local TV stations in the top 130 markets, 30 more than the company had previously anticipated.
- All of the company's multichannel programming networks must be offered to competing cable and satellite TV providers on a nonexclusive and nondiscriminatory basis.
- Satellite-industry rules requiring good-faith negotiations for broadcast-carriage rights will remain in effect for News Corp. as long as cable program-access rules are in effect, rather than expiring in 2005, as they do for the rest of the satellite industry.

The DOJ's only additional condition requires News Corp. to appoint only U.S. citizens to the Hughes Electronics/DirecTV audit committee.

On the positive impact of the merger, FCC chairman Michael Powell said, "News Corp. has a history of taking significant risks and introducing new and innovative media services. Enhanced competition will increase pressure to improve service and lower prices for both cable and satellite television subscribers" (McConnell, 2004, p. 1). On the other side, Democratic commissioners Michael Copps and Jonathan Adelstein opposed the deal, arguing that News Corp. will have unprecedented and harmful reach across nearly every major media sector—TV distribution, broadcast networks, TV stations, cable networks, major film- and TV-production studios, print, video backhaul, and electronic programming guides.

The "new" DirecTV hopes to acquire eight million new subscribers by the end of the decade (McClellan, 2004). To get there, DirecTV has proposed a number of short-term goals to be achieved:

- Standardizing set-top boxes across the company's subscriber base.
- Expanding service to 130 markets, from 60 now.
- Adapting the interactive technology of News Corp.'s European satellite operation, BSkyB, starting with news and sports programming.
- Standardizing DirecTV's electronic program guide by midyear.
- Significantly overhauling the customer-service operation.

Another factor to watch is the fallout from the January 2004 FCC auction of spectrum for a new, terrestrial, microwave-based multichannel video distribution and data Service that is seen by the FCC as a competitor to DBS, cable, and the other existing MVPD players. The auction included 214 geographic area licenses, of which 192 received winning bids. Each license consists of one 500 MHz block of unpaired spectrum in the 12.2 GHz to 12.7 GHz band used by existing DBS services. According to the FCC, permissible operations include any digital fixed non-broadcast service including one-way direct-to-home/office wireless service (FCC, 2004b).

Licensees are permitted to provide one-way video programming and data services on a non-common carrier and/or common carrier basis. Mobile and aeronautical services are not authorized. Two-way services may be provided by using other spectrum or media for the return or upstream path. Licensees are subject to a construction requirement that they provide "substantial service" in the license area within five years of receiving the license. Two of the winning bidders, DTV Norwich and South.com, are minority owned by Cablevision and EchoStar. Each company may decide to use the new spectrum to compliment or expand current services on VOOM and the DISH Network. None of the other winning bidders has yet to discuss operational plans (McConnell, 2004).

Finally, the FCC will once again head to the auction block in mid-2004 to auction the unassigned DBS channels at 157°W, 166°W, and 175°W. There are 27 channels available at 157°W, and 32 channels each at 166°W and 175°W. A number of companies have expressed interest in these orbital slots, including the incumbent DBS operators. Since they do not cover the entire continental United States, it is somewhat unlikely that a new national DBS competitor will emerge from the bidding. However, with the chance to obtain spectrum that could be used to augment current MVPD services, it will be interesting to see if any of the active cable operators, including Cablevision and Comcast, try to acquire, on their own or in partnerships, these potential profit generators.

Bibliography

Bell Canada Enterprises. (2004, March 10). *Corporate fact sheet.* Retrieved March 10, 2004 from http://www.bce.ca/en/company/corporateffactsheet/index.php.

Boyer, W. (1996, April). Across the Americas, 1996 is the year when DBS consumers benefit from more choices. *Satellite Communications*, 22-30.

DirecTV. (2002, March). *DirecTV for business*. Retrieved March 14, 2002 from http://www.directv.com/DTVAPP/buy/Business_PvtJets.jsp.

DISH Network. (2002, March). *When will I get DISH Home?* Retrieved March 14, 2002 from http://www.dishnetwork.com/content/technology/itv/dish_info/index.asp.

EchoStar. (2004, March 9). *Viacom's demands create impasse in negotiations for rights to carry channels; DISH Network to lose CBS in 16 markets.* Retrieved March 10, 2004 from http://www.corporate-ir.net/ireye/ir_site.zhtml?ticker=dish&script=410&layout=-6&item_id=503189.

Eutelsat. (2004, March). *Direct-to-home broadcasting*. Retrieved March 10, 2004 from http://www.eutelsat.org/products/2_1_1.html.

Federal Communications Commission. (1989). *Memorandum opinion and order.* MM Docket No. 86-847.

Federal Communications Commission. (1995). *Memorandum opinion and order.* MM Docket No. 95-1733.

Federal Communications Commission. (1996, February 14). *Status report.* Report No. SPB-37.

Federal Communications Commission. (1999, May 14). Dominion Video Satellite, Inc. application for minor modification of authority to construct and launch order and authorization. *Report and order.* CS Docket No. 98-102.

Federal Communications Commission. (2000a, January 14). Annual assessment of the Status of competition in the market for the delivery of video programming. *Sixth annual report.* CS Docket No. 99-230.

Federal Communications Commission. (2000b, March 14). Implementation of the Satellite Home Viewer Improvement Act of 1999/retransmission consent issues, good faith negotiation, and exclusivity. *First report and order.* FCC 00-99.

Federal Communications Commission. (2000c, December 8). In the matter of amendment of parts 2 and 25 of the commission's rules to permit operation of NGSO FSS systems co-frequency with GSO and terrestrial systems in the Ku-band frequency range. *First report and order/further notice of proposed rulemaking.* ET Docket 98-206.

Federal Communications Commission. (2001, September 4). Implementation of the Satellite Home Viewer Improvement Act of 1999 (CS Docket 00-96). *Order on reconsideration.* FCC 01-249.

Federal Communications Commission. (2002, January 14). Annual assessment of the status of competition in the market for the delivery of video programming. *Eighth annual report.* CS Docket No. 01-129.

Federal Communications Commission. (2004a, January 28). Annual assessment of the status of competition in the market for the delivery of video programming. *Tenth annual report.* MB Docket No. 03-172.

Federal Communications Commission. (2004b, January 27). *FCC's multichannel video distribution and data service concludes auction.* Retrieved March 10, 2004 from http://hraunfoss.fcc.gov/edocs_public/attachmatch/DOC-243253A1.pdf.

Frederick, H. (1993). *Global communications & international relations.* Belmont, CA: Wadsworth.

Golden Sky Systems, Inc. (2000). *Golden Sky to merge with Pegasus Communications Corporation in $1 billion transaction.* Retrieved March 10, 2000 from http://www.gssdirectv.com/ press/60X-story.html.

Hogan, M. (1995, September). U.S. DBS: The competition heats up. *Via Satellite*, 28-34.

Hogan, M. (1998a, January 19). Demand remained strong for DBS in 1997. *Multichannel News, 19* (3), 33.

Hogan, M. (1998b, February 2). Digital cable not immediate threat, says DBS. *Multichannel News, 19* (5), 12.

Hogan, M. (1998c, March 2). DBS discounts 2nd receivers. *Multichannel News, 19* (9), 3, 18.

Howes, K. (1995, November). U.S. satellite TV. *Via Satellite*, 28-34.

Hughes. (2004, February 24). *DirecTV Latin America, LLC emerges from chapter 11.* Retrieved March 10, 2004 from http://www.directvla.com/newcc/news/news.asp?February%2024,%202004.htm.

Johnson, L., & Castleman, D. (1991). *Direct broadcast satellites: A competitive alternative to cable television?* Santa Monica, CA: Rand.

JSkyB, PerfecTV to combine operations on May 1. (1998, March). *SkyREPORT.* Retrieved March 5, 1998 from http://www.skyreport.com/jskyb.htm.

Lambert, P. (1992, July 27). Satellites: The next generation. *Broadcasting & Cable, 124* (31), 55-56.

Manasco, B. (1992, April). The U.S. multichannel marketplace in the year 2000. *Via Satellite*, 44-49.

McClellan, S. (2000, May 8). Disney triumphant. *Broadcasting & Cable.* Retrieved March 10, 2004 from http://www.broadcastingcable.com/index.asp?layout=articlePrint&articleID=CA16200.

McClellan, S. (2004, January 5). *Lean, mean sat TV.* Retrieved March 10, 2004 from http://www.broadcastingcable.com/article/CA372628?display=Search+Results&text=news+corp.

McConnell, B. (2004, February 9). Bidders plunk down $119 million for new terrestrial TV service. *Broadcasting & Cable.* Retrieved March 10, 2004 from http://www.broadcastingcable.com/article/CA380259?display=Search+ Results&text=satellite+tv.

OpenTV. (2002). *The company.* Retrieved March 14, 2002 from http://www.opentv.com/company/.

Otsuka, N. (1995). Japan. In L. Gross (Ed.). *The international world of electronic media.* New York: McGraw-Hill.

Parone, M. (1994, February). Direct-to-home: Politics in a competitive marketplace. *Satellite Communications*, 28.

Pegasus Communications. (2004, March). *About Pegasus.* Retrieved March 9, 2004 from http://www.pgtv.com/about/satellite_television.asp?display=about_pegasus.

Rainbow DBS Company. (2004, March). *Cablevision's Rainbow DBS introduces "VOOM"—Nation's first television service designed to meet demand of growing, underserved HDTV market.* Retrieved March 9, 2004 from http://voom.com/util/press/press_101503.jsp.

Regional Administrative Radio Conference. (1983). *Final report and order*. Geneva: ITU.
Satellite Home Viewer Act of 1988. (1988). 17 U.S.C. § 119.
Satellite Home Viewer Improvement Act of 1999. (1999, November 19). 17 U.S.C. § 122.
SES-Astra. (2004, March). *Astra reach in Europe*. Retrieved March 10, 2004 from http://www.ses-astra.com/corporate/market-research/eutrends.shtml.
Setzer, F., Franca, B., & Cornell, N. (1980, October 2). *Policies for regulation of direct broadcast satellites*. Washington, DC: FCC Office of Plans and Policy.
Skyangel.com, Inc. (2004, March). *Welcome and vision*. Retrieved March 10, 2004 from http://skyangel.com/About/Index.asp?IdS=00101F-9294E20&x=002|000&~=.
Sky Perfect Communications, Inc. (2004, March 10). *Total registrations and DTH subscribers*. Retrieved March 10, 2004 from http://www.skyperfectv.co.jp/skycom/e/frame/fr_new_68.html.
Sky PerfectTV gets DirecTV Japan. (2000, March). *SkyREPORT*. Retrieved March 10, 2000 from http://www.skyreport.com/skyreport/mar2000/030300.htm#dtv.
SkyReport.com. (March, 2004). *Worldwide DTH platforms*. Retrieved March 10, 2004 from http://www.skyreport.com/globaldth.cfm.
STAR. (2004, March 10). *About us*. Retrieved March 10, 2004 from http://www.startv.com/eng/frame_aboutus.cfm.
Star Choice. (2004, March 10). *Learn about Star Choice*. Retrieved March 10, 2004 from http://starchoice.com/english/aboutus/learnaboutstarchoice/default.asp.
The News Corporation, Ltd. (2004, March 10). *Direct broadcast satellite television*. Retrieved March 10, 2004 from http://newscorp.com/operations/dbst.html.
TiVo, Inc. (2000). *What is TiVo?* Retrieved March 10, 2000 from http://www.tivo.com/flash.asp?page=discover_index.
Whitehouse, G. (1986). *Understanding the new technologies of the mass media*. Englewood Cliffs, NJ: Prentice-Hall.
Wold, R. N. (1996, September). U.S. DBS history: A long road to success. *Via Satellite*, 32-44.

5

Pay Television Services

Jennifer H. Meadows, Ph.D.*

Although the concept of paying for television programming seemed outrageous in the early days, pay television services have proliferated to become a successful segment of the television industry. From premium channels to video on demand, the variety and number of pay television services and the means to receive them has increased dramatically over the past few years. This chapter will discuss traditional pay television services, such as premium channels and pay-per-view (PPV), as well as newer services such as video on demand (VOD), subscription video on demand (SVOD), and near video on demand (NVOD).

Premium channels are the most visible and familiar of the pay television services. These services, such as HBO and Showtime, offer a mix of popular movies, original programming, and sports without commercial interruption. Subscribers pay a monthly fee for each channel or package of channels, usually between $10 and $15 per month, above the basic cable fee.

PPV services have been offered to cable and direct broadcast satellite (DBS) subscribers since the early 1980s. With PPV, a subscriber can order a specific program for a set price. Programming for PPV ranges from popular movies and adult programming to specialized events such as rock concerts and sporting events. Depending on the cable or DBS system, the subscriber can call and order the movie or event, or use the remote control to place an order. Movies are usually offered on several channels at staggered start times, while special events are usually offered on a one-time-only basis.

With PPV, the viewer must wait to watch an event or movie until the scheduled time the cable or DBS system airs it. However, with near video on demand or enhanced pay-per-view, the same movie

---

* Associate Professor, Department of Communication Design, California State University, Chico (Chico, California).

is scheduled on many different channels, with a different start time on each channel. This practice gives the consumer a choice of several movies starting at closely-staggered times. In the past, critics of PPV have complained that the restricted start times of movies on PPV have limited its success. NVOD is more convenient as it allows the viewer more choices of start times for movies and more movie choices, but a service provider must devote many more channels to provide NVOD than ordinary PPV.

Video on demand goes one step further by allowing viewers, using their remote controls, to order from a wide variety of entertainment choices whenever they want to. The viewer is then capable of fast forwarding, rewinding, and pausing the program. VOD puts control of the programming in the hands of the viewer instead of the video service provider, making the experience much more similar to renting a video than ordering PPV.

There are two ways to bill for VOD. The first is billing on a per-viewing basis, where subscribers pay a fee that allows them to access a single program during a specified time period. Most PPV VOD provides a 12- or 24-hour viewing window for a single fee, allowing extra time for pausing, rewinding, etc. The second is subscription VOD (SVOD), a service where subscribers pay a monthly fee for access to a slate of VOD programming.

Services such as VOD, SVOD, and NVOD are made possible through advances in technology such as digital video compression, fiber optics, digital video recorders, and new advanced set-top boxes. In order for all of these pay television services to be made available, the television household must be "addressable." Addressability means that the video service provider can communicate directly with each set-top box in every household, allowing the service provider to deliver pay television programming only to consumers who request and pay for it. The set-top box decodes the blocking signal from the video service provider and unscrambles or presents the desired programming.

The incentive for local cable systems and DBS services to carry all of these forms of premium programming is economic. As a general rule, revenues from virtually all forms of premium television are split, with about half being kept by the cable or DBS company and half being paid to the programming service.

Several key factors have emerged in the past few years that will shape the future of pay television services. This chapter will discuss the issues and technology pertaining to pay television services and will highlight factors to watch in the fast-moving and quickly-changing future of pay television.

Background

Pay television has been around almost as long as television. Zenith actually began to study the possibilities of pay television while the television was still in the research and development stage (Veith, 1976). In the 1940s, Zenith introduced Phonevision, a service that supplied movies via telephone lines (Gross, 1986). Also in the 1950s, Paramount tested the Telemeter system where customers inserted coins into a set-top decoding box in order to receive programming.

Frightened by the possibility of subscription television, broadcasters lobbied Congress and the Federal Communications Commission (FCC) against any form of pay television service. Those efforts delayed the authorization of pay TV until March 1959, and it was then allowed only on a trial basis. Over-the-air subscription television service was not authorized until 1968. Broadcast pay television systems were tested in subsequent years, but none of them ever took off. Part of this failure was due to FCC regulations intended to protect frightened broadcasters and theater owners.

The 1970s and 1980s saw the proliferation of over-the-air subscription television services that used scrambled UHF signals and set-top decoder boxes. Signal stealing was a problem with these services because the decoder boxes used to unscramble the signals were easy to make. Further deregulation in the early 1980s opened up the subscription television market and made way for 24-hour pay services. However, competition to these over-the-air pay services arrived in the form of greater cable television penetration and the availability of premium cable channels, pay-per-view services, and home video. In addition, UHF station owners began to see that they could make more money broadcasting as independent stations rather than as subscription television services (Gross, 1986).

The first major step toward success for pay television services came in 1972 with the introduction of Home Box Office. As a cable service initially delivered via microwave technology, HBO was originally provided to cable customers in Wilkes-Barre (Pennsylvania) and quickly became a success despite a large subscriber turnover problem (known as churn).

A commitment to new technology played a major role in HBO's future success. In 1975, HBO was beamed to a communications satellite, RCA's Satcom I, to become the first "national entertainment communications network" (Mair, 1988, p. 23). After HBO made its mark, other premium channels arrived on the scene, including Showtime, Cinemax, and The Movie Channel. HBO actually developed Cinemax in response to Showtime.

At the same time as HBO was established, cable companies were experimenting with pay cable television. Warner Cable deployed the Gridtronics pay cable service in several of its cable systems in 1972 (Veith, 1976). Another early pay cable service was Theatrevision, which started in 1973 in Sarasota (Florida) (Veith, 1976). Users of the service bought paper tickets to insert into a set-top decoder, and movies were offered at scheduled times. Cox Cable Communications deployed an optical system service called Channel 100 in 1973 (Veith, 1976). These first pay cable services eventually led to PPV service.

In 1977, a federal court overturned the FCC pay cable regulation that limited the service to movies less than three years old and sports broadcasts more than five years old (Baldwin & McVoy, 1983). Many industry analysts saw PPV as the real future of cable television because hit movies could be seen earlier than on network television, and VCR penetration was still quite low. In addition, cable operators recognized the market of viewers that liked movies but did not like going to movie theaters. The movie studios were thus very interested in PPV as a distribution arm for entertainment products, and they experimented with different release windows (the time it takes for a movie to go from theatrical release to pay cable, cable, home video, and broadcast television).

Over time, the release window evolved so that movie studios earned the maximum return by releasing their movies to a pre-specified sequence of markets. Because home video proved to be much more profitable than PPV, most movies are released to home video in an exclusive 45- to 90-day window before they are released to PPV. The PPV programming services have argued that the window needs to be shorter because, by the time they get a movie, most viewers have already seen it on home video. While the performance of hit movies on PPV has been disappointing, PPV has had much more success with adult services and, in the past, event programming.

Home video and DVD (digital videodisc) have had a direct impact on PPV. Until recently, the number of movies available on PPV was constrained by the number of available channels, the limited number of start times, complicated billing and ordering procedures, lack of control of the program, and lack of programming choices. PPV operators have attempted to market PPV as a convenient alternative to home video, yet, to this day, home video and DVD remain a multibillion dollar business, while PPV revenues are only about one-eighth as big.

In the past several years, new video distribution services, as well as advances in communication technologies, have made new pay television services possible. In many ways, these new opportunities overcome the limitations of services such as PPV. Advances in compression and bandwidth capacity and the upgrade to digital transmission systems have made it possible for multichannel video programming distribution (MVPD) operators to offer more channels on their systems.

In order to make their service more desirable, premium channels are multiplexing—expanding their services from one channel to several channels. For example, HBO and Cinemax began to multiplex, without an increase in subscription rates, in the early 1990s. While multiplexing essentially began with premium channels offering different feeds for different time zones (delaying the west coast schedule by three hours to allow west coast viewers to watch a program at the same scheduled time as east coast viewers), now the premium multiplexes offer different programming on each channel. Premium channels also began to offer original programming instead of just movies to attract subscribers. HBO has been very successful with its original series such as *The Sopranos*. Showtime, likewise, rolled out successful original series and movies. Other premium services, including Encore and Starz!, have followed.

DBS was the first to offer NVOD, and now the service is currently offered by most digital cable systems. Viewers can order a hit movie every 10 to 15 minutes with their remotes. While NVOD overcomes many of PPV's limitations, video on demand is viewed by the cable and satellite industry as the service with the greatest revenue potential. With these systems, subscribers can access programming immediately and have VCR functionality, meaning they can pause, fast forward, and rewind the program.

Time Warner's now defunct Full Service Network offered VOD. This service was deployed in Orlando (Florida) from 1994 to 1997 and offered switched digital interactive multimedia services using a hybrid fiber/coax network. Customers could order movies on demand, with full VCR functionality, from a library of 100 titles for about the same price as a video rental. Ultimately, the Full Service Network was a failure as customers did not use the interactive services, and thus the service could not generate enough revenue to pay for the infrastructure and the related set-top boxes.

Despite the failure of the Full Service Network, VOD and SVOD have made significant inroads. In 2001, HBO began testing HBO on Demand, an SVOD service. Subscribers pay a low monthly fee for access to a library of HBO programming and movies. Subscribers have VCR functionality to control the programming. Other networks quickly joined HBO, including Discovery, Starz!, Showtime, and the Independent Film Channel.

MVPD subscribers with digital video recorders (DVRs) such as TiVo have a "VOD-like" experience. The subscriber can record a program onto the DVR and then watch it with VCR functionality. This practice is popular with DBS subscribers who can get DVR service bundled with satellite programming service. DirecTV subscribers can elect to have TiVo service, and DISH subscribers can have DISH DVR service.

Recent Developments

There have been numerous developments in pay television services since 2002. While some aspects of pay television, such as premium channels, remain relatively unchanged, others, such as video on demand, have incurred significant change in the past few years.

HBO continues to dominate the premium channel landscape. The March 2004 season premiere of *The Sopranos* culled 12.1 million viewers (*Sopranos*, 2004). The premium channel giants HBO and Showtime, along with other services such as Starz!, continue to offer a menu of multiplexed channels. The greatest advances have come in the form of premium SVOD, such as HBO on Demand and Showtime on Demand. In order to subscribe to these services, the customer must already be a subscriber to the premium channel and digital cable. For example, to subscribe to Showtime on Demand, you must be a regular Showtime subscriber. Thus, for the convenience of on-demand services, premium channel subscribers must pay extra. The services tout themselves as more convenient than DVRs because there is no need to program a device (Clasen, 2004; Sheng, 2004).

For many years, video on demand was the ultimate, unfulfilled goal of the MVPD industry. Now, video on demand is a reality and is considered by many in the cable industry as the key tool to elevate their service above DBS. DBS does not offer VOD as of 2004 because of spectrum limitations and a lack of two-way connectivity (FCC, 2004). The two largest cable multiple system operators (MSOs)—Comcast and Time Warner Cable—both offer a full selection of VOD services in selected markets with upgraded networks. Time Warner calls its VOD service iCONTROL, and Comcast brands theirs On Demand (Time Warner, n.d.; On Demand, n.d.). The service is only available to customers who subscribe to digital cable service.

Video on demand service can be divided into three general areas: SVOD, VOD movies, and free VOD. Subscription video on demand is described above. VOD movies are similar to traditional pay-per-view, except that the customer gets access to the movie for a 24-hour period, the movie is available immediately, and the customer can rewind, fast forward, and pause. These video on demand movies are making a dent in traditional PPV and NVOD services. PPV service providers such as iN DEMAND and TVN now offer VOD services, along with traditional PPV services (iNside iN DEMAND, n.d.; TVN Entertainment Corporation, n.d.). In February 2004, TVN announced that it was converting 10 channels from PPV to VOD, leaving some small cable operators in a bind. Because the cost of upgrading their systems to offer VOD is too great, many of the small operators will see a revenue loss because they will lose those PPV channels (Haugsted, 2004).

The third category of VOD is free VOD. In this service, the cable company offers a select group of programming from traditional cable channels. The big issue here is to generate revenue from this service. Certainly, offering free VOD will entice some subscribers to upgrade to digital cable services but, in the long run, there has to be a way to make the service profitable for both the MSO and the programming provider. This is where advertising and audience ratings must be considered. A limiting factor for VOD is the lack of reliable ratings, an important factor when considering advertising.

Programming providers are worried that, if they provide programming for VOD, then the ratings for their linear programming will be reduced. Initial research has shown that this does not happen; ratings grow or stay the same. For example, the Cartoon Network has offered two shows via free VOD, *Samurai Jack* and *Ed, Edd n Eddy*, and found that both shows were watched more overall with the combination of VOD and linear availability (Stump, 2004a; Stump, 2004b; Stump, 2004c). What VOD providers need now is a way to measure viewership to ensure that the programs offered replace lost linear viewers or increase overall total viewership. Only when programming providers can be ensured that they will not lose viewers with VOD will they begin to expand the amount of programming on VOD. On the other hand, the increasing use of DVRs might spur programming providers to increase involvement with VOD. VOD can potentially offer more control and more reliable measurement of advertising than DVRs (Sheng, 2004).

The NBA and the NFL are two programming providers not afraid of VOD. Both professional sports leagues are looking at using free VOD to promote their new television channels: NFL Network and NBA TV. The NFL plans to offer game highlights, out-of-market games, and library programming. The NBA has not announced their plans yet, but is reportedly looking at highlight packages (Umstead, 2004). In addition, NASCAR offers VOD called NASCAR IN CAR with iN Demand (iNside iN Demand, n.d.).

Regardless of whether the service is PPV, NVOD, or VOD, one of the most important limiting factors is the home video release window—the time from home video release on DVD and VHS to when the film makes it to pay television services. Because most films go to home video weeks before pay television, many people have already rented or purchased it before they have the chance to watch it on PPV. The pay television industry has pleaded with the movie industry to shorten the release window. A study by In Demand found that, when release windows between home video and VOD are less than 30 days, buy rates increase 50%. The average home video release window is 51 days (Umstead, 2003a; Umstead, 2003c). In 2002, DVD and home video generated $8.2 billion compared with $1.47 billion for PPV/VOD. The numbers clearly show why the movie industry is hesitant to change release windows and jeopardize any profit.

Another issue of concern is copyright. The FCC issued its digital television (DTV) plug-and-play rules in 2003. The rules classify programming into different groups according to whether it can be copied or not. The Motion Picture Association of America (MPAA) wants SVOD programming to be classified as "copy never," and has filed comments to the FCC arguing such. The FCC classified SVOD as an "undefined business model" as was promoted by members of the SVOD industry such as Starz Encore Group. Starz Encore argued that putting SVOD into "copy never" would "dampen consumer interest in SVOD and perhaps doom the service as a business model" (Hearn, 2004). Cable MSOs are in the middle of the argument and can negotiate copy-never. The FCC rulings allow copyright holders to challenge cable operators' SVOD copyright status.

Finally, it is no surprise that one of the biggest revenue generators for PPV and VOD is adult programming. Adult programming is big business for cable companies, PPV and VOD programming suppliers, DBS, and hotels. In hotels, adult VOD titles make up 80% of in-room entertainment profits (Dirty business, 2004). DirecTV generated $150 million from adult programming in 2003 (Dirty business, 2004). However, this lucrative category has been under scrutiny by the FCC and the Justice Department, especially since the "wardrobe malfunction" during the Super Bowl half-time show in January 2004. FCC Chairman Powell (2004) and FCC Commissioner Copps (2004) both made speeches at the National Association of Broadcasters (NAB) Summit on Responsible Programming that strongly suggested that the industry adopt stronger voluntary procedures, technologies, and codes of conduct to limit the exposure of obscene and indecent material to children.

Targeting more than broadcasters, Copps (2004) said, "Successful resolution of the indecency issues must, in the end, include cable and satellite." The cable industry has responded to concerns about indecency with a major campaign called "Control Your TV," which educates parents on how cable and television technologies can be used to control exposure to objectionable programming (NCTA's Sachs, 2004). The industry also pledged to make channel blocking technology available to any subscriber who requests it, apply ratings and content labels on programming, put ratings icons on programming, and encode the ratings for the V-Chip.

Current Status

HBO continues to lead in the number of subscribers to premium channels with 27 million subscribers, earning $2.2 billion in 2003. Showtime is second with 22 million subscribers and $1.08 billion in revenue. Table 5.1 shows the number of premium *cable* television subscribers from 1994 to June 2003. Note that the number of cable subscribers to these pay services has decreased slightly. This may be due to the migration to DBS. Table 5.1 also shows the increase in pay units (premium cable subscriptions) per household from 1994 to June 2003. Units continue to increase per year, although at a slower rate (FCC, 2004).

VOD continues to grow. The FCC's *10th Annual Assessment of the Status of Competition in the Market for the Delivery of Video Programming* (2004) reported that there were 6.5 million VOD households in the United States at the end of 2002, and this number was forecast to grow to 12.8 million by the end of 2003. There were 700,000 SVOD households by the end of 2002, and this was expected to grow to 3 million by year-end 2003. A study by the Cable and Telecommunications Association for Marketing found that:

- 74% of digital cable subscribers knew about VOD.
- 22% of cable subscribers had ordered VOD.
- 34% of premium subscribers and 35% of digital subscribers had ordered VOD (Stump, 2004a).

In another hopeful note, a study by Horowitz Associates found that 36% of digital cable subscribers are willing to pay more for advanced services such as SVOD (Reynolds, 2004). Comcast found that orders jumped 700% a month when Showtime on Demand and HBO on Demand were launched. With HBO on Demand, in October 2003, Comcast found that 45% of subscribers with access to On Demand ordered it (Stump, 2004c).

VOD and SVOD are not possible without digital cable. The number of digital cable customers continues to grow each year. The National Cable & Telecommunications Association reports that there were 21.5 million digital cable customers as of third quarter 2003 (see Figure 5.1). Churn is around 5%, high for digital cable. Although this number may appear small, it is twice the rate of regular cable. Most subscribers switch to DBS or adopt DVR services (McKinsey Quarterly, 2003).

Finally, Showtime Event Television reported that PPV/VOD generated $2.45 billion in 2002, a significant increase of 19.5% over 2001. Gross PPV movie revenues were up 18.5% to $1.47 billion. Adult PPV grew 15.1% to $609 million. Events were up 26.9% to $363 million, $106 million due to the Lennox Lewis-Mike Tyson fight. DBS generated $1.4 billion in PPV revenue, while cable generated $1.1 billion. That breaks down to $73.11 per DBS household and $34.57 per cable household (Umstead, 2003b).

Table 5.1
Premium Cable Services: 1994-June 2003 (in millions)

Year	Premium Cable Service Subscribers (Pay HH)		Premium Cable Service Subscriptions (Pay Units)	
	Total	% Change	Total	% Change
1994	27.7	4.9%	46.5	7.6%
1998	32.9	3.5%	58.6	6.0%
1999	34.3	4.3%	60.2	2.7%
2000	35.7	4.1%	65.9	9.5%
2001	36.0	0.8%	75.4	14.4%
2002	35.3	-1.9%	81.1	7.6%
June 2003	35.4	0.3%	82.0	1.1%

Source: FCC (2004)

Figure 5.1
Digital Cable Customers: 2000-2003

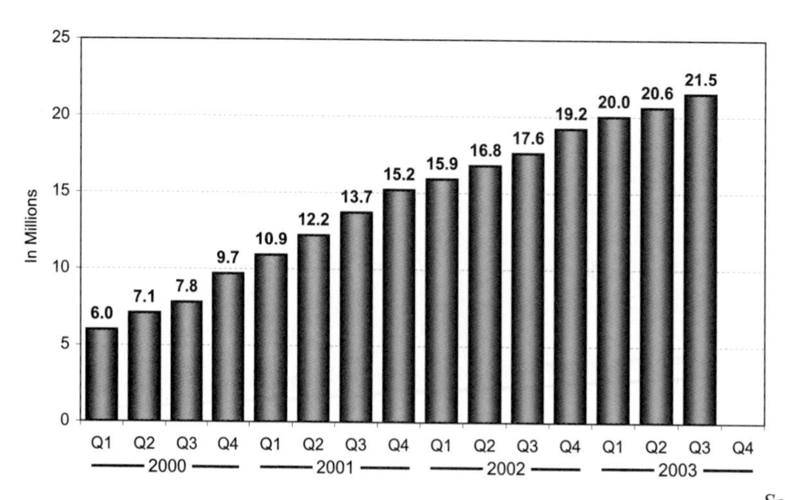

Source: NCTA (2004)

Factors to Watch

As the number of digital cable households continues to grow and as cable MSOs continue to upgrade to digital, look for an increase in the availability of VOD and SVOD services. Look for DBS providers to begin to provide VOD and SVOD services to customers, as cable companies aggressively defend their MVPD market share.

Premium channels should look to SVOD services to increase revenue, as multiplexed channels appear to be maxed out. If programming can be recycled into SVOD while subscribers pay extra to view these programs, there is a win/win situation for the premium channels and cable operators.

It is unclear how the DVR will affect the future of pay television. On one level, DVRs are a threat to SVOD and VOD because users can have a similar experience. On the other hand, some programming providers and MVPDs are experimenting with sending programming straight to DVRs and calling it "VOD." DVRs are discussed in more detail in Chapter 16.

Finally look for competition to come from Internet movie services and mail order DVD services. Movielink and CinemaNow both use broadband to allow users to download full-length movies to their computers (Olsen, 2004a and b). Both services work with Microsoft Windows XP Media Center. MGM, Sony Pictures, Paramount, Warner Bros., and Universal Studios back Movielink (Olsen, 2004b). Mail order DVD services, such as Netflix, allow subscribers to rent an unlimited number of DVDs for a monthly fee. Netflix subscribers can have up to three DVDs at a time for a $19.95 fee per month. The service touts a huge collection of DVDs including hard-to-find titles. Netflix has almost two million subscribers as of March 2004 and adds 125,000 new subscribers per month (Liedtke, 2004). The company expects to have $1 billion in revenue in 2006 (Netflix CEO, 2004).

Bibliography

Baldwin, T., & McVoy, D. S. (1983). *Cable communication*. Englewood Cliffs, NJ: Prentice-Hall.

Clasen, R. (2004). Getting out in front of SVOD's issues. *Multichannel News*. Retrieved February 10, 2004 from http://www.multichannel.com/index.asp?layout=print&articleID=CA374077.

Copps, M. (2004). Remarks of FCC Commissioner Michael J. Copps. *NAB Summit on Responsible Programming*. Washington, DC. Retrieved April 7, 2004 at http://www.fcc.gov.

Dirty business. (2004). *ABC News*. Retrieved April 7, 2004 from http://printerfriendly.abcnews.com/printerfriendly/Pring?ftechFromGLUE=true&GLUEService=ABCNewsCom.

Federal Communications Commission. (2004). *10th annual assessment of the status of competition in the market for the delivery of video programming*, MB Docket No. 03-172.

Gross, L. S. (1986). *The new television technologies*. Dubuque, IA: Wm. C. Brown Publishers.

Haugsted, L. (2004). TVN shift pinches small ops. *Multichannel News*. Retrieved February 10, 2004 from http://www.multichannel.com/index.asp?layout=print&articleID=CA380338.

Hearn, T. (2004). Studios want FCC to re-think SVOD copying rule. *Multichannel News*. Retrieved February 10, 2004 from http://www.multichannel.com/index.asp?layout=print&articleID=CA737900.

iNside iN DEMAND. (n.d.). *iN DEMAND*. Retrieved March 22, 2004 from http://www.indemand.com/about/who.jsp.

Liedtke, M. (2004). Netflix has a "Blockbuster" plan. *The San Francisco Examiner*. Retrieved April 7, 2004 from http://www.examiner.com/templates/print.cfm?storyname=03010bu_netflix.

Mair, G. (1988). *Inside HBO*. New York: Dodd, Mead & Co.

McKinsey Quarterly. (2004). Fast-forwarding digital cable. *C/Net News*. Retrieved April 7, 2004 at http://news.com.com2009-1069-994354.html

NCTA's Sachs tells families "cable puts you in control" in speech to cable public affairs conference. (2004). *NCTA*. Retrieved April 7, 2004 from http://www.ncta.cion.docs/pfriendsy.cfm?prid=463&pPRess=ok.

Netflix CEO sees $1 billion in revenue in 2006. (2004). *Forbes*. Retrieved April 7, 2004 from http://www.forbes.com/reuters/newswire/2004/02/26/rtrl277224.html.

Olsen, S. (2004a). CinemaNow debuts download-to-own movies. *C/Net News*. Retrieved April 7, 2004 from http://news.com.com/2102-1026_3-5141683.html?tag=st.util.print.

Olsen, S. (2004b). Intel, Movielink co-star in Web movie push. *C/Net News*. Retrieved April 7, 2004 from http://news.com.com/2102-1026_3-5160223.html?tag=st.util.print.

On Demand. (n.d.). *Comcast*. Retrieved April 7, 2004 from http://www.comcast.com/Benefits/CableDetails/Slot2PageOne.asp

Powell, M. (2004, March 31). Remarks of FCC Chairman Michael Powell. *NAB Summit on Responsible Programming*. Washington, DC. Retrieved April 7, 2004 from http:www.fcc.gov.

Reynolds, M. (2004). Studies differ on cable, DBS sub-approval levels. *Multichannel News*. Retrieved February 10, 2004 from http://www.multichannel.com/index.asp?layout=print&articleID=CA378342.

Sopranos premiere earns big ratings (2004). *CNN*. Retrieved April 7, 2004 from http://edition.cnn.com/2004/SHOWBIZ/TV/03/10/neilsens.ap/.

Sheng, E. (2004). Concurrent down 8%; video on demand rev view disappoints. *The Wall Street Journal*. Retrieved February 10, 2004 from http://www.wsj.com/article_print/0,,BT_CO_20040130_003855,00.html.

Stump, M. (2004a). CTAM: People know VOD. *Multichannel News*. Retrieved February 10, 2004 from http://www.multichannel.com/index.asp?layout=print&articleID=CA379423.

Stump, M. (2004b). Looking for VOD's guardian angel. *Multichannel News*. Retrieved February 10, 2004 from http://www.multichannel.com/index.asp?layout=print&articleID=CA380145.

Stump, M. (2004c). Demanding need: Ratings, soon. *Multichannel News*. Retrieved February 10, 2004 from http://www.multichannel.com/index.asp?layout=print&articleID=CA375254.

Time Warner Cable. (n.d.). Retrieved March 22, 2004 from http://www.timewarnercable.com.

Top 25 TV Networks. (2003). *Broadcasting & Cable*. Retrieved March 24, 2004 from http://www.broadcastingcable.com/article/CA338554?doc_id=128888&promocode=SUPP&display=Features&pubdate=12%2F01%2F2003.

TVN Entertainment Corporation. (n.d.). *Who we are*. Retrieved March 22, 2004 from http://www.tvn.com/about/.

Umstead, R. T. (2003a). Narrowing cable's VOD window. *Multichannel News*. Retrieved February 10, 2004 from http://www.multichannel.com/index.asp?layout=print&articleID=CA303420.

Umstead, R. T. (2003b). '02 was kind to PPV, VOD. *Multichannel News*. Retrieved February 10, 2004 from http://www.multichannel.com/index.asp?layout=print&articleID=CA277343.

Umstead, R. T. (2003c). Research: Early windows pay off. *Multichannel News*. Retrieved Febrary 10, 2004 from http://www.multichannel.com/index.asp?layout=print&articleID=CA338633.

Umstead, R. T. (2004). NBA, NFL pitch free VOD product. *Multichannel News*. Retrieved February 10, 2004 from http://www.multichannel.com/index.asp?layout=print&articleID=CA375255.

Veith, R. (1976). *Talk-back TV: Two-way cable television*. Blue Ridge Summit, PA: TAB Books.

6

Interactive Television

Paul J. Traudt, Ph.D.*

Interactive television (ITV) was seen in the 1970s as the future be-all and end-all video and information technology for everything from electronic civic forums to consumer marketing. By the mid-1990s, different software and hardware platforms were married to more traditional television technologies and programming, often with Web-based functions. More recent developments under the ITV umbrella have included the emergence of prominent hardware and interactive applications, such as video on demand (VOD), interactive program guides (IPGs), and digital video recorders (DVRs), while interactive gaming, Internet tools, and commerce applications continue to show less promise.

Finding profitable economic models, as well as the one true killer application for ITV, continue to be elusive. The 2001 National Association of Television Programming Executives (NAPTE) convention saw numerous vendors pushing the "interactive" features of their hardware, middleware, interactive platforms, and user-level applications. This push was followed by retrenchment to mostly after-production game enhancements for reality programming in 2002 (Broadcasters grab, 2003). The current perspective finds industry observers pointing to ongoing failures in viable streaming video and Web-based delivery of television services (Arlen, 2004).

The ITV industry remains handcuffed by a range of problems in the areas of technology "incompatibility, manufacturing, copyright, creative, and economic problems" (Merli, 2002, p. 10). Central to these problems is viewer/consumer inertia. Because the history of television is dominated by one-way analog communication, audiences are, by tradition, habitually passive, not interactive users. Such "lean back" behaviors confront any efforts on the part of the ITV industry to compel a critical mass of viewer/consumers to become "lean in" and *paying* users of a range of television services. The result is

---

* Associate Professor of Media Studies, Hank Greenspun School of Journalism and Media Studies, University of Nevada Las Vegas (Las Vegas, Nevada).

an ITV industry unable to answer two fundamental questions, "Do tens of millions of American homes even want [ITV]? Will televiewers ... pay extra for iTV and broadband services?" (Merli, 2002, p. 10).

ITV is typically some combination of traditional linear video and interactive services. It may be as simple as providing viewers with program selection options from entertainment programming guides, to home shopping and Web-based information features, to VOD, DVR, and Internet-type tools on television (Jarvis, 2001; ITV: How we, 2001; Merli, 2002, p. 10). The range of configurations can include linear analog or digital video accompanied by layered, transparent, or semi-transparent graphics or text. Some ITV providers distribute HTML (hypertext markup language) information to analog or digital television receivers, digital set-top boxes, or personal computers (Swedlow, 2002; Interactive television, 2002).

The industry flow chart (Figure 6.1) shows how the ITV industry can be broken down into two major groups, with numerous players in each group. One group includes interactive production services, addressable technologies, interactive advertising and programming content providers, and broadband delivery systems. The second group includes digital set-top box hardware manufacturers, middleware network architecture developers, interactive application platform providers, and user-level applications. This last category—user-level applications—carves out the landscape for viewer-consumer services.

Interactive games. This service typically shares the common feature of content provided by programmers or advertisers without Web browsing capability. For example, ACTV's software located in set-top boxes allows end-user control of picture selection of live sporting events via remote control. NTN, diego, Static, and Vivendi Universal are among prominent service providers.

Interactive program guides. This service is typically provided by means of satellite, cable, or other set-top box system. The typical configuration provides programming schedules and information sent via the vertical blanking interval (VBI). Advanced digital set-top boxes now include increasing memory capability for individual users and navigational preferences for more than one household user. Revenues are based on some combination of a percentage of monthly cable service fees and advertising. Decisionmark, Gemstar, Gist, iSurfTV, TMS, TV Gateway, and Wave Systems vie for dominance in this area, seen by many industry observers as the most competitive area of current ITV development.

Internet tools. This service provides window or browser-type interaction by means of graphical or text enhancement for features such as e-mail, Web surfing, and chat over a broadcast video background. Internet-on-television services are now offered through direct broadcast satellite services or broadcast television. The typical configuration includes proprietary software housed in a receiver, set-top box, tuner card, or cable headend. Revenues are generated by some combination of set-top box purchases or rentals, subscriptions, or monthly fees. HyperTV, ICTV, and MSN TV compete in this area of the ITV arena.

Figure 6.1
Interactive Television Industry Boundaries Map

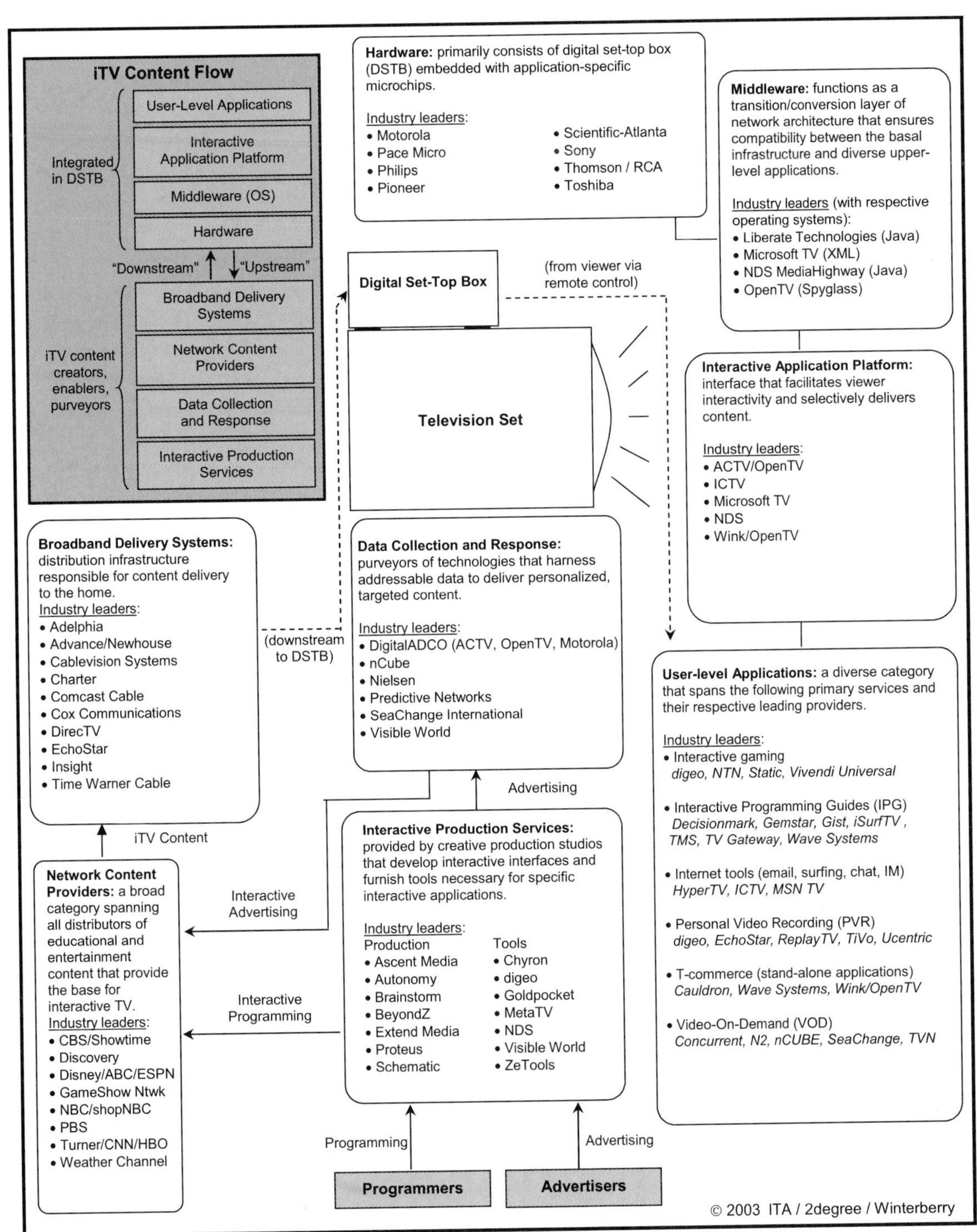

Digital video recorders. DVRs allow audience time shifting with a set-top digital hard drive that records video programming. Users can record, pause, resume, and play original program feeds. Users can now record hundreds of hours of programming as memory features for such devices continue to expand. Revenues are generated by some combination of DVR purchases, rentals, and monthly subscription fees. EchoStar, diego, ReplayTV, TiVo, and Ucentric are major players. (DVRs are discussed in more detail in Chapter 16.)

T-commerce. Television commerce provides an opportunity for television viewers to request more information about or even purchase advertised products or services. Revenues are generated through a combination of cable subscriptions and advertising. Stand-alone applications include those provided by Cauldron, Wave Systems, and Wink/OpenTV.

Video on demand. VOD users access programming by means of two-way cable or digital set-top boxes. VOD requires considerable bandwidth and can provide information whenever the end-user chooses, either by means of real-time interactivity or previously delivered programming stored on hard drives at the user's end. Revenues are generated by some combination of subscriptions and advertising. Concurrent, N2, nCube, SeaChange, and TVN are among those competing in this area of ITV. (For more on VOD, see Chapter 5.)

Major players continue to converge two or more of these groupings into bundled services. Some industry observers refer to all of these services as *enhanced television.* A convenient way to understand and evaluate these services includes a look at ITV's troubled history.

Background

History illustrates an ongoing series of ups and downs in the ITV industry—one characterized by many trials and failures reflecting a cycle of full-scale financial investments, converged technologies, frenzied marketplace speculation, and false promises of lucrative returns on capital investments. Early ITV efforts included CBS's *Winky Dink* in 1953, an animated series designed to encourage child-audience interaction with program characters. The program included screen prompts encouraging children at home to apply a transparent sheet to their home television screens and use crayons to connect dots and reveal "secret messages." The program aired in some markets as late as the early 1970s (Swedlow, 2002).

A major effort was mounted in 1977 by Warner Amex's QUBE TV, an analog, two-way cable experiment that provided interactive programming and banking services in addition to standard programming. The Columbus (Ohio) set-top box service, offering 36 channels and a wired remote, soon expanded to the Dallas and Pittsburgh markets. End users could specify programming preferences during programs specifically designed for the system. Unreliable technology and declining investor support contributed to the system's demise (Rosenstein, 1994; Swedlow, 2002).

Successful European efforts in teletext and videotex services were also tested in the United States in the late 1970s and early 1980s. GTE's Viewdata and Knight-Ridder's Viewtron provided electronic newspapers, weather, and agribusiness information via telephones. Flawed technology and limited programming resulted in end-user indifference. GTE later offered interactive home shopping, banking, video games, and VOD via cable television to residents of Cerritos (California) in the late 1980s and early 1990s, with less than 5% household penetration (Rosenstein, 1994).

Time Warner launched their Full Service Network to 4,000 Orlando test households in 1994, and projected that over 750,000 households would subscribe to the system by 1998. Test households could use VOD, IPGs, postal services, interactive shopping, and video games on a pay-per-view basis. Time Warner announced the end of its $250 million dollar experiment in 1997, citing high costs and user disinterest (Full Service Network, 2002).

Microsoft Corporation fueled wild speculation regarding the potential of interactive television in 1997 with its acquisition of WebTV. WebTV grew to 200,000 subscribers by November 1997, ultimately attracting an estimated one million subscribers before growth leveled off in 2000 (Marriage of, 1997; Desmond, 1997). In related business alliances, in 1998, Microsoft agreed to provide ITV technologies to Tele-Communications, Inc. (TCI) for the production of Internet-ready set-top boxes and invested $1 billion in Comcast and $5 billion in AT&T Broadband. This endorsement of cable system bandwidth as a digital distribution medium was used to establish the corporation's operating system as the standard for use in digital set-top boxes (Lesly, et al., 1977; Devin, 2002). Microsoft rival, Oracle Corporation, reacted by expanding into enhanced TV (Clark, 1997).

The new millennium saw continued ITV activity. ACTV joined with the National Basketball Association in a 2002 launch of live fantasy basketball games to accompany actual game coverage on TBS and TNT networks (NBA Entertainment, 2002). Gemstar, a prominent IPG provider, boasted strong earnings in 2001 with its GUIDE Plus+ or TV Guide Interactive (Gemstar-TV Guide, 2001). Microsoft's WebTV was merged with MSN to become MSN TV in spring 2001. The same corporation's Ultimate TV was also folded into MSN TV in 2002, after attracting only 100,000 subscribers. At the time, Ultimate TV was a satellite receiver digital set-top box with 35 hours of DVR capability and Internet-on-television features offered via DirecTV (Johnston, 2002).

DVR growth continued slowly during 2001 and 2002. Industry experts predicted significant growth in this area as word-of-mouth communication spread from early end-users to mainstream television users. Wink Communications competed with RespondTV and WorldGate Communications in 2001 for clients intent on developing interactive forms of television advertising, during which time Ford Motor Corporation, GlaxoSmithKline, and Lands' End all produced and aired interactive commercials (Jarvis, 2001). Less than 1% of all commercials aired were enhanced for interactive commerce, in part because of incompatibility between competing proprietary platforms (Wink tops, 2000). SeaChange initiated a significant undertaking in VOD development in 2001 with Comcast, Cablevision, Time Warner Cable, and Rogers of Canada, fueled in large part by a $10 million investment by Comcast.

Estimates of ITV's economic future were extremely optimistic. In 2000, Forrester Research predicted ITV revenues in 2004 would exceed $40 billion, with advertising the major revenue source (Meeting focuses, 2000). USA Warburg Bank was more conservative in their estimate, suggesting a tripling of revenues from 2002 to 2004 to $11.2 billion (Entertain me, 2002). Following suit, Forrester soon adjusted its prediction of more than $11 billion in ITV advertising to $1.5 billion (Pomerantz, 2002). How accurate were these predictions? A sample of more recent developments in the ITV arena suggests an even wider gap between optimistic projections and market realities.

Recent Developments

The ITV Production Standards Initiative, a U.S. effort to standardize specifications for content production was introduced in May 2002. Members of the initiative included cable industry leaders such as Cablevision Systems and Charter; content providers including Warner Bros., ESPN, and NBC; and interactive production services such as Goldpocket and nCube. The initiative was designed to help distribution of content over middleware and digital set-top box applications (Kerschbaumer, 2002a). On a similar front, the Advanced Television Systems Committee approved the standard for a DTV Application Software Environment, thus allowing, on paper, for over-the-air television broadcasters to provide interactive content as part of digital television (DTV) and high-definition television (HDTV) transmissions. This set the middleware standards for set-top boxes, allowing for such things as HTML graphics and JAVA-based advertising. Broadcasters were not expected to race forward into actual programming applications (Kerschbaumer, 2002c).

VOD's future dominated the 2002 summer National Cable & Telecommunications Association convention, relegating other ITV applications to a lounge act of sorts because of ongoing speculation, mergers, and litigations. Microsoft downsized its interest in ITV by focusing on low-end IPGs. Liberty Media Corporation entered into ITV by buying controlling interests in OpenTV, a middleware producer, and by purchasing ACTV, a leading platform in interactive applications (Arlen, 2002). Liberty also acquired Wink Communications, a one-way applications platform deployed to both cable and satellite operators that offers 30 interactive channels to more than one million subscribers via News Corp.'s DirecTV. Experts questioned the long-range economic viability of the service (Kerschbaumer, 2002b). By fall 2002, Liberty announced it would reorganize its holdings by "selling" Wink and ACTV to OpenTV while retaining controlling interests. The move was an effort to consolidate ITV applications and hasten profitability (Kerschbaumer, 2002d). By November 2003, DirecTV announced it was pulling out of the arrangement (Kerschbaumer, 2003), and News Corp. announced plans to make the satellite service more interactive in an effort to reduce subscriber churn (Pasztor & Goldsmith, 2003).

Meanwhile, a number of European ventures continued to explore low-end ITV applications, including the use of cellular telephone text messaging as a form of audience feedback for reality television genres. British, German, Belgian, Netherlands, and Spanish mobile operators, broadcasters, and program producers enjoyed reasonable economic successes (Texting, 2002). On the T-commerce front, Johnson & Johnson sponsored a multi-brand interactive advertising experiment as part of the ABC Network's airing of *His/Her Body Test* in June 2003. Results showed over 54,000 of those viewers who used an interactive platform "spent more time watching the show, and paid more attention to and had better brand recall of the commercials" (McClellan, 2003).

Current Status

Europe's early success with text-based ITV services continues and outpaces U.S. developments. For example, some form of ITV penetration in U.S. households was estimated at 8% in 2002, compared with significantly higher percentages in Great Britain. Teletext, for example, is found in over 80% of British households, with other forms of ITV exceeding 20% (Towler, 2002). Reasons include Britain's later adoption of cable technologies, thus allowing for more digital platforms. Home computer penetration is also lower, resulting in less time spent surfing the Internet. British Sky Broadcasting, partially owned by Rupert Murdoch's News Corp., is the largest provider of digital programming in Great Britain, and includes opportunities for viewers to gamble on sports and horse races. Together, these factors have aided ITV's success in Great Britain (Chen, 2002; Grant, 2004).

Figure 6.2
Percentages of U.K. Households with Selected Television Equipment (n = 1,191 Adults Aged 16+)

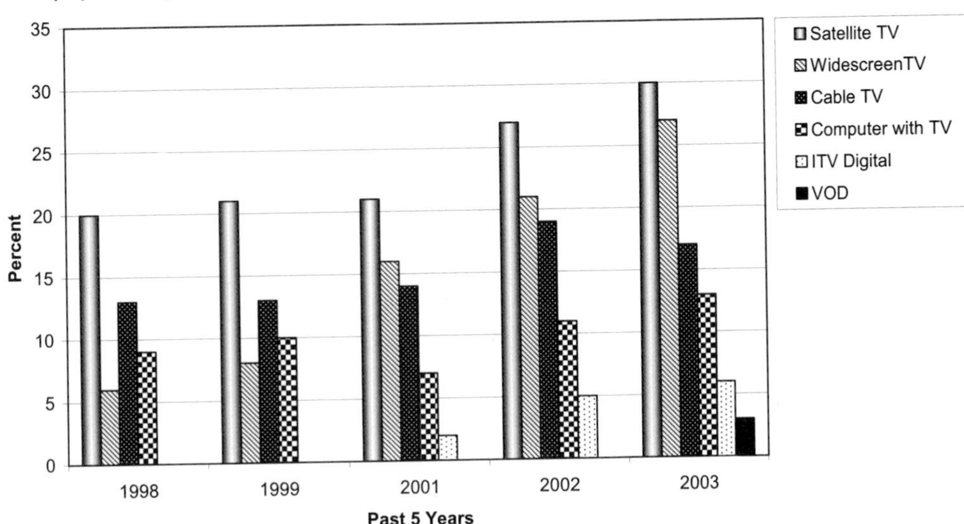

Source: Towler (2002)

In the United States, DVRs continue to lead ITV growth. There were 1.5 million DVRs sold in 2002, a market projected to expand to 5.8 million homes by the end of 2004 and nearly double to 11 million in 2005 (Taub, 2003; Having it, 2004). One of the more recent and aggressive ITV programming efforts was *Hockey Night in Canada Plus*, a service offered coast-to-coast to 1.3 million direct broadcast satellite (DBS) subscribers for an additional fee. Among the programming features were 45-second instant replay and viewer camera control options to focus on full-rink, goal, or individual player perspectives. The service accompanied standard television program coverage and was targeted to affluent demographics (Merli, 2004).

Factors to Watch

The one ITV killer application remains elusive, and VODs and DVRs are now seen as the industry's "savior apps" (Fritz, 2002). In the United States, ITV continues to suffer from a legacy of lofty projections and few successes, fueled in large part by audience confusion regarding what ITV actually means and how it differentiates from the services they enjoy via Web surfing. Viable bundling of services and price considerations further complicate the scene. Results from one consumer survey showed 75% of U.S. respondents were familiar with "several" ITV applications, but few had ever used any of these services (Fritz, 2002). Despite these realities, ITV continues to have its proponents.

Ben Mendelson of the Interactive Television Alliance argues that significant differences characterize the ITV landscape compared to only a few years ago. Now, the "technology is here, … there is a digital infrastructure … [and] the cost of equipment is no longer prohibitive." Only three years ago the focus … was essentially directed around the operating system in the set-top box. Now, we are finding out that is not all that important" (Shaw, 2003, p. 4A).

The future of cable-based expansion of ITV services is also contingent on roles played by the consumer electronics industry. "The deluge of connectivity-based consumer products—including home network gateways, digital video recorders/media centers and related converged products … has been looking for opportunities to sidestep the cable supply duopoly" (Brown, 2003, p. 8). Sony, Panasonic, Thomson, Philips, Samsung, and others may well win future ITV battles for U.S. audiences.

Bibliography

Arlen, G. (2002, June 9). Cable 2002: Waiting for the killer app. *TV Technology, 20* (11), 1, 10.
Arlen, G. (2004). Tomorrow's television remains elusive. *TV Technology, 22* (2), 35.
Broadcasters grab text messaging. (2003, April). *TV Technology, 21* (8), 8.
Brown, P. (2003, February 19). Passage: Cable's options expand. *TV Technology, 21* (4), 8.
Chen, C. (2002, April 1). I want my ITV. *Fortune, 145* (7), 124.
Clark, D. (1997, August 13). Oracle plans to integrate TV programs with data from the World Wide Web. *The Wall Street Journal,* B7.
Desmond, E. (1997, November 10). Set-top boxing. *Fortune, 136,* 91-93.
Devin, L. (2002, April 1). The most valuable square foot in America. *Fortune, 145,* 118.
Entertain me. (2002, April 13). *Economist, 8,268* (363), 4.
Fritz, M. (2002, April). VOD primes the ITV pump. *EMedia Magazine, 15* (4), 12.
Full Service Network (FSN) in Orlando, Florida. (n.d.). *Hong Kong University of Science & Technology.* Retrieved April 26, 2002 from http://www.ust.hk/~webiway/content/USA/Trial/fsn.html.
Gemstar-TV Guide International, Inc. reports financial results for the quarter ended September 30, 2001. (2001). *Gemstar-TV Guide.* Retrieved November 11, 2001, from http://gemstartvguide.com/pressroom/display_pr.asp?prId=95.
Grant, P. (2004, January 22). Cable operators gear up to offer interactive TV. *The Wall Street Journal, 243* (15), B1.
Having it your way with DVRs. (2004, January). *PC World, 22* (1), 141.
Interactive television industry boundaries map. (2002). *ITV Alliance.* Retrieved March 15, 2004, from http://www.itvalliance.org/indexprgITVa.htm.
ITV: How we define it. (2001, June 9). *Broadcasting & Cable, 131* (29), 34.
Jarvis, S. (2001, August 27). ITV finally comes home. *Marketing News, 1,* 19, 20.
Johnston, C. (2002, February 20). Microsoft keeping ITV. *TV Technology, 20* (4), 1, 17.
Kerschbaumer, K. (2002a, May 13). ITV gets first standard. *Broadcasting & Cable, 132* (20), 39.
Kerschbaumer, K. (2002b, July 1). More than a Wink. *Broadcasting & Cable, 132* (27), 25.
Kerschbaumer, K. (2002c, September 9). ATSC approves DASE. *Broadcasting & Cable, 132* (40), 40.
Kerschbaumer, K. (2002d, September 30). Merger in ITV space. *Broadcasting & Cable, 132* (40), 15.
Lesly, E., Cortese, A., Reinhardt, A., & Hamm, S. (1997, November 24). Let the set-top wars begin. *Business Week, 3,554,* 74-75.
Marriage of convenience. (1997, November). *Time Digital*, 60-64.
McClellan, S. (2003, September 15). Report: No more baby steps for interactive TV. *Broadcasting & Cable, 37* (23), 23.
Merli, J. (2002, April 3). In search of the "active" in ITV. *TV Technology, 20* (7), 1, 10.

Merli, J. (2004, February 4). NHL plays the interactive "angles." *TV Technology, 22* (3), 16.

Meeting focuses on interactive TV. (2000, May 13). *Las Vegas Review Journal,* D6.

NBA Entertainment teams with TBS Superstation, TNT and Sprite to launch first ever live fantasy game on NBA.com. (2002). *ACTV, Inc.* Retrieved January 10, 2002, from http://www.actv.com/actvfinal/january10.html.

Pasztor, A., & Goldsmith, C. (2003, December 19). Murdoch plans bells, whistles for DirecTV. *The Wall Street Journal, 121* (242), B1.

Pomerantz, D. (2002, April 15). Spray and pray. *Forbes, 169* (9), 48.

Rosenstein, A. (1994). Interactive television. In A. Grant (Ed.). *Communication technology update, 3rd edition.* Newton, MA: Focal Press.

Shaw, R. (2003, March 31). The interactive living revolution. *Broadcasting & Cable, 133* (13), 3A-4A, 6A, 8A, 10A-13A.

Swedlow, T. (2002), April 2). 2000 interactive enhanced television: A historical and critical perspective. *Interactive TV Today.* Retrieved March 15, 2002, from http://www.itvt.com/etvwhitepaper.html.

Taub, E. (2003, December 29). Commercial skippers, network's enemy no. 1. *The New York Times*, C6.

Texting the television. (2002, October 19). *Economist, 8,295* (36), 59.

Towler, R. (2002). The public's view 2002: An ITC/BSC research publication. *Office of Communications.* Retrieved March 15, 2004, from http://www.ofcom.org.uk/research/consumer_audience_research/tv_audiencereports/?a=87101.

Wink tops the ITV pecking order. (2000, June 28). *TV Technology, 18* (3), 10, 12.

7

Digital Television

Peter B. Seel, Ph.D. & Michel Dupagne, Ph.D.*

The period between 2004 and 2006 will usher in a crucial stage in the diffusion of digital television (DTV) in North America, Europe, and Japan. As DTV standards and transition policies have been implemented over the past decade, much of the debate has now shifted to receiver costs, innovative display technologies, and new programming choices. In the United States, December 31, 2006 will mark the end of the 10-year window defined by the Federal Communications Commission (FCC) (1997) for the national conversion from analog to digital television broadcasting. While it is doubtful that this deadline will be met due to the lack of critical mass in consumer adoption, all U.S. full-power terrestrial television stations will be simulcasting their programming in both analog and digital formats by 2006. The rate of consumer adoption will be driven by the declining price of DTV sets and the perceived relative advantage of both digital display technology and DTV programming (Rogers, 2003). To date, the U.S. conversion to digital broadcasting has cost $4.5 billion, with an expected eventual investment of $16 billion (McConnell, 2004b). Japan and Europe are experiencing the same transition challenges as the United States, although European broadcasters have chosen to focus their digital migration efforts on standard-definition television (SDTV) instead of high-definition television (HDTV).

This very expensive conversion from analog to digital television broadcasting is the most significant change in global broadcast standards since color images were added in the 1950s and 1960s. For better or for worse, television is a ubiquitous global medium. The glow of the TV screen can be found in rustic cabins near the Arctic Circle and in open-air shelters on remote Pacific islands. It is a primary source of news and entertainment for viewers from New Delhi to New York. The diffusion of

* Peter B. Seel is Associate Professor, Department of Journalism and Technical Communication, Colorado State University (Fort Collins, Colorado). Michel Dupagne is Associate Professor, School of Communication, University of Miami (Coral Gables, Florida).

digital television will provide a movie-friendly display with a wider image, improved sound quality, and the higher resolution needed for projecting a sharp image on a screen. However, the most significant change may be that improvements in digital compression technology will facilitate the simultaneous transmission of multiple streams of HDTV and SDTV programming from a single broadcast antenna. Every broadcast station in the United States will have the capability of becoming a multichannel program provider. These programs can be combined with data supplied by broadcasters or from the Internet for enhanced entertainment or educational viewing options.

As discussed in this chapter, digital television refers primarily to native digital (ATSC, or Advanced Television Systems Committee) programming aired by terrestrial broadcasters, even though it may be retransmitted into a majority of homes by cable or satellite operators. The FCC (1998) defines DTV as "any technology that uses digital techniques to provide advanced television services such as high-definition TV (HDTV), multiple standard definition TV (SDTV), and other advanced features and services" (p. 7,420). Therefore, digital cable and standard direct broadcast satellite (DBS) services, which deliver digitized and compressed NTSC (National Television System Committee) signals in MPEG2 format, are not covered in this chapter. (For more on digital cable, see Chapter 3. Digital signals via DBS are discussed in Chapter 4.)

A key attribute of digital technology is "scalability"—the ability to produce audio/visual quality as good (or as bad) as the viewer desires (or will tolerate). This does not refer to program content quality—that factor will still depend on the creative ability of the writers and producers. Within the constraints of available transmission bandwidth, digital television facilitates the dynamic assignment of sound and image fidelity in a given transmission channel. The two common options are high-definition television and standard-definition television.

HDTV represents the highest pictorial and aural quality that can be transmitted through the air. It is defined by the FCC in the United States as a system that provides image quality approaching that of 35mm motion picture film, that has an image resolution of approximately twice (1080I or 720P) that of conventional NTSC television, and has a picture aspect ratio of 16:9 (FCC, 1990). At this aspect ratio of 1.78:1 (16 divided by 9), the television screen is wider in relation to its height than the 1.33:1 (4 divided by 3) of NTSC. It is closer in aspect ratio to the images seen in movie theaters that are 1.85:1 or even wider. Figure 7.1 compares a 16:9 HDTV set with a 4:3 NTSC display—note that the higher resolution of the HDTV set permits the viewer to sit closer to the set, which results in a wider angle of view.

SDTV is another type of DTV technology that can be transmitted *along with,* or *instead of,* HDTV. SDTV transmissions offer lower resolution (480P or 480I) than HDTV, and they are available in both narrow- and widescreen formats. Using digital video compression technology, it is feasible for U.S. broadcasters to transmit up to five SDTV signals instead of one HDTV signal within the allocated 6 MHz digital channel. The development of multichannel SDTV broadcasting, called "multicasting," is an approach that several broadcast networks are actively investigating. A local television station can now deliver a network daytime soap opera while simultaneously broadcasting children's programming and three additional dedicated news, sports, and weather channels in SDTV. Most stations will reserve true HDTV programming for evening prime-time hours, and some networks are working on compression schemes that will allow the simultaneous transmission of one HD and one SD program in prime time.

Figure 7.1
Wider Viewing Angle with HDTV

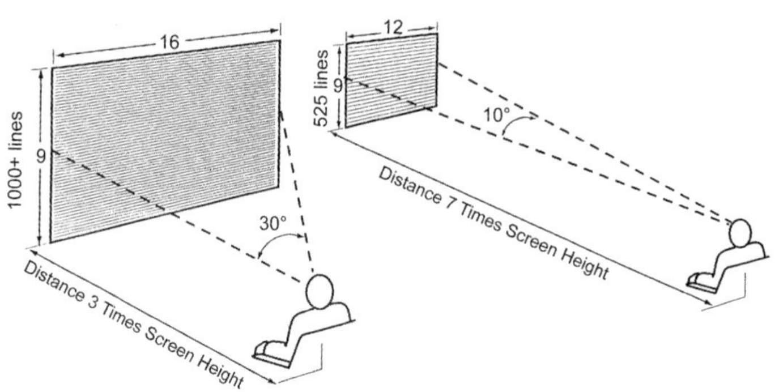

Source: M. Dupagne & P. B. Seel

Background

In the 1970s and 1980s, Japanese researchers at NHK developed two related analog HDTV systems: an analog "Hi-Vision" *production* standard with 1,125 scanning lines and 60 fields (30 frames) per second, and an analog "MUSE" *transmission* system with an original bandwidth of 9 MHz designed for DBS distribution throughout Japan. Japanese HDTV transmissions began in 1989 and steadily increased to a full schedule of 17 hours a day by October 1997 (Nippon Hoso Kyokai, 1998).

The decade between 1986 and 1996 was a significant era in the diffusion of HDTV technology in Japan, Europe, and the United States. There were a number of key events during this period that shaped advanced television technology and related industrial policies:

- In 1986, the Japanese Hi-Vision system was rejected as a world HDTV production standard by the CCIR, a subgroup of the International Telecommunications Union (ITU), at a Plenary Assembly in Dubrovnik, Yugoslavia. European delegates successfully lobbied for a postponement of this initiative that effectively resulted in a *de facto* rejection of the Japanese technology. European high-technology industries were still recovering from Japanese dominance of their VCR markets, and resolved to create their own distinctive HDTV standard that would be intentionally incompatible with Hi-Vision/ MUSE (Dupagne & Seel, 1998).

- By 1988, a European research and development consortium, EUREKA EU-95, had created a system known as HD-MAC that featured 1,250 widescreen scanning lines and 50 fields (25 frames) displayed per second. This analog 1,250/50 system was used to transmit many European cultural and sporting events, such as the 1992 summer and winter Olympic Games in Barcelona, Spain, and Albertville, France.

- In 1987, the FCC in the United States began a series of policy initiatives that led to the creation of the Advisory Committee on Advanced Television Service (ACATS). This committee was charged with investigating the policies, standards, and regulations that would facilitate the introduction of advanced television (ATV) services in the United States (FCC, 1987).

- U.S. testing of analog ATV systems by ACATS was about to begin in 1990 when the General Instrument Corporation announced it had perfected a method of digitally transmitting a high-definition signal. This announcement had a bombshell impact since many broadcast engineers were convinced that digital television transmission would be a technical impossibility until well into the 21st century (Brinkley, 1997). The other participants in the ACATS competition quickly developed digital systems that were submitted for testing. Ultimately, the three competitors in the testing process with digital systems (AT&T/Zenith, General Instrument/MIT, and Philips/Thomson/Sarnoff) decided to merge into a consortium known as the Grand Alliance. With the active encouragement of the Advisory Committee, they combined elements of each of their ATV proponent systems in 1993 into a single digital Grand Alliance system for ACATS evaluation.

The FCC adopted a number of key decisions during the ATV testing process that defined a national transition process from NTSC to an advanced broadcast television system:

- In August 1990, the commission outlined a *simulcast* strategy for the transition to an ATV standard (FCC, 1990). This strategy required that U.S. broadcasters transmit *both* the new ATV signal and the existing NTSC signal concurrently for a period of time, at the end of which all NTSC transmitters would be turned off. Rather than try and deal with the inherent flaws of NTSC, the FCC decided to create an entirely new television system that would be incompatible with the existing one. This was a decision with multi-billion dollar implications for broadcasters and consumers since it meant that all existing production, transmission, and reception hardware would have to be replaced with new equipment capable of processing the ATV signal.

- The FCC originally proposed a transition window of 15 years (now set at 10 years) from the adoption of a national ATV standard to the shutdown of NTSC broadcasting (FCC, 1992). This order caused consternation on the part of the television broadcast industry at what they perceived was too short a transition period, involving very high costs. The transition cost for a single television station was estimated to range from $2 million for simply retransmitting the network signal ("pass-through") to $10 million for complete digital production facilities (Ashworth, 1998).

- In summer 1995, the Grand Alliance system was successfully tested, and a digital television standard based on that technology was recommended to the FCC by the Advisory Committee on November 28 (ACATS, 1995).

- In May 1996, the FCC proposed the adoption of the ATSC Digital Television (DTV) Standard based upon the work accomplished by ATSC in documenting the technology developed by the Grand Alliance consortium (FCC, 1996a). The proposed ATSC DTV standard specified 18 digital transmission variations as outlined in Table 7.1 below. Stations would be able to choose whether to transmit one channel of high-resolution, widescreen HDTV programming, or four to six channels of standard-definition programs during various day parts.

Table 7.1
U.S. Advanced Television Systems Committee (ATSC)
DTV Formats

Format	Active Lines	Horizontal Pixels	Aspect Ratio	Picture Rate*	U.S. Adopter
HDTV	1,080 lines	1,920 pixels/line	16:9	60I, 30P, 24P	CBS & NBC (60I)
HDTV	720 lines	1,280 pixels/line	16:9	60P, 30P, 24P	ABC (30P)
SDTV	480 lines	704 pixels/line	16:9 or 4:3	60I, 60P, 30P, 24P	Fox (30P)
SDTV	480 lines	640 pixels/line	4:3	60I, 60P, 30P, 24P	None

* In the Picture Rate column, "I" indicates interlaced scan in *fields*/second and "P" means progressive scan in *frames*/second.

Source: ATSC

Note that the standard allows for both interlaced and progressive scanning. Interlaced scanning is a form of signal compression that first scans the odd lines of a television image onto the screen, and then fills in the even lines to create a full video frame every 1/30th of a second. The present NTSC standard uses interlaced scanning to conserve the bandwidth needed for broadcast transmission and reduce the screen flicker that would result if the lines were scanned sequentially. While interlaced scanning is spectrum-efficient, it creates unwanted visual artifacts that can degrade image quality. Progressive scanning—where each complete frame is scanned on the screen in only one pass—is utilized in computer displays because it produces fewer image artifacts than interlaced scanning. Progressive-scan DTV receivers are capable of displaying small-font text (e.g., from a Web site) that would be illegible on a conventional NTSC television set.

In December 1996, the FCC finally approved a DTV standard that deleted *any* requirement to transmit any of the 18 transmission video formats listed in Table 7.1 (FCC, 1996b). The commission resolved a potential controversy over the image aspect ratio and scanning structure by leaving these decisions up to broadcasters. They were free to transmit digital programming in either the 4:3 or 16:9 aspect ratio and in progressive or interlaced format as they wished. The commission also declined to mandate any requirement that broadcasters must transmit true HDTV on their digital channel.

The ATSC standard did specify the adoption of the Dolby AC-3 (Dolby Digital) multichannel audio system. The AC-3 specifications call for a surround-sound, six-channel system that will approximate a motion picture theatrical configuration with speakers front-left, front-center, front-right, rear-left, rear-right, and a subwoofer for bass effects. These audio systems are enhancing the diffusion of "home theater" television systems with multiple speakers and either a video projection screen, a direct-view CRT monitor, or a flat-panel display that can be mounted on the wall like a painting.

In April 1997, the FCC defined how the United States would make the transition to DTV broadcasting and set December 31, 2006 as the target date for the phase-out of NTSC broadcasting (FCC, 1997). However, the U.S. Congress passed a bill in 1997 that would allow television stations to continue operating their NTSC transmitters as long as more than 15% of the television households in a market cannot receive digital broadcasts through cable or DBS *and* do not own a DTV set or a digital-to-analog converter box capable of displaying digital broadcasts on their older analog television sets (Balanced Budget Act, 1997). This law gave broadcasters some breathing room if consumers do not adopt DTV technology as fast as set manufacturers would like.

To demonstrate their good faith in an expeditious DTV conversion, the four largest commercial television networks in the United States (ABC, CBS, NBC, and Fox) made a voluntary commitment to the FCC to have 24 of their affiliates in the top 10 markets on the air with a DTV signal by November 1, 1998 (Fedele, 1997). All of their affiliates in the top 30 markets were expected to broadcast digital signals by November 1, 1999, and, except for delays in a few markets caused by tower construction, this deadline was met. Table 7.2 outlines the rest of the rollout.

Table 7.2
U.S. Digital Television Broadcasting Phase-in Schedule

Phase	# of Stations	Market Size	Type of Station	DTV Transmission Deadline	% of U.S. HHs*
1	24	Top 10	Voluntary	November 1, 1998	--
2	40	Top 10	Network Affiliates	May 1, 1999	30%
3	80	Top 30	Network Affiliates	November 1, 1999	53%
4	1,315	All	All Commercial	May 1, 2002	~100%
5	373	All	Non-Commercial	May 1, 2003	--
6	1,688**	All	All	December 31, 2006	Planned NTSC Reversion Date

* Television households capable of receiving at least one local DTV broadcast signal.
** Station statistics as of April 6, 2004.

Sources: FCC and *TV Technology*

Recent Developments

United States

Receiver sales. In 1998, the first HDTV receivers went on sale in the United States at prices ranging from $5,000 to $10,000 or more (Brinkley, 1998). Since then, average prices of DTV sets have declined by 50% (see Table 7.3). Nearly nine million DTV sets and displays were sold to dealers between 1998 and 2003. Rapidly rising sales in 2003 of widescreen television displays with HDTV decoders or set-top boxes indicate that consumer adoption is finally taking off in the United States. The Consumer Electronics Association (CEA) has forecast that digital TV shipments will reach 5.8 million units in 2004, 8.3 million units in 2005, and 11.9 million units in 2006 (Arlen, 2004). If these predictions hold true, about 30% of U.S. households could own a DTV receiver by 2006.

Table 7.3
Actual and Predicted Sales to Dealers of Digital TV Sets and Displays, 1998-2006*

Year	Units Sales in Thousands	Dollar Sales in Millions	Average Unit Price in Dollars
1998	14	$44	$3,147
1999	121	$294	$2,433
2000	648	$1,426	$2,200
2001	1,460	$2,645	$1,812
2002	2,535	$4,279	$1,688
2003	4,100	$6,384	$1,557
2004§	5,800	$7,510	$1,295
2005§	8,300	$10,093	$1,216
2006§	11,900	$13,495	$1,134

* This category includes (1) DTV-capable sets (direct view, rear projection, DLP), (2) integrated DTV sets (direct view, rear projection, DLP), and (3) LCD and plasma TV sets (EDTV and HDTV).
§ = predicted.

Source: Consumer Electronics Association

However, these rosy statistics must be tempered with an important caveat. According to the CEA, only about 12% of these nine million sets were fully integrated (i.e., were equipped with a built-in HDTV tuner) or were matched with a separate external HDTV decoder (Wargo, 2004). Thus, while most of the DTV receivers are HDTV capable, few can actually decode and display images in HDTV format. The FCC has addressed this issue with two key rulings that require television display manufacturers to address the "tuner gap" and the compatibility of DTV sets with cable systems. In 2002, the FCC issued a Report and Order that mandated set manufacturers include DTV tuners in *all* new television receivers 36 inches or larger by July 1, 2005, and set benchmarks for smaller sets to meet this standard over the following two years (FCC, 2002). The CEA argued that the FCC did not have the statutory authority to issue such a mandate and challenged the decision in court. In October 2003, the U.S. Court of Appeals for the D.C. circuit upheld the FCC's Digital Tuner Order as "a reasonable exercise of the Commission's authority under the All Channel Receiver Act [of 1962]" (*Consumer Electronics Ass'n v. F.C.C.*, 2003, p. 293).

The second key action of the Commission was the approval of a "plug-and-play" standard to allow cable subscribers to receive digital basic and premium (excluding video on demand) cable programming on their DTV sets *without* the need of a set-top box (FCC, 2003a). In essence, this new compatibility rule is the digital equivalent of the familiar cable-ready set specification. The standard was the result of negotiations between Cable Television Laboratories (CableLabs) and the Consumer Electronics Association. Since almost 70% of U.S. television households access programming via cable, the resolution of the interface between digital sets and cable television service was a key benchmark in the diffusion of DTV. Digital cable-ready sets will also include digital over-the-air reception functionality and may be available in retail stores as early as the second half of 2004.

Display types. As of March 2003, rear-projection models accounted for 73% of all DTV sets sold since 1998 (CEA, 2003). Yet, it is premature to declare which display technology will eventually dominate the HDTV marketplace because there are new contenders.

- *Direct-view CRTs*—These sets feature the traditional cathode ray tube (CRT) technology that has been the standard display since the invention of television. As sets have become larger with the advent of HDTV, the CRT has become very bulky because the volume of glass used has also increased. Direct-view sets have traditionally been brighter and displayed better overall image quality than projection models (Schiesel, 2003), but this is no longer true compared with the emerging technologies described below.

- *Plasma displays*—In these sets, plasma gas is used as a medium in which tiny color elements are switched off and on in milliseconds. These displays can be quite large compared with CRT models—Samsung exhibited a model at the 2004 Consumer Electronics Show that was 80-inches diagonally and only three inches deep (Taub, 2004a). The images displayed on plasma sets are sharp and offer vivid colors, but are vulnerable to image burn-in. These displays are still very costly (e.g., the 80-inch Samsung model retails for $70,000), thereby limiting their adoption to the affluent (Taub, 2004a). Plasma TV sets are available in both 480p enhanced-definition television (EDTV) and HDTV resolutions.

- *Liquid crystal display (LCD) models*—LCDs work by rapidly switching color crystals off and on. They can be used in direct-view or rear-projection models and can create very sharp and colorful images. LCD technology is commonly used to create laptop and flat-panel computer displays. In 2004, Samsung Electronics introduced the largest LCD model yet manufactured—a 57" display retailing for $30,000 (Taub, 2004a). LCD displays are lighter weight than other technologies, but their circuitry blocks some of the light transmitted to the screen.

- *Digital light processing (DLP) projectors*—Developed by Texas Instruments, DLP technology utilizes hundreds of thousands of tiny micro-mirrors mounted on a 1-inch chip that can project a very bright and sharp color image (Texas Instruments, 2004). This technology is used in a three-chip system to project digital "films" in movie theaters. For under $3,000, a consumer can create an impressive digital home theater with a single-chip DLP projector, a glass-beaded movie screen, and a six-channel Dolby Digital surround-sound system. By early 2004, consumer demand for DLP receivers had increased significantly because of the perceived value of the technology relative to its cost (Austen, 2003; Kerschbaumer, 2004).

- *Liquid crystal on silicon displays*—LCoS high-definition televisions are a new entrant in the DTV marketplace. Utilizing a liquid crystal technology similar to that of LCDs, LCoS offers high-resolution displays that proponents say are superior to the others. Unlike LCDs, all the circuitry on an LCoS chip is located behind a silicon layer, and more light is transmitted to the screen in higher resolution. Unlike a DLP chip, there are no moving parts that might fail, making the LCoS technology more durable (Taub, 2004b).

Figure 7.2
Plasma and LCD TV Sets on Display at a Retail Store

Source: M. Dupagne

Consumer awareness and interest. Consumer awareness of HDTV in the United States continues to rise, with 2003 polls indicating that 73% to 78% of surveyed respondents knew or heard about HDTV (Arlen, 2004; Ipsos-Insight, 2003; Yankee Group, 2004). However, it is not clear whether consumers really understand the specifics of DTV because marketing studies generally do not measure actual knowledge. Chan-Olmsted and Chang (2003) found that only 34% of their sample knew the difference between digital cable and DTV, and almost 60% did not know or erroneously answered a series of DTV questions. Yet, in consumer surveys, a majority of respondents appeared to be relatively familiar with the technological benefits that HDTV could offer (e.g., superior image quality, widescreen aspect ratio, and surround-sound). They also expressed interest in watching movies and sports in HDTV format (Arlen, 2004; Ipsos-Insight, 2003; Yankee Group, 2004). According to the latest 2003 CEA study, nearly seven out of 10 respondents had seen HDTV, primarily in a retail store. Not surprisingly, the typical early DTV buyer is male, between 35 and 54 years old, and has an annual household income of more than $100,000 (CEA, 2001).

Programming. Consumer electronics companies, such as Zenith, Panasonic, and Hitachi, have underwritten production and analog-to-digital conversion costs on all broadcast networks in an effort to increase DTV set sales. Since 2003, the major television networks have dramatically increased the amount of DTV programming that they simulcast.

Table 7.4 outlines the sources of HDTV programming available in mid-2004. The Big Six broadcast networks are all simulcasting HD programs in prime time, and an increasing number of cable networks are transmitting selected programs in high definition, led by HBO, ESPN HD, Showtime, and the HDNet programming service co-founded in 2001 by Mark Cuban. Most major sporting events, such as the Super Bowl, the Final Four NCAA basketball tournament, the World Series (baseball), and the Masters (golf), are now routinely simulcast in HDTV. In the next two years as the 2006 conversion deadline approaches, the amount of simulcast programming will continue to increase.

In October 2003, Cablevision Systems headed by cable pioneer Charles Dolan announced the launch of a new DBS service called VOOM that would carry only HDTV programming (Higgins, 2003b). The company said that it would provide 39 channels of HD content, 21 of them created by VOOM. The new company is a $450 million gamble by Dolan that the service would attract affluent early adopters seeking unique HDTV programming. (For more details on VOOM, see Chapter 4.)

Multicasting. As of mid-2004, more than 190 U.S. television stations aired multiple SDTV programs at certain times of the day, and many other stations are making plans to follow suit (Eggerton & Kerschbaumer, 2003). Public television stations have been at the forefront of multicasting experiments to fulfill their educational mandate to children and adult learners, while providing a variety of programming choices for other audiences at the same time (Expanding PBS', 2004).

Commercial stations have different operational philosophies and contemplate using multicasting to sell advertising spots on several SDTV channels transmitted at the same time. NBC Universal announced in October 2003 that it was studying multicasting technology for the potential daytime transmission of five channels of diverse program content in SDTV: the regular network feed, a movie channel (with films provided by Universal), a program preview channel, a locally-produced news and traffic channel, and a local "alert channel" of weather and safety information (Higgins, 2003a). The network and its affiliates would revert to single-channel HDTV programming in the evening hours. The proposal hinges on the willingness of local affiliates to provide the news and weather content, and on the cooperation of local cable companies to carry all five SDTV channels. In December 2003, CBS signed a deal with Comcast, the largest U.S. multiple system operator (MSO), that assures that the cable company will carry a CBS affiliate's network feed and their multicast SDTV channels in the markets that Comcast serves (Higgins, 2003c).

Broadcasters are also studying multicasting as a means of providing a wireless "cable" service to paying subscribers. In January 2004, U.S. Digital Television began offering the HDTV feeds of local stations plus 11 channels of typical cable programming to subscribers for $19.95 a month (Kerchbaumer & Higgins, 2004). The system works by pooling unused DTV spectrum capacity from local broadcasters. Each local station was awarded enough spectrum for HDTV transmission of 19.4 megabits per second (Mb/s), but new compression technologies allow HDTV broadcasting in the 12 Mb/s to 14 Mb/s range. The unused 5 Mb/s to 7 Mb/s bandwidth is sufficient for the transmission of up to two SDTV channels. Similar multicasting bandwidth pooling in other large cities would facilitate the creation of wireless subscription services that could compete with cable providers on a limited basis.

Station conversion. As of February 25, 2004, 97% (1,632) of television stations in all U.S. markets have been granted a DTV construction permit or license by the FCC. Of these stations, 85% (1,385) are transmitting a DTV signal, but many are operating at low power, especially in smaller markets (FCC, 2004b). Most stations in the top 30 markets have been on the air in DTV since November 1, 1999. As noted in Table 7.2, all *commercial* stations were supposed to air a DTV signal by May 1, 2002, but as of February 2004, 104 stations have petitioned the FCC for a third extension of the deadline (FCC, 2004b). A total of 214 noncommercial public television stations missed the May 1, 2003 deadline, and 129 have asked the FCC for a second extension. It is not surprising that many public television stations have experienced difficulty financing the conversion to DTV. The number of extensions requested confirms this expectation.

Table 7.4
Current and Planned HDTV Programming in the United States (as of January 26, 2004)

Network	Type of Programming
ABC	Sitcoms, dramas, sports, and special events
Bravo HD+	Dramas, music concerts, and arts
CBS	Sitcoms, dramas, sports, soap opera, and special events
Cinemax HD	Movie
Comcast SportsNet HDTV	Sports
Discovery HD Theater	Nature shows, documentaries, and special events
ESPN HD	Sports
Fox	Spring 2004
Fox Sports Net HD	Sports
HBO HDTV	Movies and dramas
HDNet	Sitcoms, dramas, action series, documentaries, travel programs, music concerts, special events, and news features
HDNet Movies	Movies
INHD	Movies, sports, and special events
INHD2	Movies, sports, and special events
Movie Channel HD	Movies
MSG Network HDTV/FSNY HD	Sports
NBC	Sitcoms, dramas, sports, and late-night show
NBATV HD	Sports
NFL Network HD	Sports
New England Sports Net HD	Sports
PBS	HD Loop, documentaries, music concerts
Showtime HD	Movies and dramas
STARZ! HD	Movies
TNT HD	May 2004
UPN	Movies and dramas
WB	Dramas

Source: Broadcasting & Cable

Cable deployment. Because nearly 70% of U.S. television households receive their signals from a cable provider, digital conversion of cable facilities is essential to pass along DTV programming from broadcasters to consumers. However, the cable industry has been ambivalent about providing HDTV/SDTV service for two fundamental reasons:

1) They would have to provide an extra DTV channel for every existing analog channel, constraining system capacity without increasing revenue.

2) SDTV technology would permit terrestrial broadcasters to transmit four to six of these channels simultaneously over the air—in effect transforming them into multichannel pro-

viders in direct competition with cable news and sports channels, as the proposed NBC multicast system illustrates.

Only in the past two years has the cable industry embraced HDTV transmission. This may be due to the urgent prodding by FCC Chairman Michael Powell, but it may equally be attributed to the transmission of HD programs by DBS providers. The FCC declined to require must-carry status for multicast DTV channels in a January 2001 decision, but left the door open for future action if foot-dragging by cable companies inhibited the DTV transition (FCC, 2001). HDTV has since become a factor in the intense competition between cable and satellite providers. In March 2004, the National Cable & Telecommunications Association (2004) reported that 84 million TV homes could receive ("were passed") HDTV programming through a cable system and that cable operators carried a total of 382 local DTV stations.

"Broadcast flag." An interesting and related issue to the DTV conversion in the United States is the perceived notion by copyright holders (especially the Motion Picture Association of America) that piracy of digital television programming on the Internet is going to become a problem as significant as that of music piracy. As a result, the program production industries have insisted that the FCC adopt regulations that require the inclusion of "broadcast flag" technology in DTV decoders, displays, and recorders that would prevent copying of "flagged" programs. The "flag" is a 16- or 24-bit label inserted in a DTV signal that would instruct a digital device, such as a digital video recorder, as to whether a broadcast or cablecast program could be copied and how many times (Boutin, 2004). The FCC adopted such a provision in a November 4, 2003 order that gives DTV set manufacturers until July 1, 2005 to include copy protection technology in their products (FCC, 2003b).

In March 2004, the FCC was sued by several public interest groups (including the American Library Association and the Electronic Frontier Foundation) who allege that "broadcast flag" technology will unfairly restrict the copying of televised programs (Singel, 2004). The FCC has yet to clarify exactly what type of program copying would be inhibited, but the public interest groups are concerned that simply recording an evening network newscast would be impossible under the new system. Fair use of such broadcast content by educators (especially for distance education classes) would be blocked by the "flag" digitally embedded in the programs as they are transmitted. This will be a contentious issue for the broadcasting industry and the FCC through 2006.

Japan

As noted earlier, Japan has been a primary innovator in HDTV technology. When General Instrument announced in June 1990 that it would design an all-digital HDTV system, Japan was caught off guard and was forced to reconsider its entire HDTV program. After some initial resistance, the Japanese forged ahead with their digital satellite and terrestrial plans to recapture their leadership position in HDTV deployment (Dupagne & Seel, 1998). In December 2000, digital broadcasting transmissions began from a high-power satellite (MPHPT , 2002). As of August 2003, 19 broadcasters offered six HDTV channels, 19 SDTV channels, 23 radio channels, and nine data channels. There were about 4.4 million subscriptions to the BS digital broadcasting service at that time. In March 2002, digital broadcasting transmissions started from a new low-power satellite. As of August 2003, 18 broadcasters supplied two HDTV channels, 61 SDTV channels, 20 radio channels, and 10 data channels from this satellite (MPHPT, 2003).

In December 2003, NHK and 16 private broadcasters launched digital terrestrial television service in Tokyo, Osaka, and Nagoya, although apparently not at full power to avoid interference with analog frequencies (Nervous energy, 2003). The service will debut in the other cities by the end of 2006. The technical specifications of the Japanese ISDB-T (Integrated Services Digital Broadcasting-Terrestrial) standard, a variant of the European DVB-T standard, are summarized in Table 7.5. Analog NTSC broadcasting is scheduled to end in July 2011.

In addition, the MPHPT has specified controversial numerical targets for the diffusion of digital TV receivers: 12 million sets or 21% household penetration by 2006, 36 million sets or 50% household penetration by 2008, and 100 million sets that would be bought by all 48 million households by 2011 (Third action plan, 2003). The Ministry has acknowledged that diffusing 100 million DTV sets in seven-and-a-half years represents "a difficult hurdle" (p. 5), but has remained committed to meeting these goals. Industry insiders, however, question the feasibility of even producing such a large output of DTV sets for domestic consumption—13 million sets every year between 2004 and 2011 (Nervous energy, 2003). They are also quick to point out that the conversion from analog to digital terrestrial transmissions will cost between ¥400 billion ($3.5 billion) and ¥800 billion ($7.5 billion). The estimated investment for each private broadcaster would average ¥6 billion ($56 million), a burdensome proposition for some cash-strapped local stations (Suzuki, 2003).

Moreover, Japanese consumers do not seem much aware of DTV or enthusiastic at the prospect of spending between $2,300 and $3,500 for a DTV set. One commercial network executive summarized the situation as follows: "To be honest, it is absolute chaos. We really are not sure whether all of this makes any business sense" (Schwarzacher, 2003, p. 36). Like its U.S. counterpart, the Japanese DTV migration is likely to be a challenging ride for all parties involved.

Europe

While U.S. and Japanese broadcasters have opted for a dual HDTV/SDTV track, their European counterparts have decided to embrace SDTV exclusively using the DVB-T standard (see Table 7.5). The following reasons are often cited for Europe's lack of interest in HDTV broadcasting (Commission of the European Communities, 2004):

1) European broadcasters still perceive that HDTV was a resounding market failure in the 1990s.

2) European broadcasters believe that HDTV is too expensive and that there is no business model for offering HDTV programming to viewers.

3) European broadcasters prefer to use their limited spectrum to provide more viewing options with multiple SDTV channels rather than a single HDTV source.

One recent exception to this mindset has been Belgian-based Alfacam's satellite-delivered Euro1080. Launched in January 2004, this first European HDTV channel represents a surprising turnabout on a continent where the availability of HDTV programming to viewers has been virtually nonexistent in the last 10 years. The programming content, which focuses on sports, music, and cultural events, is now available at no charge to satellite households that are equipped with an HDTV receiver and decoder. It will also be distributed to cable operators and movie theaters outfitted with digital projection systems (Alleyne, 2003; Van Overstraeten, 2003). Whether the presence of a new European HDTV programming outlet will renew viewer and broadcaster interest in the technology is, of course, an open question.

Table 7.5
International Terrestrial DTV Standards

System	ISDB-T	DVB-T	ATSC DTV
Region	Japan	Europe	North America
Modulation	OFDM	COFDM	8-VSB
Aspect Ratio	1.33:1, 1.78:1	1.33:1, 1.78:1, 2.21:1	1.33:1, 1.78:1
Active Lines	480, 720, 1080	480, 576, 720, 1080, 1152	480, 720, 1080*
Pixels/Line	720, 1280, 1920	varies	640, 704, 1280, 1920*
Scanning	1:1 progressive, 2:1 interlaced	1:1 progressive, 2:1 interlaced	1:1 progressive, 2:1 interlaced*
Bandwidth	6-8 MHz	6-8 MHz	6 MHz
Frame Rate	30, 60 fps	24, 25, 30 fps	24, 30, 60 fps*
Field Rate	60 Hz	30, 50 Hz	60 Hz
Audio Encoding	MPEG-2 AAC	MUSICAM/Dolby AC-3	Dolby AC-3

* As adopted by the FCC on December 24, 1996, the ATSC DTV image parameters, scanning options, and aspect ratios were not mandated, but were left to the discretion of display manufacturers and television broadcasters (FCC, 1996b).

Source: P. B. Seel & M. Dupagne

By the end of 2003, six European countries had formally rolled out digital terrestrial television (DTTV or DTT) service, the equivalent of DTV service in the United States (see Table 7.6). Other countries are at different stages of their analog-to-digital transition plans. Two European experiences have attracted a great deal of interest from the National Association of Broadcasters (NAB) and the U.S. Congress: the free-to-air (or free-to-view) model of Freeview in the United Kingdom and the digital switchover model in Berlin-Brandenburg, Germany.

Rising from the ashes of the defunct premium ITV digital platform, Freeview began transmitting its 30 digital channels to 75% of U.K. households in October 2002. U.K. viewers can receive these signals at no cost provided they own an integrated digital television (iDTV) set or a £80 ($150) digital-to-analog converter box *and* a standard rooftop antenna (Freeview, 2003). Like other European DTTV services, Freeview uses a video encoding resolution of 576 vertical (active) lines by 704 horizontal pixels per active line (Gardiner, 2003). The scanning structure is interlaced at 50 fields per second.

By the end of 2003, more than three million U.K. homes were watching Freeview channels (Ofcom, 2004). The success of this free DTTV service has prompted other European countries to reexamine their strategies for DTTV deployment. With the demise of ITV digital and Spain's Quiero

TV in mid-2002, the future of European DTTV as a pay-TV platform was seriously compromised. European Broadcasting Union senior analyst Alexander Shulzycki noted that, "A free model is the best way to drive penetration, and with rising penetration more and more business models can work" (Wynn & Wales, 2003). The National Association of Broadcasters is exploring whether the Freeview model could be a viable business model in the United States and help turn stations into multichannel operations (Eggerton, 2004a).

Table 7.6
Existing and Planned Digital Terrestrial Television (DTTV) Services in Selected Western European Countries (2003)

Country	Launch Date	Number of Multiplexes*	Number of Channels	Analog Shut-off Date	Number of DTTV users
Austria	2004	3	16	2012	
Finland	2001	3	15	2006	100,000
France	2004	6	33	?	
Germany	2002	?	26	2010	170,000
Ireland	2007	6	?	?	
Italy	2006	18	90	2006	
Netherlands	2003	5	24	2005	NA
Portugal	2004	?	5	2007	
Spain	2000	11	17	2012	130,000
Sweden	1999	4	21	2008	140,000
United Kingdom	1998	6	28	2010	1,600,000

*Multiplexes (MUX) are assigned frequencies to public service and commercial broadcasters for providing DTTV services. Each multiplex/frequency can support multiple DTTV channels.

Source: IDATE

The other major European development is the August 2003 decision of the Medienanstalt Berlin-Brandenburg (MABB), in cooperation with public and private broadcasters, to turn off analog PAL transmitters and frequencies in the Berlin and Brandenburg states (MABB, 2003). This German region became the world's first example of a complete and successful conversion from analog to digital terrestrial television. However, when deciding to apply the same switchover approach to other areas or countries, caution must be taken to recognize the auspicious conditions under which the Berlin experience took place. Of the 1.8 million households living in Berlin and Brandenburg, more than 1.6 million were cable subscribers or received analog or digital satellite signals. Only 160,000 households (9%) watched television over the air on their primary TV sets, and another 90,000 cable or satellite households watched television over the air on their second or third sets (MABB, 2003). Interestingly, cable operators had to convert DTTV signals to analog signals for their subscribers because cable television service in that region was primarily provided in analog format (Bakarinow, 2004).

In addition, several incentives, including public subsidies, influenced the pace of the switchover:

1) The DTTV service offered 28 channels instead of 10 channels for the old analog television service.

2) Set-top boxes were widely available from manufacturers and cost as little as €88 ($107).

3) Broadcasters and MABB launched an aggressive communication campaign, budgeted at €1.2 million ($1.5 million), to inform all viewers about the conversion and provide them with Web and telephone support.

4) The MABB earmarked €1 million ($1.2 million) to purchase 6,000 set-top boxes for low-income households.

5) Private broadcasters received €60,000 to €70,000 ($74,000 to $86,000) annually to defray the technical cost of the digital switchover (Cole, 2003; MABB, 2003).

MABB's DTV project leader, Sascha Bakarinow, insisted that the primary motivation for shutting off analog television transmissions was economic: "Simulcasts are expensive. In Germany, it costs around €800,000 [$976,000] ... a day for an analog channel and €200,000 [$244,000] ... a day for a digital channel" (Cole, 2003, p. 12). The U.S. General Accounting Office is expected to release a report drawing upon the lessons of Berlin's analog shut-off in 2004.

Factors to Watch

The global diffusion of DTV technology will evolve over the first decade of the 21st century. Led by the United States, Japan, and the nations of the European Union, digital television will gradually replace analog transmission in technologically-advanced nations. It is reasonable to expect that many of these countries will have converted their cable, satellite, and terrestrial facilities to digital technology by 2010. In the process, this digital technology will influence what people watch, especially as it will offer easy access to the Internet and other forms of entertainment that are more interactive than traditional television.

The global development of digital television broadcasting is entering a vital stage in the coming decade as terrestrial and satellite DTV transmitters are turned on, and consumers get their first look at HDTV and SDTV programs. Among the issues that are likely to emerge in 2004 and 2005 in the United States:

Receiver prices—Since 1998, about nine million DTV sets have been sold in the United States, and retail prices have declined by 50%. How much further do prices of DTV sets have to drop before stimulating consumer demand beyond affluent innovators and early adopters? In 2003, the price of a DTV set in the United States still averaged more than $1,500. As the Japanese experiment with Hi-Vision demonstrated, high retail prices can thwart the successful diffusion of a new technology despite programming availability and consumer interest (Dupagne & Seel, 1998). After more than 10 years of diffusion, sales of analog HDTV sets in Japan still only numbered 831,000 by January 2000.

Indoor reception—In 2001, *New York Times* columnist Eric Taub (2001) summarized his experience with antenna reception of HDTV broadcasts with wry humor: "If you are the kind of person who would have loved owning a car in 1910, believing that the new worlds a vehicle would open outweighed the need to change a tire every 10 miles or crank the engine by hand, then HDTV is for you" (p. G7). Despite improvements in next-generation ATSC (8-VSB) receivers and decoders, satisfactory over-the-air reception of DTV signals can still pose a challenge in 2004, especially with indoor antennas. Causes include multipath propagation (i.e., ghosting), interference with analog frequencies, and impulse noise (e.g., energy from electrical appliances) (Kutzner, 2001).

For instance, if a DTV receiver cannot handle multipath problems effectively, it could lock up and display a black or frozen image. Unlike NTSC reception, which may produce degraded but viewable pictures in less-than-ideal conditions, DTV reception shows either perfect pictures or nothing (i.e., the cliff effect). External rooftop antennas generally provide better DTV reception than indoor "rabbit ear" antennas (Dupagne, 2003). Given that 85% of TV households subscribe to a multichannel video programming distributor (MVPD), such as a cable or DBS operator (FCC, 2004a), most viewers will eventually access local DTV channels via their MVPD.

Deadline date—Although the target date for the completion of the digital conversion in the United States was codified into law (Balanced Budget Act, 1997), it is quite doubtful that local broadcasters will shut down their NTSC transmitters on December 31, 2006. In fact, FCC Chairman Michael Powell recently acknowledged that such deadline was "very unlikely" (Johnston, 2004, p. 19). By the end of 2006, only one-third, instead of the required 85%, of TV households in local television markets might be capable of receiving DTV signals either over-the-air or through an MVPD. Therefore, an extension of NTSC broadcasting beyond 2006 is almost guaranteed. Of course, it is anyone's guess when the United States will reach this 85% saturation level and shut off analog frequencies and transmissions. It appears quite probable that broadcasters will need to operate two transmitters to simulcast in both analog and digital formats past 2010 if present DTV diffusion rates remain unchanged.

In March 2004, the FCC proposed a plan to speed up the pace of the conversion toward the requisite 85% DTV penetration level by having cable and DBS operators converting local stations' digital signals to analog and delivering them in that format to subscribers (McConnell, 2004a). This approach exhibits some similarities to the Berlin experience described earlier. The commission would be able to recover and auction the analog television spectrum, valued at $6.1 billion (Congressional Budget Office, 1999) and complete the switchover as early as 2007 (McConnell, 2004b). Over the years, spectrum auctions have become an important source of income for the federal government. However, the FCC's draft immediately angered broadcasters and some members of Congress because

1) DTV set owners who subscribe to cable or DBS might not be able to receive local stations in the expected HDTV quality.

2) The 16 million TV households who receive local stations over-the-air would be forced to purchase a digital-to-analog converter within the next few years.

As of this writing, it is not clear in what shape or form this plan will be adopted.

Multicasting and cable carriage—The use of multiple SDTV channels will remain on the economic agenda of many broadcast stations and networks in the next two years. Media Global TV analyst Brian Wieser views "multicasting as an opportunity for advertisers to more narrowly target consumers and to invest directly in the new programs needed to fill the digital channels" (Eggerton, 2004b, p. 14). On paper, multicasting sounds like an attractive strategy to yield additional revenue streams, but business models have yet to be developed to make it a viable option. In addition, the 2001 FCC Cable Carriage Order, which limited the post-2006 digital must-carry obligations to the *primary* video signal or channel, would have to be amended to grant broadcasters digital must-carry status for their multicast channels. As of this writing, cable operators continue to firmly oppose such expansion on the basis that it would violate their First and Fifth Amendment rights (Salvos continue, 2003). The commission itself is divided on the subject and is still reviewing the constitutionality of multicast must-carry. In February 2004, CBS affiliates applied more pressure on the FCC by threatening to forego multicasting altogether if they could not secure expanded DTV carriage rights (McConnell, 2004c).

In the United States, the volume of simulcast HDTV programming is increasing each year. This programming will entice consumers to consider paying a premium for new digital television sets to access this HDTV content. Since nearly 70% of U.S. television viewers are cable subscribers, they will start to implore their cable operators to dedicate additional channel capacity for DTV programming. The speed of the transition to DTV broadcasting in the United States is now in the hands of television manufacturers who must find innovative ways to reduce the cost of digital sets, and cable operators who must balance the need to serve a minority of DTV subscribers with the demands of their NTSC customers seeking additional analog channels. There is a venerable Chinese saying that states, "may you live in interesting times." This will certainly apply during the next two years of the transition from analog to digital television broadcasting.

Bibliography

Advisory Committee on Advanced Television Service. (1995). *Advisory Committee final report and recommendation*. Washington, DC: Author.

Alleyne, P. (2003, September 8). High-def crosses the pond. *Broadcasting & Cable, 133,* 17.

Arlen, G. (2004, Winter). HDTV awareness on the rise. *HDTV Guide*, 8, 10. Retrieved March 20, 2004 from http://www.ce.org/publications/books_references/dtv_guide/HDTV_Guide_Winter_04.pdf.

Ashworth, S. (1998, March 23). Finding funds for the transition. *TV Technology, 16*, 10, 12.

Austen, I. (2003, December 24). Consumers want rear-projection TVs, and now. *The New York Times,* C2.

Balanced Budget Act of 1997. (1997). Pub. L. No. 105-33, § 3003, 111 Stat. 251, 265.

Bakarinow, S. (2004, March 24). Personal communication.

Boutin, P. (2004, February). Why the broadcast flag won't fly. *Wired,* 32.

Brinkley, J. (1997). *Defining vision: The battle for the future of television.* New York: Harcourt Brace.

Brinkley, J. (1998, January 12). They're big. They're expensive. They're the first high-definition TV sets. *The New York Times,* C3.

Chan-Olmsted, S., & Chang, B. (2003, August). *Consumer awareness and adoption of digital television: Exploring the audience knowledge, perceptions, and factors affecting the adoption of terrestrial DTV.* Paper presented at the annual meeting of the Association for Education in Journalism and Mass Communication, Kansas City, MO.

Cole, G. (2003, September 9). Berlin's big bang. *Financial Times,* 12. Retrieved January 13, 2004 from LexisNexis Academic.

Commission of the European Communities. (2004, January 13). *The contribution of wide-screen and high definition to the global rollout of digital television* (Commission staff working paper). SEC(2004) 46. Brussels: Author.

Congressional Budget Office. (1999). *Completing the transition to digital television.* Washington, DC: Author.
Consumer Electronics Ass'n v. F.C.C., 347 F.3d 291 (D.C. Cir. 2003).
Consumer Electronics Association, eBrain Market Research. (2001). *DTV owner satisfaction II.* Arlington, VA: Author.
Consumer Electronics Association. (2003). *Digital America 2003.* Arlington, VA: Author.
Dupagne, M. (2003, February). Review of the HiPix digital television computer card. *Feedback, 44* (1) 56-66. Retrieved March 27, 2004 from http://www.beaweb.org/feedback/feed44v1.pdf.
Dupagne, M., & Seel, P. (1998). *High-definition television: A global perspective.* Ames: Iowa State University Press.
Eggerton, J. (2004a, January 5). NAB explores UK's Freeview DTV success. *Broadcasting & Cable, 134,* 1, 14.
Eggerton, J. (2004b, February 2). Magna Global pro touts a multicasting future. *Broadcasting & Cable, 134,* 14.
Eggerton, J., & Kerschbaumer, K. (2003, December 8). Suddenly it's hip to spectrum-split. *Broadcasting & Cable, 133,* 1, 30-31.
Expanding PBS' reach and mission. (2004). *PBS digital television.* Retrieved March 28, 2004 from http://www.pbs.org/digitaltv/multiNS.html.
Fedele, J. (1997, September 25). DTV schedule breeds apprehension. *TV Technology, 15,* 16.
Federal Communications Commission. (1987). *Formation of advisory committee on advanced television service and announcement of first meeting*, 52 Fed. Reg. 38523.
Federal Communications Commission. (1990). Advanced television systems and their impact upon the existing television broadcast service. *First report and order*, 5 FCC Rcd. 5627.
Federal Communications Commission. (1992). Advanced television systems and their impact upon the existing television broadcast service. *Second report and order/Further notice of proposed rulemaking*, 7 FCC Rcd. 3340.
Federal Communications Commission. (1996a). Advanced television systems and their impact upon the existing television broadcast service. *Fifth further notice of proposed rulemaking*, 11 FCC Rcd. 6235.
Federal Communications Commission. (1996b). Advanced television systems and their impact upon the existing television broadcast service. *Fourth report and order*, 11 FCC Rcd. 17771.
Federal Communications Commission. (1997). Advanced television systems and their impact upon the existing television broadcast service. *Fifth report and order*, 12 FCC Rcd. 12809.
Federal Communications Commission. (1998). Advanced television systems and their impact upon the existing television broadcast service. *Memorandum opinion and order on reconsideration of the sixth report and order*, 13 FCC Rcd. 7418.
Federal Communication Commission. (2001). Carriage of digital television broadcast signals. *First report and order and further notice of proposed rulemaking*, 16 FCC Rcd. 2598.
Federal Communications Commission. (2002). Review of the commission's rules and policies affecting the conversion to digital television. *Second report and order and second memorandum opinion and order*, 17 FCC Rcd. 15978.
Federal Communications Commission. (2003a). Compatibility between cable systems and consumer electronics equipment. *Second report and order and second further notice of proposed rulemaking*, 18 FCC Rcd. 20885.
Federal Communications Commission. (2003b). Digital broadcast content protection. *Report and order and further notice of proposed rulemaking*, 18 FCC Rcd. 23550.
Federal Communications Commission. (2004a). Annual assessment of the status of competition in the market for the delivery of video competition. *Tenth annual report*, MB Docket No. 03-172. Retrieved March 27, 2004 from http://hraunfoss.fcc.gov/edocs_public/attachmatch/FCC-04-5A1.pdf.
Federal Communications Commission. (2004b). Summary of DTV applications filed and DTV build out status. Retrieved March 25, 2004 from http://www.fcc.gov/mb/video/files/dtvsum.htm.
Freeview. (2003). *FREEVIEW consumer press pack.* London: Author. Retrieved March 21, 2004 from http://www.freeview.co.uk/files/818_040_CONSUMER_PRESS_PACK.pdf.
Gardiner, P. (2003, December 11). Personal communication.

Higgins, J. (2003a, October 13). NBC eyes big ideas: VOD, DTV multicast. *Broadcasting & Cable, 133*, 1, 44.
Higgins, J. (2003b, October 20). Voom launches with promises and big gaps. *Broadcasting & Cable, 133*, 29.
Higgins, J. (2003c, December 22). Comcast makes deal for CBS multicasting. *Broadcasting & Cable, 133*, 2.
Ipsos-Insight. (2003, December 9). *High-definition TV: Is the signal getting stronger?* Retrieved March 21, 2004 from http://www.ipsos-na.com/news/pdf/media/mr031209-2.pdf.
Johnston, C. (2004, February 4). Powell: 2006? Not likely. *TV Technology, 22*, 19.
Kerschbaumer, K. (2004, January 19). HD sets slim down, are DLP-based. *Broadcasting & Cable, 134*, 39.
Kerschbaumer, K., & Higgins, J. (2004, January 12). Utah's uncable surprise. *Broadcasting & Cable, 134*, 1, 59.
Kutzner, J. A. (2001, April). *The challenges of indoor reception.* Paper presented at the annual meeting of the Broadcast Engineering Conference, National Association of Broadcasters Convention, Las Vegas, NV.
Medienanstalt Berlin-Brandenburg. (2003). *Berlin goes digital. The switchover of terrestrial television from analogue to digital transmission in Berlin-Brandenburg: Experiences and perspectives.* Berlin: Author. Retrieved March 21, 2004 from http://www.mabb.de/bilder/Projektbericht_engl.pdf.
McConnell, B. (2004a, February 27). CBS affils threaten to abandon multicasting. *Broadcasting & Cable.* Retrieved March 27, 2004, from http://www.broadcastingcable.com/index.asp?layout=articlePrint&articleID=CA387058.
McConnell, B. (2004b, March 15). Millions could be without TV. *Broadcasting & Cable, 134*, 5, 18.
McConnell, B. (2004c, March 26). More reps, pan Powell plan. *Broadcasting & Cable.* Retrieved March 27, 2004 from http://www.broadcastingcable.com/index.asp?layout=articlePrint&articleID=CA406122.
Ministry of Public Management, Home Affairs, Posts, and Telecommunications. (2002). *Information and communications policy in Japan: 2002 annual report.* Tokyo: Author. Retrieved March 21, 2004 from http://www.soumu.go.jp/joho_tsusin/eng/Resources/AR2002/ar2002.pdf.
Ministry of Public Management, Home Affairs, Posts and Telecommunications. (2003, September). *Major aspects of Japan's broadcasting policy.* Tokyo: Author. Retrieved March 21, 2004 from http://www.soumu.go.jp/joho_tsusin/eng/major_aspects.html.
National Cable & Telecommunications Association. (2004, March 29). *Cable HDTV now available in 99 of top 100 U.S. markets; service reaches 84 million homes.* Retrieved March 29, 2004 from http://www.ncta.com/press/press.cfm?PRid=465&showArticles=ok.
Nervous energy: Digital terrestrial television is nigh in Japan. (2003, November). *Television Asia*, 36. Retrieved March 20, 2004 from LexisNexis Academic.
Nippon Hoso Kyokai. (1998). *NHK factsheet '98.* Tokyo: Author.
Ofcom. (2004). Digital television update Q4 2003. Retrieved March 21, 2004 from http://www.ofcom.org.uk/research/industry_market_research/m_i_index/dtv/uptake/?a=87101.
Rogers, E. (2003). *Diffusion of innovations* (5th ed.). New York: Free Press.
Salvos continue over digital must-carry. (2003, November 28). *TV Technology.* Retrieved November 28, 2003 from http://www.tvtechnology.com/dailynews/one.php?id=1591.
Schiesel, S. (2003, November 27). TV maze: A survival guide. *The New York Times*, E1, E7.
Schwarzacher, L. (2003, December 1-7). Japan gets a yen to go digital. *Variety, 393*, 32.
Singel, R. (2004, March 11). Calls to burn the broadcast flag. *Wired News.* Retrieved March 11, 2004 from http://www.wired.com/news/politics/0,1283,62619,00.html.
Suzuki, Y. (2003, December 2). Local TV may lose in digital shift. *The Daily Yomiuri*, 4. Retrieved January 13, 2004 from LexisNexis Academic.
Taub, E. (2001, February 15). High-definition TV: All or nothing at all. *The New York Times*, G7. Retrieved March 21, 2004 from LexisNexis Academic.
Taub, E. (2004a, January 8). Flat-panel sets to enhance the visibility of Samsung. *The New York Times*, C1, C4.
Taub, E. (2004b, February 5). For better HDTV displays, it's all about the chip. *The New York Times*, E5.
Texas Instruments, Inc. (2004). *How DLP technology works.* Retrieved March 11, 2004 from http://www.dlp.com/dlp_technology/dlp_technology_overview.asp.

Third action plan for the promotion of digital broadcasting formulated. (2003, May 14). *MPHPT Communications News, 14* (3), 4-7. Retrieved March 21, 2004 from http://www.soumu.go.jp/joho_tsusin/eng/Releases/NewsLetter/Vol14/Vol14_03/Vol14_03.pdf.

Van Overstraeten, M. (2003, December 30). La TV haute définition démarre jeudi. *La Libre Belgique.* Retrieved January 1, 2004 from http://www.lalibre.be/article_print.phtml?art_id=147803.

Wargo, S. (2004, January 20). Personal communication.

Wynn, C., & Wales, A. (2003, November 14). Europe looks to UK's Freeview model for DTT. *New Media Markets.* Retrieved March 21, 2004 from LexisNexis Academic.

Yankee Group. (2004). *2003 digital home entertainment survey confirms HDTV momentum.* Retrieved March 12, 2004 from http://www.yankeegroup.com/public/home/daily_viewpoint.jsp?ID=11241.

8

Streaming Media

Jeffrey Wilkinson, Ph.D.*

The technology of "*streaming*" media over the Internet is, at its simplest, the transmission of audio and video from one computer to another (McEvoy, 2001, p. 118). Early predictions were that streaming was the new broadcasting, and, in the late 1990s, hundreds of radio stations offered their signals online. Because the quality of the streamed content has always been dependent on bandwidth, live audio/radio has been possible since the mid-1990s, and continues to be an extremely popular application. Video content (because of its size) has been slower to develop. Compression technologies and widespread broadband Internet connections, however, have enabled greater use, and demand for streaming video is rising quickly. Research from Accustream iMedia Research (http://www.accustreamresearch.com), reported that demand for video streaming essentially doubled from 2002 to 2003, up 104% to 7.8 billion (Accustream, 2004). Research from Arbitron also confirms that broadband users consume significantly more streamed content than dial-up users, and demand for both audio and video continues to rise dramatically (Arbitron/Edison, 2003).

Streamed content is delivered either live or on-demand. Dial-up Internet connections were sufficient to sustain live music broadcasts, and it is not surprising that Arbitron research finds users see little difference between broadcast and Webcast clips (for listings of both types of live Webcasters, see, for example, http://radioshowlinks.com).

* Associate Professor, Communication Studies Department, Hong Kong Baptist University (Kowloon Tong, Hong Kong, China).

Background

Streaming technology was introduced in 1995 by Progressive Networks (now called RealNetworks). Their initial product, RealAudio, enabled people to listen to high-quality audio signals over the Internet through a computer (using 14.4 Kb/s and 28.8 Kb/s modems at the time). Streaming video was introduced in early 1997, and global interest took off. Microsoft began offering Windows Media Player shortly thereafter, and, in 1999, Apple launched its streaming version of QuickTime. Since then, all three have become common.

Types of Streaming

There are a number of variations on streaming. One distinction is between "true streaming" and "progressive download." Both are common, and both have their uses. True streaming typically sends audio or video content across a network where it is displayed in real time. Typically, a few seconds are needed to buffer the file before it is played back on the receiving computer using a suitable player. True streaming employs a special streaming server (such as Real's Helix 10 server, Apple's Darwin server, and Microsoft's Windows Media Server). The streaming server uses the appropriate protocols (such as RTSP [real-time streaming protocol] and RTP [real-time transport protocol]) so that the content is not downloaded onto a client hard-drive. This differs from what is called "progressive streaming" that uses a regular HTTP (hypertext transfer protocol). Web server. In this case, when the user clicks on the file, the material is actually downloaded onto the user's hard drive before playback. Given the size of video files, this is normally not the first choice for users, but it is the easiest (and maybe the only) way around firewalls.

The main advantages of true streaming are speed, (host) control, and flexibility. Streamed material is played back quickly—almost instantaneously. Control is maximized for the host, because the original content remains on the server, and access can be controlled using password, registration, or some other security feature. Finally, since streamed segments are placed individually on a host server, and can be updated or changed as needed, there is tremendous flexibility. The major advantages for progressive download are that playback quality may be higher (because it is now on the user's computer) and the user has a digital copy of the material (user control).

Another streaming distinction is between "live" versus "on demand." To offer on-demand streaming, a streaming server with enough capacity to hold the content that can be configured to enable a sufficient number of simultaneous users—whatever the need—must be used. To do live streaming, all of the above are needed, but with the addition of a dedicated computer (called a "broadcaster" or "encoder") to create the streams from the live inputs (such as a microphone or camera) by digitizing, compressing, and sending out the content to the streaming server as an RTP stream.

Architectures & Codecs

There are two primary tasks in streaming:

1) Synchronizing, managing, and playing media.

2) Making audio and video components small enough to play (on computer/through network).

The technologies developed to accomplish this are architectures and codecs. Architectures handle and synchronize digital files. The most common are familiar names: *QuickTime*, *Real*, and *Windows Media Player*. *MPEG* is both an architecture and a codec. Architectures are often called formats, but this is misleading because they are more than just formats. Some architectures work better on the Web, some for CD-ROM.

Codec is short for compression/decompression (or) coder/decoder. A codec will compress the audio or video file so it is small enough to be streamed through a network connection and be played back on a computer. The codec is the actual encoding/decoding algorithm; they are not always interchangeable or work cross-architecture. One example is *Sorenson Video,* which is a QT Web codec, and *Cinepak* is a cross-architecture codec often used for CD-ROM projects. Codecs are needed because, for example, one second of uncompressed NTSC video can take up to 27 MB of disk space.

Playback: Streaming Platforms

To play streamed material, the user has to install a compatible player on his or her computer. Several types of proprietary systems have come and gone, and many new ones continue to be launched. By far, the three most common platforms for providing and playing streamed content are Real (RealNetworks), Windows Media Player (Microsoft), and QuickTime (Apple). Another type, MPEG-4 players, is becoming increasingly popular, and Macromedia Flash is now getting into video streaming as well. Each has some technical differences, each has its merits, and all are regularly upgraded. As far as Real, WMP, and QT, each has strengths and weaknesses. Critical reviews and comparisons typically summarize by saying that it is a matter of taste, and recommend users install all three on their computer (see Table 8.1).

RealNetworks/RealOne. The early dominance of RealNetworks has been attributed to it having been the first on the scene, and having the foresight to give away the players rather than sell them. Since 1995, several generations of improvements have been made in its products. The overall pricing structure for Real services has remained stable, and the basic player is a free download (new models with tech support are offered for a price).

Real offers a wide variety of servers and other products, as streaming applications/needs expand. Real has allowed developers a free "starter" streaming server capable of allowing 25 concurrent Internet viewers. At this time, RealNetworks offers the Helix 10 streaming server and RealProducer 10 encoder to clients who wish to "offer any media format from any point of origin across any network transport, running any OS to any person on any Internet-enabled device anywhere in the world" (RealNetworks.com, 2004a).

Microsoft/Windows Media Player. Shortly after RealNetworks began streaming audio, Microsoft became interested and, in the late 1990s, the two enjoyed a technology-sharing relationship. After the partnership dissolved, Windows Media Player was offered as a free plug-in to users. Since then, improved versions of Windows Media Player have been regularly released, and WMP is now the most-installed player in the world. The platform has gained the upper hand by virtue of being a good player that was also self-compatible (operating system, browser, media player, and various tools).

Because streaming is just one of the many things Microsoft does, it is easy to get lost in the vast array of products and services. There is no extra cost for the streaming server because it is just one part of the larger whole. As you buy into the full range of services Microsoft provides, costs begin to accumulate. Once you have purchased a package, you can efficiently control content and Web transactions, including pay-per-view (e.g., pay-per-download or pay-per-stream), registration, subscription, and digital rights management (DRM).

Table 8.1
Comparison of Streaming Platforms

Player	Pros	Cons
Real	- Supports HD content - Broadcast support for portable devices - Linux support	- Intrusive messages and nag screens
Windows	- Supports HD content - Leading pay-for-player VOD - Media player can read DVDs, create custom music playlists, and burn CDs - Supports numerous codecs - Comes with Windows - Best at real-time streaming	- Comes with Windows - Default settings may override competing players or may be incompatible with others
QuickTime	- Supports HD content - Open standard, open architecture (compatible) Supports many codecs (even non-Apple) - Bundled with portable devices (cams)	- No standard DRM solution to prevent piracy - No full-length movies, must register for FS playback - Messages can nag (re: register) - Not very good at real-time streaming - Some problems with firewalls

Source: *USA Today* (2003)

Apple/QuickTime. In 1999, Apple began offering streaming with the QuickTime 4 platform. QuickTime has been quite popular because it plays on both PC and Macintosh computers and can be delivered from any kind of file or Web server. The QuickTime file format is an open standard and was adopted as part of the MPEG family of standards.

Apple has an open-source code approach where they allow vendors to configure the QuickTime format for their use, then send the changes back to Apple for incorporation into new versions of the server software. QuickTime has worked out many bugs in earlier versions and is highly regarded.

MPEG. MPEG is a standards-based format specifically designed to handle digital audio and video. Most of the available Internet video content is MPEG1. MPEG2 is DVD/broadcast-quality video that has been too demanding and unstable for popular streaming applications. More recently, there are more material and devices configured for MPEG4. MPEG4 combines all media objects (text, video, images, etc.) into a synchronized, interactive presentation, and it is expected to gain acceptance and use.

For Internet audio (and music in general), the most important and by far the most common format has been MP3. MP3 (MPEG1 audio layer 3) is part of the MPEG standard, and is a highly compressed format, which means the files are extremely small—typically 128 Kb/s for two-channel stereo (Halsall, 2001). Despite the small size, the quality is extremely high and the sound of MP3 files can be indistinguishable from uncompressed CD-audio. A wide variety of players handle MP3 files. (For more on MP3, see Chapter 17.)

Recent Developments

Technical Advancements

For over a decade, MP3 has been the industry standard for audio files. The ability to compress a 1.4 Mb/s audio stream into high-quality 128 Kb/s stream has made it the standard for digital music distribution. This will be changing. Now, as MPEG4 is emerging as a new standard, the MPEG4 AAC audio codec is rated as being superior to MP3. The AAC audio codec is the root of the MP4 file type, recently popularized by Apples iTunes, among others (Bouthillier, 2004). The result is even smaller files, but with higher audio quality.

A number of groups have endorsed MPEG4 as the technical standard for streaming media, and QuickTime, Real, and Windows Media all now offer some degree of MPEG4 compatibility. Still, there are some problems, and there remain a number of incompatibilities between devices. The Internet Streaming Media Association (ISMA) continues to work to bring about universal interoperability among products and services. One company specializing in MPEG4 products is Envivio (http://www.envivio.com). MPEG4 encoders and servers are being adopted by broadcasters, and in 2003, Indian broadcaster New Delhi Television Ltd., launched what is said to be the world's first digital satellite news gathering network based entirely on MPEG4 encoding (Envivio, 2003).

Applications

The Internet continues to expand and change, reflecting and anticipating the way people will use it. It has reached a level of maturity that suggests stability in both content and use (Pew Internet, 2002). Multi-year trends have now been compiled, and confirm that the Net both complements and supplements traditional media use. For example, as time spent viewing traditional TV goes down, time spent consuming Internet video viewing is going up.

Streaming Audience Measurement. Measures of the most popular streaming content are becoming more available, but Internet audience measurement is still in its infancy. Just as in the early days of broadcasting, several firms were initially involved in the measurement of Internet streaming audiences, but that number is dwindling, and familiar faces are taking top spots. Perhaps the best-known providers of research are Arbitron (http://www.arbitron.com) and Nielsen (http://www.nielsen-netratings.com/). Two other well-known producers of Internet audience research are comScore Media Metrix (http://www.comscore.com/metrix/gs.asp), and Jupiter Research (http://www.jupiterresearch.com/).

Significantly for Arbitron, in November 2002, they acquired a license to the streaming audio audience measurement system and related assets from MeasureCast. MeasureCast continues to develop new technologies for strategic partner Nielsen Media Research. All the firms mentioned here provide regular research reports and measurements of Internet radio, video, marketing, and advertising services.

In the realm of enterprise streaming, Keynote Systems (http://www.Keynote.com) is emerging as a notable provider of streaming measurement and diagnostics. Keynote acquired Streamcheck in July 2003 and is now focusing entirely on streaming media performance (Meserve, 2003). The Keynote System supports Microsoft Windows Media Player, RealNetwork's RealPlayer, and Apple's QuickTime, and allows customers to gauge the effectiveness of their streaming system with statistics such as initial connect time, buffer time, and whether any rebuffering occurs. The services begin at $250 per month, and diagnostic testing starts at $650 per URL (uniform resource locator) monitored up to 75 kilobits per second (Meserve, 2003).

Popular Types of Streamed Content

According to figures from Arbitron, in July 2003, the number of "people over age 12 who have ever tried Internet audio or video" was 108 million Americans (Arbitron/Edison, 2003). Further, an estimated 50 million Americans watched video online or listened to audio that month. The number of broadband home connections tripled between 2001 and 2003 (from 7% to 21% overall), and audio and video streaming demand has grown accordingly.

Video Streaming. Because of limitations on bandwidth and content, video streaming popularity has lagged behind audio/radio. According to Arbitron, from 2000 through 2002, an average of 8% of all Americans watched some form of Internet video, but that leapt to 12% in 2003. This has been attributed to broadband adoption, the war in Iraq, and brands such as ESPN introducing "compelling" video content to consumers. Both Arbitron and Accustream Research have identified that the most popular types of video content are film trailers and music videos (see Figure 8.1). Other popular content includes information (news and sporting events) and entertainment (short or full-length movies).

According to Accustream Research, most of the streaming video growth has been occurring on "few selected big aggregator platforms" such as Real Networks, ESPN, Yahoo!, and AOL. Broadband streams made up 78% of total video streams accessed in 2003 (Accustream Research, 2004).

Figure 8.1
Most Watched Online Video Programming

Source: Arbitron

Internet Radio. Radio stations and online music continue to grow in acceptance and popularity. The 2002 decision by CARP (Copyright Arbitration Royalty Panel) to establish royalty payment fee rates resulted in the decision by a number of stations to stop Webcasting (Arbitron/Edison, 2003; Miles, 2003). Still, in April 2003, there were reported to be more than 4,000 Internet radio stations and Internet-only stations worldwide, and more than 8,000 not-for-profit Webcasters (Miles, 2003). All types of formats are available and listings are easily found (see for example, the World Radio Network site at http://www.wrn.org). Online radio listening overall continues to trend upward dramatically (see Figure 8.2).

Figure 8.2
Listeners to Online Radio

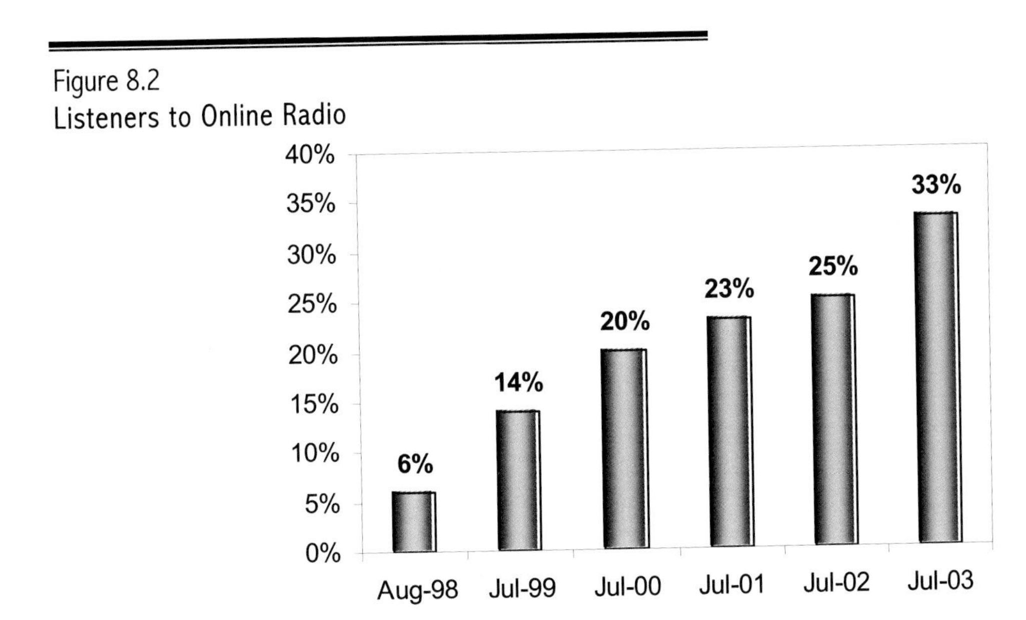

Source: Arbitron

According to Arbitron, the weekly Internet broadcast audience grew from 16 million in July 2000 to 30 million in July 2003. Internet-only broadcasters have also taken advantage of online music's popularity. In 2002, AOL introduced Radio@AOL, and Yahoo! strengthened its Internet broadcast capabilities under its LAUNCH brand. By mid-2003, Internet-only audio services and Internet radio listenership was nearly equal (Arbitron, 2004). However, the Internet music audience is dominated by Internet-only brands, as shown by Table 8.2

Table 8.2
Internet Broadcasters and Sales Networks Weekly Top 15—February 2-8, 2004

Rank	Company	Type	TTSL*	Cumulative Audience
1	AOL Radio@Network	Commercial	6,124,864	1,547,903
2	LAUNCH	Commercial	4,692,881	987,898
3	MUSICMATCH	Non-commercial	2,442,561	571,527
4	The Adsertion Network	sales network	724,086	107,673
5	Virgin Radio	Commercial	484,195	85,331
6	Educational Media Foundation	Non-commercial	409,712	56,124
7	AccuRadio	Commercial	210,047	75,707
8	KillerOldies.com	Commercial	134,847	26,082
9	KPLU	Non-commercial	105,148	14,561
10	Emap	Commercial	97,065	39,441
11	Mak Radio	Commercial	94,051	25,928
12	WOXY-FM	Commercial	93,459	16,273
13	WBUR	Non-commercial	91,411	25,879
14	WXPN-FM	Non-commercial	83,039	12,057
15	WKSU-FM	Non-commercial	78,884	20,802

* TTSL = Total Time Spent Listening

Source: http://www.arbitron.com/newsroom/home.htm

Non-Broadcast Streaming—Music. Perhaps the greatest competition to Internet broadcasters comes from the music industry. Non-broadcast streaming partnerships abound, allowing users to access vast music libraries for a fee. This has been a boon for the music industry, which had been wracked with losses due to piracy and illegal downloads. Offering a combination of "digital jukebox" with music file downloading has been advantageous to service and content providers alike.

RealNetworks has reported success with the Rhapsody music jukebox service. Rhapsody was purchased from Listen.com in 2003 for an estimated $36 million in cash and stock. In its 2003 annual report, RealNetworks said that more than 70% of their revenue was derived from consumer-delivered services (mostly videogames). Still, a sizable portion of cash came from the 350,000 music subscribers who pay under $10 per month for those memberships (RealNetworks, 2004b). The following month, after cutting the price of a song purchase to $.79, Real announced that subscriber use of Rhapsody rose by 10% (RealNetworks, 2004b).

Although Real has offered a monthly-fee jukebox since 1999, the current trend of offering both can be traced to the launch of Apple's iTunes music store in 2003. Apple announced the then-unheard-of price of $.99 for each song file. Although initially critical, the music industry took note when consumers downloaded one million files the first week (Olsen & Yamamoto, 2004). Other businesses have followed suit on the per-song download charge, and it has become the industry standard.

In March 2004, Virgin announced it would develop its own digital jukebox and online music store with MusicNet, to be available in August 2004 (C/Net News.com, 2004). The service, Virgin Digital, would follow the standard $.99 per file. The monthly charge had not been announced, but would reportedly be "hyper-competitive." The service would work on mobile phones, hand-held devices, and other consumer electronics gear. The downloaded files would be in WMA format, supported by Windows and Real, but not QuickTime. Later that month, Wal-Mart officially launched their own music download service, offering "88 cents, every song, every day." Using only the WMA format, this move by the world's largest retailer is seen as a "direct attack on Apple iTunes" (Swint, 2003). (For more on digital audio services, see Chapter 17.)

Education and Training. Businesses continue to employ streaming for education, training, and corporate communications. Research from the Yankee Group indicated that more than one-third of companies with more than 100 employees use streaming media for training and education, and over one-quarter use streaming video for corporate marketing, advertising, and branding programs (Miles, 2003). As businesses invest in developing in-house streaming operations, outsourcing is still available and competitive. In April 2003, MCI began offering such a video streaming service. Teaming with Yahoo Broadcast Solutions, MCI's Global Streaming Video service was aimed at enterprise streaming applications such as employee training, executive meetings, and live or on-demand events (Pappalardo, 2003). Customers could set up video-streaming sessions with up to 500 simultaneous users for a flat fee of $7,500 for three hours of content.

Public schools and universities also continue to embrace streaming. While a number of teachers and professors experiment with streaming lecture materials for individual classes, more formal structures have also been launched to make streaming an integral part of modern higher education. Many online course development software packages, such as WebCT, include the ability to incorporate streaming with other materials. For example, the Public Broadcasting System (PBS) now has launched *PBS Campus* (http://www.pbs.org/campus), a joint effort by PBS, local public television stations, and colleges nationwide (PBS joins, 2004). PBS provides the course materials (including text, audio, and video content), and the colleges provide the instructors, registration, and award the credit (courses are not available directly from PBS).

Current Status

As of March 2004, the latest versions of the big three are RealPlayer 10, Windows Media Player 9 series, and Apple QuickTime 6.5. Since all three platforms offer free versions, it is best to make sure all three are installed. Windows Media Player and Real are still the most popular in terms of numbers; the Apple Website notes that QuickTime 6 has been downloaded over 175 million times (Apple.com, 2003).

Competition, Conflict

The field of streaming continues to be highly competitive and contentious, especially between RealNetworks and Microsoft. In March 2004, RealNetworks filed a lawsuit against Major League Baseball claiming the league violated a contract that required it to use the streaming media company's software. MLB.com's subscription service offers live audio and video streaming, and had been encoding only in Microsoft's Windows Media format. The three-year contract was coming to a close, and MLB had begun approaching a number of Web portals for signing a new online distribution deal (Hu, 2004).

If having to choose among four formats is not enough (Real, WM, QT, MPEG), consumers can now consider a fifth. In March 2004, Macromedia (http://www.macromedia.com/) announced it is now aggressively pushing into streaming video. Macromedia's Flash Video Streaming Service will compete with Apple, Microsoft, and Real (and those using MPEG4). The service will stream Flash video using VitalStream's network in a way similar to Apple's use of the Akamai Technologies network to stream its QuickTime content (Dalrymple, 2003). This effort will make it easier for companies to use Flash video, but educating users will be a key barrier Macromedia must overcome.

Factors to Watch

Privacy

There seems to be a zero-sum conflict between privacy and standardization, and corporate concerns may ultimately win from one of two directions. Perhaps the more innocent initiative comes from those working on MPEG21. MPEG21 would solve a number of compatibility issues, but would also enable content distributors to have complete control over all aspects of content (Bouthillier, 2004). Ironically, even as it resists adopting MPEG as a standard, this is precisely what Microsoft is aggressively implementing through DRM (digital rights/restrictions management). Also called TC (trusted computing by proponents; treacherous computing by opponents), wielding control over the home environment has long been the goal of corporate content creators such as Disney and the music industry. In early 2004, Disney entered into a non-exclusive agreement to allow Microsoft to handle DRM issues for their content (Microsoft and Disney, 2004).

Longstanding problems with piracy make it no surprise that the music industry also favors greater control over content. Whichever effort becomes the standard, the consumer will probably lose more than win. According to Bouthillier (2004), the benefits of DRM are standardization and "no more hassles from incompatible devices." But the consequence is that control shifts, perhaps irrevocably, from the user to the content owner. This control would extend ownership all the way to the user's device, affecting the way the media is consumed.

Mobile Streaming

Streaming is also moving into the domain of wireless (Miles, 2003). Delivering content to mobile phones and portable devices is being aggressively pursued by businesses. RealNetworks has entered into agreements with major mobile market leaders (Nokia, Ericcson, Sprint, AT&T Wireless, Vodafone, and Motorola). They hope to grow digital media services on mobile devices. Microsoft has also entered into the mobile services realm. In January 2004, Microsoft announced their "Windows Media Connect" technology that aims to connect music, video, and photos from the PC to consumer electronics devices throughout the home (and linked to the Windows Media Center, also to link content throughout the home).

Real has launched a mobile wireless streaming system, and PacketVideo has begun testing a system in France, China, Korea, and Switzerland. In addition, Nextel announced a deal with XSVoice to allow live wireless audio. Billed as "better than streaming," Nextel's service brings content immediately to consumers through their wireless phones.

A new standard has been issued for high-speed wireless streaming by IEEE (Institute of Electrical and Electronics Engineers). The 802.15.3 standard enables home and small-businesses to connect audio and video devices to high-speed wireless networks at data rates of up to 55 Mb/s at distances of up to 300 feet (Blau, 2003).

Streaming Becomes Mainstream

Streaming requires content, but what that content is and who creates it are not as clear as we would like. Media convergence enables us to combine all forms of content (text, audio, video) and deliver it across the Web. The full impact of this convergence is still only hinted at, as the tools and skills of content creation are diffused throughout society.

Like computing, streaming will become ubiquitous. Now, even AOL Instant Messenger users can engage in live video exchanges (Perez, 2004). Version 5.5 of AIM allows live video streaming if the users have a Webcam, a microphone, and a PC running Windows XP. The feature is compatible with Apple's iChat AV 2.1 instant messaging service, so Mac and PC users can interact across platforms.

This development underscores the way convergence is changing what it means to be an Internet user. Recent figures noted that 44% of American Internet users have contributed their thoughts and digital content to the online world (Pew Internet, 2004). Admittedly the numbers are small—only 7% use Webcams, and only 3% have contributed video files to Websites. But these numbers translate to roughly nine million Webcams, and even more staggering—around four million people have placed video files on the Web. As the number of streaming content-creators increases, so also will the potential availability of free media-rich content.

Figure 8.3
Streaming Media Screen Shot

This screen shot includes the three most popular players: QuickTime 4.0, RealOne, and Windows Media Player (left to right). In addition to the standard look of each player, a user can modify the appearance of most players by applying a "skin" that changes the layout of the player, graphics, colors, etc. Technology does not yet, however, allow us to control the appearance of the people in the videos.

Source: J. S. Wilkinson

Bibliography

Accustream iMedia Research. (2004, February 18). *Accustream research data shows Internet video streams grew by 104% in '03 to over 7.8 billion.* Retrieved March 12, 2004 from http://www.accustreamresearch.com/news/feb1804.html.

Apple.com. (2003, December 18). *Apple expands 3GPP support with QuickTime 6.5.* Retrieved March 12, 2004 from http://www.apple.com/pr/library/2003/dec/18quicktime.html.

Arbitron. (2004, February 24). *AOL Radio@Network is the top Internet broadcaster according to Arbitron Internet broadcast ratings.* Retrieved February 25, 2004 from http://arbitron.com/newsroom/archive/WCR02_24_04.htm.

Arbitron/Edison Media Research. (2003, July). *Internet and multimedia 11: New media enters the mainstream.* Retrieved February 25, 2004 from http://www.arbitron.com/webcast_ratings.

Blau, J. (2003, August 8). IEEE issues new spec for high-speed wireless streaming. *Computerworld.* Retrieved March 12, 2004 from http://www.computerworld.com/printthis/2003/0,4814,83837,00.html.

Bouthillier, L. (2004, February 18). *The MPEG video standards—From 1 to 21.* Retrieved March 1, 2004 from http://www.streamingmedia.com/.

C/Net News.com. (2004, March 7). Virgin to launch music jukebox, Net music store. *C/Net News.* Retrieved March 10, 2004 from http://www.news.com.com/2100-1027-5171229.html.

Comparing the Big 3 players. (2003, November 17). *USA Today.* Retrieved February 6, 2004 from http://www.usatoday.com/tech/news/techinnovations/2003-11-17-players-side_x.htm.

Dalrymple, J. (2003, December 16). Flash streaming video hits the Web. *Macworld.* Retrieved February 12, 2004 from http://maccentral.macworld.com/news/2003/12/15/flashvideo/.

Envivio.com. (2003, October 29). *NDTV uses Envivio 4Caster encoders to launch the world's first MPEG4 digital satellite news gathering network.* Retrieved March 12, 2004 from http://www.envivio.com/news/news/031029_ndtv.html.

Halsall, F. (2001). *Multimedia communications: Applications, networks, protocols and standards.* Harlow, England: Addison-Wesley.

Hu, J. (2004, March 9). Real hits Major League Baseball with lawsuit. *C/Net News.* Retrieved March 10, 2004 from http://news.com.com/2102-1023_3-5171852.html.

McEvoy, S. (2001). *Microsoft Windows Media Player for Windows XP handbook.* Redmond WA: Microsoft Press.

Meserve, J. (2003, October 14). Keynote upgrades streaming media monitoring service. *Network World Fusion.* Retrieved March 12, 2004 from http://www.nwfusion.com/news/2003/1014keystream.html.

Microsoft and Disney announce multiyear agreement to cooperate on digital media initiatives and for Disney to license Windows Media digital rights management software. (2004, February 9).*Microsoft.com.* Retrieved March, 3, 2004 from http://www.microsoft.com/presspass/press/2004/Feb0402.

Miles, P. (2003, April). The state of the industry. *Streaming Magazine.* 12, 14-15.

Olsen, S., & Yamamoto, M. (2004, March 4). Real's reality: Microsoft just one factor in Net pioneer's chaotic history. *C/Net News.* Retrieved March 9, 2004 from http://news.com.com/2009-1023_3-5168648.html.

Pappalardo, D. (2003, April 22). MCI rolls out video streaming. *Network World Fusion.* Retrieved March 9, 2004 from http://www.nwfusion.com/edge/news/2003/0422yahoomci.html.

PBS joins with five higher education institutions to offer "PBS Campus." (2004, March 10). *Streaming Magazine.* Retrieved March 10, 2004 from http://www.streamingmagaziine.com/.

Perez, J. C. (2004, February 5). AOL launches AIM 5.5 with streaming video. *The Industry Standard.* Retrieved March 12, 2004 from http://thestandard.com/article.php?story=2004020521478888.

Pew Internet & American Life Project. (2002, March 3). *Getting serious online: As Americans gain experience, they use the Web more at work, write e-mails with more significant content, perform more online transactions, and pursue more serious activities.* Retrieved March 3, 2002 from http://www.pewinternet.org/releases/index.asp.

Pew Internet & American Life Project. (2004, February 29). *Content creation online: 44% of American Internet users have contributed their thoughts and digital content to the online world.* Retrieved March 3, 2004 from http://www.pewinternet.org/reports/toc.asp?Report=113.

RealNetworks.com. (2004a, January 29). *RealNetworks reports fourth quarter and 2003 results.* Retrieved March 3, 2004 from http://www.realnetworks.com/company/press/releases/2004/q403results.html.

RealNetworks.com (2004b, February 3). *RealNetworks Rhapsody usage jumps 10% in January.* Retrieved February 20, 2004 from http://www.realnetworks.com/company/press/releases/2004/rhapsody_usage.html.

Swint, K. (2003, December 18). Wal-mart.com faces the music. *Wired.* Retrieved April 14, 2004 from http://www.wired.com/news/digiwood/0,1412,61659,00.html.

9

Radio Broadcasting

Gregory Pitts, Ph.D.*

> The NRSC therefore recommends that the iBiquity FM IBOC [DAB] system ... should be authorized by the FCC as an enhancement to FM broadcasting in the United States, charting the course for an efficient transition to digital broadcasting with minimal impact on existing FM reception and no new spectrum requirements (NRSC, 2001, p. 9).

> We ... want to enable terrestrial radio broadcasters to better compete with satellite radio services now in operation. As such, we seek comment on what changes in our rules would likely encourage radio stations to convert to a hybrid or an all-digital format (FCC, 2004c, April 20, p. 8).

The word "radio" fails to deliver an image of sexy technology capable of competing with the latest digital communications innovations. Radio is, however, a big part of the lives of most people. Each week, the more than 13,500 FM and AM radio stations in the United States are heard by more than 223 million people, and advertisers spent $19.4 *billion* on radio advertising in 2002 (FCC, 2004b; RAB, 2003). Deregulation has increased the number of radio stations a company or individual may own, pushing operators to own clusters of stations in individual cities, or to own hundreds or even a thousand stations around the United States. Radio may not be viewed as sexy, but the changes envisioned by the Federal Communications Commission (FCC) and endorsed by the National Association of Broadcasters (NAB) to firmly establish digital audio broadcasting may reinvent this sleepy medium.

The next step in radio is to re-invent the technology to enable stodgy analog radio transmissions to compete in the digital present and future. Ownership groups have an economic incentive to pursue

* Associate Professor, Department of Communication, Bradley University (Peoria, Illinois).

new technology. Whether consumers will view these technological offerings as worthy of adoption when they already possess the ability to download digital music to various portable music players will determine the success of new radio technology.

This chapter examines the factors that have redirected the technological path of radio broadcasting. Among these are radio station ownership consolidation, competition from two digital satellite radio services, and, most importantly, technological improvements to promote the implementation of digital terrestrial audio broadcasting capable of delivering near CD-quality audio and a variety of new data services, from song/artist identification to local traffic and weather to subscription services yet to be imagined.

Background

The history of radio is rooted in the earliest wired communications—the telegraph and the telephone—although no single person can be credited with inventing radio. Most of radio's "inventors" refined an idea put forth by someone else (Lewis, 1991). Although the technology may seem mundane today, until radio was invented, it was impossible to simultaneously transmit entertainment or information to millions of people. The radio experimenters of 1900 or 1910 were as enthused about their technology as are the employees of the latest tech startup. Today, the Internet allows us to travel around the world without leaving our seats. For the listener in the 1920s, 1930s, or 1940s, radio was the only way to hear live reports from around the world.

Probably the most widely known radio inventor-innovator was Italian Guglielmo Marconi, who recognized its commercial value and improved the operation of early wireless equipment. The one person who made the most lasting contributions to radio and electronics technology was Edwin Howard Armstrong. Armstrong discovered regeneration, the principle behind signal amplification. He invented the superheterodyne tuner, leading to a high-performance receiver that could be sold at a moderate price, thus increasing home penetration of radios. In 1933, Armstrong was awarded five patents for frequency modulation (FM) technology (Albarran & Pitts, 2000).

The two traditional radio transmission technologies are amplitude modulation (AM) and frequency modulation. AM varies (modulates) signal strength (amplitude) and FM varies the frequency of the signal.

The oldest commercial radio station began broadcasting in AM in 1920. Although AM technology had the advantage of being able to broadcast over a wide coverage area (an important factor when the number of licensed stations numbered just a few dozen), the AM signal was of low fidelity and subject to electrical interference. FM, which provides superior sound, is of limited range. Commercial FM took nearly 20 years from the first Armstrong patents in the 1930s to begin significant service, and did not reach listener parity with AM until 1978 when FM listenership finally exceeded AM's.

FM radio's technological add-on of stereo broadcasting, authorized by the FCC in 1961, along with an end to program simulcasting (airing the same program on both AM and FM stations) in 1964, expanded FM listenership (Sterling & Kittross, 1990). Other attempts, such as Quad-FM (quadraphonic sound), ended with disappointing results. AM stereo, touted in the early 1980s as the savior in AM's competitive battle with FM, languished for lack of a technical standard because the marketplace and government failed to adopt an AM stereo system (FCC, n.d.-c; Huff, 1992).

Table 9.1
Radio in the United States at a Glance

Households with Radios	99%
Average Number of Radios per Household	5.6
Number of radios in U.S. homes, autos, commercial vehicles and commercial establishments	800 million
	Source: U.S. Bureau of the Census (2003) and FCC (2004b)
Radio Station Totals	
AM Stations	4,794
FM Commercial Stations	6,217
FM Educational Stations	2,552
Total	13,563
FM Translators and Boosters	3,383
LPFM Stations	237
	Source: FCC (2004b)
Radio Audiences	
Persons Age 12 and Older Reached by Radio:	
Each week:	94.1% (About 223 million people)
Each day:	75% (About 178 million people)
Persons Age 12 and Older Time Spent Listening to the Radio:	
Each week:	20 hours
Each weekday:	3 hours
Each weekend:	5 hours
Where Persons Age 12 and Older Listen to the Radio:	
At home:	36.5% of their listening time
In car:	45.5% of their listening time
At work or other places:	18.0% of their listening time
Daily Share of Time Spent With Various Media:	
Radio	44%
TV/Cable	41%
Newspapers	10%
Magazines	5%
	Source: Radio Advertising Bureau (2003)
Satellite Subscribers	
XM Satellite Radio	1,680,000
Sirius Satellite Radio	351,663
	Source: XM Satellite Radio (2004b) and Sirius Satellite Radio (2004a)

Why have technological improvements in radio been slow in coming? One obvious answer is that the marketplace did not want the improvements. Station owners were unwilling to invest in changes; instead, they shifted music programming from the AM band to the FM band. AM attracted listeners by becoming the home of low-cost talk programming. Listeners were satisfied with the commercially supported and noncommercial radio programming offered by AM and FM stations. Consumers, either wanting something new or tiring of so many radio commercials, first bought tape players, then CD players, and today MP3 players to provide improved audio quality and music choice. Government regulators, primarily the FCC, were unable to support and institute new radio technologies. The consumer electronics industry focused on other technological opportunities, including video recording and computer technology.

The Changing Radio Marketplace

The Federal Communications Commission (FCC) elimination of ownership caps mandated by the Telecommunications Act of 1996 fueled many of the changes that have taken place in radio broadcasting in the last decade. Before the ownership limits were eliminated, there were few incentives for broadcasters, equipment manufacturers, or consumer electronics manufacturers to upgrade the technology. Analog radio, within the technical limits of a system developed more than 80 years ago, worked just fine. Station owners did not have the market force to push technological initiatives. (At one time, station owners were limited to seven stations of each service. Later, it was increased to 18 stations of each service, before further deregulation completely removed ownership limits.) The fractured station ownership system ensured station owner opposition to FCC initiatives and manufacturer efforts to pursue technological innovation.

Ownership consolidation, along with station automation and networking, reflect new management and operational philosophies that have enabled radio owners to establish station groups consisting of 100 or more stations. The behemoth of the radio industry is Clear Channel Communications. The San Antonio-based company owns more than 1,200 radio stations that reach 54% of all U.S. residents ages 18 to 49 on a daily basis in more than 300 markets (Clear Channel, n.d.). Cumulus Media, the second largest station owner, owns a comparatively small 266 stations in 56 mid-sized markets (Cumulus Media, n.d.)

Recent Developments

There are five areas where technology is affecting radio broadcasting:

1) Enhancements to improve present-day, on-air transmissions.

2) Delivery competition from satellite digital audio radio services (SDARS).

3) New digital audio broadcasting transmission modes that are compatible with existing AM and FM radio.

4) New voices for communities: low-power FM service.

5) New technologies offer substitutes for radio.

Enhancements—Stations Install Digital Audio Equipment

Most radio stations have upgraded their on-air and production capabilities to meet the digital standards of computer-based audio systems. Virtually all portions of the audio chain are digital or are capable of handling a digital signal. Music and commercial playback is digital through compact discs, computer hard-drive systems, MiniDiscs, or other digital media. Most new audio consoles are digitally capable. The program signal delivered to a station's transmitter is digitally processed and travels to a digital exciter in the station's transmitter where the audio is added to the carrier wave. The only part of the process that is still analog is the final transmission of the over-the-air FM or AM signal.

Computer systems have also allowed stations to create "walk-away" operational technology, where live announcers and engineers are not needed at the station facility. Announcer comments, along with music and commercials, are stored on a computer hard drive system and played back. Some radio owners, including Clear Channel, have even used the computer system and computer networks to create virtual radio stations where announcers are neither local nor live (Mathews, 2002). The system, called voice-tracking, lets an announcer prepare an on-air show for listeners in another city. The announcer's comments, complete with locale-specific commentary, are sent through a network connection to the receiving station's on-air computer.

Competition from SDARS

The single biggest competitive challenge for free, over-the-air radio broadcasting in the United States has been the introduction of competing subscriber-based satellite radio service, a form of out-of-band digital "radio." The service was launched in the United States in 2001 by XM Satellite Radio, followed by Sirius Satellite Radio in 2002. The service was authorized by the FCC in 1995 and, strictly speaking, is not a radio service. Rather than delivering programming through terrestrial (land-based) transmission systems, each service utilizes of pair of geosynchronous satellites to deliver its programming (see Figure 9.1). Although listener reception is over-the-air and electromagnetic spectrum is utilized, the service is national instead of local, it requires users to pay a monthly subscription fee of between $10 and $13, and it requires a proprietary receiver to decode the transmissions (Sirius homepage, n.d.; XM homepage, n.d.). XM Satellite Radio and Sirius Satellite Radio, both publicly traded companies, offer about 100 channels of music and talk programming.

A key attraction to the services is that the music programming is supplied commercial-free, providing content not available from existing radio stations including show tunes, blues, folk, bluegrass, a variety of narrowcast rock and country channels, American standards, and commercial supported content from sources such as Fox News, BBC, CNBC, CNN, Discovery, Weather Channel, and ESPN (Sirius homepage, n.d.; XM homepage, n.d.).

The initial question, of course, was whether consumers would pay for an audio product they have traditionally received for free. While both companies continue to lose money as of mid-2004, subscriber growth has been strong. XM, by the end of the first quarter of 2004, had more than 1.68 million subscribers; Sirius had more than 351,663 subscribers (XM, 2004b; Sirius, 2004a). Each service needs between 3.5 million and 4.3 million subscribers to become profitable by 2004 or 2005 (Elstein, 2002; Stimson, 2002; O'Dell, 2004). While subscriber growth has been strong, startup losses for each company have been significant. XM reported a loss in 2003 of $584.5 million, and Sirius reported a $226.2 million loss (O'Dell, 2004). Given the development costs of each service, satellite radio technology faces an uncertain future. The cost to attract and add each new subscriber remains high; XM calculates its cost per gross addition (including subscriber acquisition costs, advertising, and marketing expenses) at $125 per subscriber. This essentially means that the company generates no positive revenue from a subscriber's first-year customer subscription fees. If the subscriber discontinues the service (referred to as listener churn) at that point, the company makes no money.

Figure 9.1
Satellite Radio

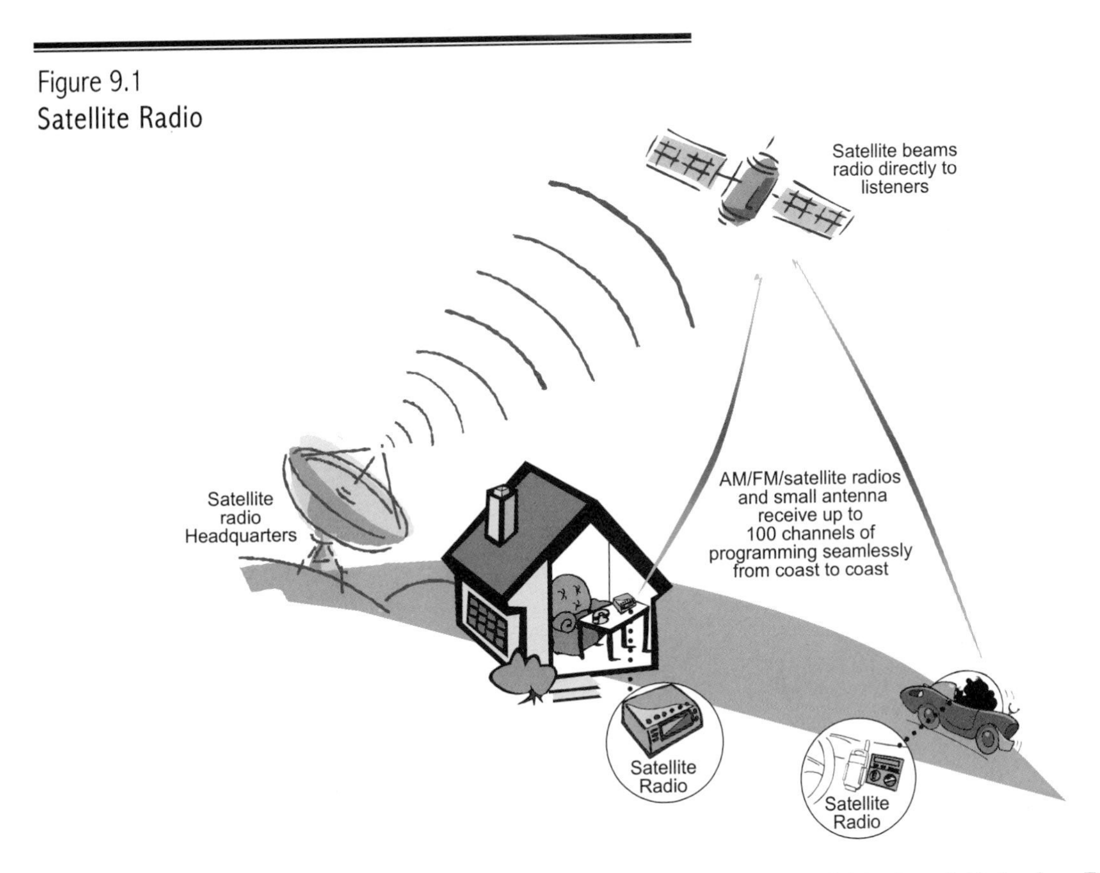

Source: J. Meadows & Technology Futures, Inc.

Helping the growth of both companies has been an array of savvy partnerships and investments. Both companies have alliances with various automobile manufacturers. General Motors and Clear Channel are investors in XM Satellite Radio. DaimlerChrysler is an investor in Sirius. Sirius and XM have a technology-sharing agreement that allows for production of receivers that work with either service (XM, 2000). Both receiver manufacturers and major electronics retailers offer automobile, home, and portable receivers; Sirius has recently signed a deal to sell receivers at Wal-Mart stores, the world's largest retailer (Sirius, 2004c). Programming rights, such as exclusive sports deals with the NFL and NHL, provide branded content and marketing opportunities. Both services also offer market-specific traffic and weather information in about 20 cities to further enhance the appeal of the service and compete head-to-head with local radio stations. Blunders by radio station group owners have encouraged a curious public to investigate the services, as public concern increases regarding radio station playlists that restrict music diversity (O'Dell, 2004).

New Digital Audio Broadcasting Transmission Modes

Free, over-the-air radio broadcasting is responding to satellite radio with terrestrial digital audio broadcasting for existing FM and AM stations. The digital signal will eliminate many of the external environmental factors that often degrade a conventional FM or AM station's signal (Morgan, 2002). The new system will not require any new spectrum space, as stations will continue to broadcast on the existing analog channel but will also use a new digital system to broadcast on the same frequency. In December 2001, the National Radio System Committee (NRSC) submitted a report to the FCC recommending adoption of a digital system developed by iBiquity Digital (NRSC, 2001). The FCC approved the new system a few months later. This led the commission to apply a new label to FM and AM radio technology. Today, the FCC speaks of establishing a terrestrial digital audio broadcasting service (FCC, 2004c).

As illustrated in Figure 9.2, this digital audio broadcasting (DAB) system uses a hybrid in-band, on-channel (IBOC) system that allows simultaneous broadcast of analog and digital signals by existing FM stations through the use of compression technology, without disrupting the existing analog coverage. The FM IBOC system is capable of delivering near-CD-quality audio and new data services including song titles, traffic, and weather bulletins. A similar system for AM stations has also been approved for daytime use, although there are concerns about IBOC's impact on nighttime AM signals. The AM IBOC will provide FM stereo quality signals from AM stations.

Digital audio broadcasting is more than just a new broadcast technology—it is a new means of delivering a variety of forms of broadcast content and requires a new set of programming standards to accompany the new technology. On April 15, 2004, the FCC launched a Further Notice of Proposed Rulemaking and Notice of Inquiry (MM Docket 99-325) to seek comments on what rule changes would be necessary to "... foster the development of a vibrant terrestrial digital radio service ..." and thus encourage stations to convert to IBOC while, at the same time, ensuring some measure of continued free over-the-air broadcasting (FCC, 2004c).

DAB will also allow wireless data transmission similar to the radio broadcast data system (RBDS or RDS) technology that allows analog FM stations to send traffic and weather information, and programming and promotional material from the station for delivery to smart receivers, telephones, or personal digital assistants (PDAs). Data streaming or datacasting could become a significant second revenue stream for radio broadcasters.

iBiquity Digital was formed in August 2000 by the merger of the two leading IBOC companies, USA Digital Radio and Lucent Digital Radio (iBiquity, n.d.-c). iBiquity's success reflects the changing radio industry; Clear Channel, Viacom/Infinity Radio, Disney/ABC, Susquehanna Radio, Cox Radio, and Hispanic Broadcasting—some of the largest radio groups—are investors in the company and support the conversion to IBOC digital. iBiquity Digital's IBOC technology consists of audio compression technology called Perceptual Audio Coder (PAC) that allows the analog and digital content to be combined on existing radio bands, and digital broadcast technology that allows transmission of music and text while reducing the noise and static associated with current reception.

The establishment of terrestrial digital audio broadcasting involves not only a regulatory procedure by the FCC, but also the marketing of the technology to radio station owners, broadcast equipment manufacturers, consumer and automotive electronics manufacturers and retailers, and most important, to the public. iBiquity Digital markets the new technology to consumers as HD Radio, a static-free service without hiss, fading, or pops, and available without a monthly subscription fee (iBiquity, n.d.-a). According to iBiquity, the cost for a station to implement hybrid IBOC broadcasts is about $75,000 (FCC, 2004c). As of mid-2004, 100 stations broadcast full-time HD Radio signals, and over 300 stations had licensed the IBOC digital technology (iBiquity, 2004b). The Corporation for Public Broadcasting has designated more than $5 million for 76 public radio stations to fund HD Radio broadcasts (iBiquity, 2004a). As with satellite radio's earliest efforts to attract subscribers, receiver availability is a consumer barrier. Furthermore, receivers are expensive; as of mid-2004, an HD Radio tuner built by Kenwood costs about $400, and a receiver/tuner from Panasonic retails for about $1,000 (Crutchfield.com, n.d.-a; Crutchfield.com, n.d.-b).

Figure 9.2
Hybrid and All-Digital AM & FM IBOC Modes

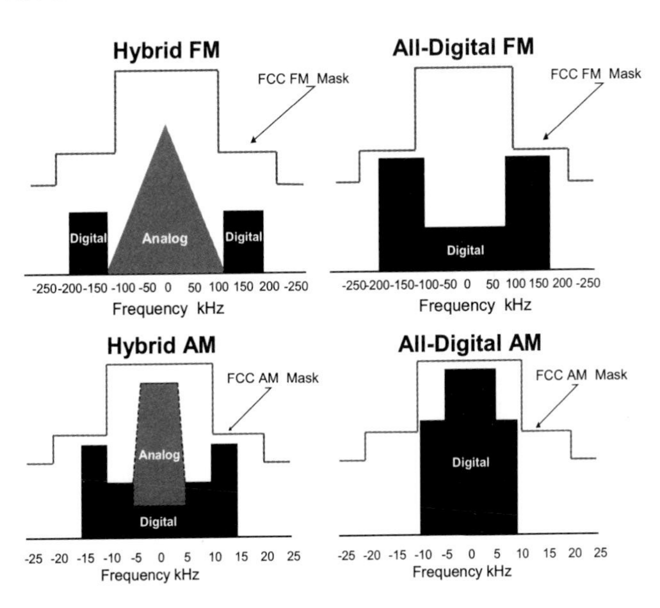

Source: iBiquity

The FCC estimates there are nearly 800 million radios in use in U.S. households, private and commercial vehicles, and commercial establishments (FCC, 2004c). All of these radios will continue to receive signals from radio stations even after the stations convert to HD Radio broadcasts. An important marketing challenge for HD Radio will be to inform listeners about the new service, while not creating the false impression that listeners receiving a simulcast signal from an analog FM station, mistakenly think they are listening to an HD Radio broadcast.

Unlike digital television, there has been no discussion as to when or if stations should ever be required to cease their analog broadcasts in favor of DAB-only transmissions. Because the DAB system uses no new spectrum, it is highly unlikely that broadcasters will be compelled to end analog broadcasting. Thus, there are questions as to whether consumers will want the new service, given the potential expense of new receivers and the availability of other technologies including subscriber-based satellite delivered audio services, MP3 audio transfer and playback, and competition from digital television. The availability of reasonably priced, factory-installed DAB automobile receivers will be an important determinant of the success of the iBiquity system. Another factor will be radio stations' ability to offer content of interest to listeners—often, this means local content (news, sports, and weather) that cannot readily be provided by national audio services and certainly not from recorded music. To that end, the FCC has launched a Localism Task Force to examine and promote broadcaster commitments to local service (FCC, n.d.-b).

A compelling motive for broadcasters to adopt DAB technology is the ability to engage in multi-casting by scaling the digital portion of the hybrid FM broadcast. IBOC provides for a 96 Kb/s (kilo-bits per second) digital data rate, but this can be scaled to 84 Kb/s or 64 Kb/s to allow 12 Kb/s or 32 Kb/s for other services, including non-broadcast services such as subscription services. Creating a broadcast marketplace demand for auxiliary services becomes a challenge for radio broadcasters. Previous efforts to excite consumers about radio data options have failed. A 1999 Consumer Electronics Association survey of 603 U.S. radio stations concluded that as many as 20% of all FM stations had the technological capability to broadcast in RDS, but stations cited "a lack of consumer demand" as the primary reason for not transmitting an RDS signal (CEMA, 1999).

A different form of DAB service is already in operation in a number of countries outside the United States. The Eureka 147 system broadcasts digital signals on the L-band (1452-1492 MHz), or a part of the spectrum known as Band III (around 221 MHz), and is in operation or experimental testing in Canada, the United Kingdom, Sweden, Germany, France, and about 40 other countries. Because of differences in the Eureka system's technological approach, it is not designed to work with existing AM and FM frequencies. Broadcasters in the United States rejected the Eureka 147 system in favor of a "backward and forward" compatible digital technology that would allow listeners to receive analog signals without having to purchase a new receiver for the DAB system (FCC, 2004c).

The World DAB Forum (www.worlddab.org), an international, non-government organization to promote the Eureka 147 DAB system, reports that more than 300 million people can receive DAB signals and that there are more than 600 different DAB services available for listeners (World DAB, n.d.-b). The Eureka numbers seem impressive until put into perspective: 300 million people can potentially receive the signal, but only if they have purchased one of the required receivers. Eureka 147 receivers have been on the market since the summer of 1998; about 100 models of commercial receivers are currently available and range in price from around $150 to several thousand (World DAB, n.d.-a).

Germany has been identified as a key country for the success of DAB. The German market contains more than 80 million people, 38 million households, and 42 million cars. About 150 stations provide coverage, reaching 80% of the population. About 300 local, regional, and national stations provide DAB coverage to about 80% of the population of the United Kingdom. Canada is another important DAB market: 73 stations broadcast digital signals in Toronto, Vancouver, Montreal, Ottawa, and Windsor, providing service to 11 million people or 35% of the country's population. Eureka DAB in Canada will face eventual competition from the U.S. IBOC system. Signal spillover from the United States, a flood of receivers from the U.S. marketplace, and the sheer size of the U.S. market could limit the success of Eureka DAB in Canada. The Canadian Radio-Television and Telecommunications Commission is set to review additional digital audio proposals from U.S.-based

SDARS providers XM Satellite Radio and Sirius Satellite Radio, and a terrestrial system proposed by Canadian broadcaster CHUM, Ltd. (Goldman, 2004).

New Voices for Communities: Low-Power FM Service

The FCC approved the creation of a controversial new classification of noncommercial FM station on January 20, 2000 (Chen, 2000). LPFM, or low-power FM, service limits stations to a power level of either 100 watts or 10 watts (FCC, n.d.-a). Although the service range of a 100-watt LPFM stations is about a 3.5-mile radius, full-power commercial and noncommercial stations feared interference. A little more than a year after approving the service, and before any stations were licensed, the commission acquiesced to Congressional pressure on behalf of the broadcast industry and revised the LPFM order. To prevent encroachment on existing stations' signals, Congress slipped the Radio Broadcasting Preservation Act of 2000 into a broad spending bill, which was reluctantly signed by President Clinton (McConnell, 2001; Stavitsky, et al., 2001).

The congressionally-mandated revision required LPFM stations to provide a third adjacent channel separation/protection for existing stations. Practically speaking, this meant that a currently licensed station operating on 95.5 MHz would not have an LPFM competitor on a frequency any closer than 94.7 MHz or 96.3 MHz. This revision, though minor in appearance, either killed or severely reduced the opportunity for service in most major metropolitan areas (McConnell, 2001). By February 2004, *The MITRE Study and Report*, commissioned by the FCC, found no substantive interference to stations at the third adjacent channel (FCC, 2004a). The FCC forwarded the report to Congress with a recommendation to eliminate the existing third adjacent channel separations.

As of April 2004, 237 LPFM stations have been licensed by the FCC and an additional 683 construction permits had been granted from the more than 3,100 applications reviewed (FCC, n.d.-b). If Congress were to accept the FCC's third adjacent channel recommendation, a second wave of station applications could be filed as new frequencies became available. Additional civic groups, ranging from church groups to community organizations to schools, could file applications for service. As part of the FCC's proposed examination of DAB, the commission has also requested comments regarding future DAB obligations by LPFM stations.

New Competition—Internet Radio

Internet radio, audio streaming, and legal music downloading offer consumers alternatives to over-the-air broadcasting. According to Arbitron, 20 million Americans aged 12 and older listen to Webcasts at least once a week, and about 39 million listen at least once a month (Tedeschi, 2004). Much of Internet radio's growth, both streaming of radio station audio and custom content, came to a screeching halt in 2002 when a panel established by the U.S. Copyright Office proposed a copyright payment plan deemed excessive by Internet audio providers (Harmon, 2002). A 2004 royalty agreement sets a rate of $0.117 cents per aggregate hour tuned in for free, advertising-supported services (Zeidler, 2004). Streaming media is discussed in more detail in Chapter 8.

New Competition—MP3s

Downloading music, while still viewed suspiciously by the recording industry, has become mainstream. In the United States, the sale of portable MP3 players nearly doubled in 2003 to 3.5 million units (McMullen, 2004). Apple's iPod remains the favorite player for many consumers; the iPod Mini weighs just 3.6 ounces and holds 1,000 songs (Garrity, 2004). (For more on digital audio, see Chapter 17.) The opportunity to control music listening options and to catalog those options in digital form as MP3 files presents not just a technological threat to radio listening but also a lifestyle threat of greater magnitude than listening to tapes or CDs. Listeners have at their fingertips thousands of songs that can be programmed for replay according to listener mood, and the playback will always be commercial-free and in high fidelity. As MP3 playback technology migrates from portable players to automotive units, the threat to radio will only increase.

Cellular telephones are seen as the leading delivery vehicle in the future for mobile entertainment (Borzo, 2002). Consumers in Europe, Japan, and Australia already use cell phones to receive music files, video clips, and video games. All of these offerings take consumers away from radio. When cell phone manufacturers and service providers to the U.S. market begin to pursue music downloading as a revenue stream, cell phones equipped with MP3 players could become pervasive here as well. The iBiquity DAB system might even become a means for radio stations to compete with themselves as DAB datacasting streams MP3 audio to cell phones.

Factors to Watch

Radio stations have been in the business of delivering music and information to listeners for more than 80 years. Listenership, more than any particular technological aspect, has enabled radio to succeed. Stations have been able to sell advertising time based on the number and perceived value of their audience to advertising clients. Technology, when utilized by radio stations, focused on improving the sound of the existing AM or FM signal or reducing operating costs.

DAB technology has the greatest potential to shift the entire radio industry and consumers into a new mode of operation. Digital broadcasting offers the possibility of streaming data content that might hold added value for consumers. Satellite audio itself holds the promise to create multiple revenue streams—the sale of the audio content, sale of commercial content on some programming channels, and possibly delivery of other forms of data. Regulatory barriers for these new technologies are not the issue. Consumer interest in the technologies, perfecting the technology so that it is as easy to use as traditional radio broadcasting has always been, marketing receivers at affordable prices, and delivering content that offers value will determine the success of these new radio technologies.

Consumer ability to easily store and transfer digital audio files to and from a variety of small personal players that have a pricing advantage over soon-to-be-introduced DAB receivers will be a determining factor in the success of DAB. Listeners desiring only entertainment will find little compelling advantage to purchasing a DAB receiver. Localism, the ability of stations to market not only near-CD quality audio content but also valuable local information will be the content that will retain listeners. The importance of local content, though, begs the question of whether listeners will desire a new, expensive receiver when existing stereo radios will continue to work.

Bibliography

Albarran, A., & Pitts, G. (2000). *The radio broadcasting industry*. Boston: Allyn and Bacon.

Borzo, J. (March 5, 2002). Phone fun. *The Wall Street Journal*, R8.

Chen, K. (2000, January 17). FCC is set to open airwaves to low-power radio. *The Wall Street Journal*, B12.

Clear Channel Communications. (n.d.). *Clear Channel radio fact sheet*. Retrieved April 25, 2004 from http://www.clearchannel.com/fact_sheets/radio_factsheet.pdf.

Consumer Electronics Manufacturers Association. (1999, February 4). *CEMA survey shows growing acceptance for radio data system among broadcasters*. Retrieved March 11, 2002 from http://www.ce.org/newsroom/newsloader.asp?newsfile=5139.

Crutchfield.com. (n.d.-a). *Kenwood KTC-HR100*. Retrieved April 12, 2004, from http://www.crutchfield.com/S-3J8kVQ1rvIf/cgi-bin/prodview.asp?s=O&cc=01&i=113KTCHR10&g=186650.

Crutchfield.com. (n.d.-b). *Panasonic CQ-CB9900U*. Retrieved April 12, 2004, from http://www.crutchfield.com/S-3J8kVQ1rvIf/cgi-bin/prodview.asp?s=O&cc=01&i=113CB9900&g=186850.

Cumulus Media. (n.d.). *Company profile*. Retrieved April 26, 2004 from http://www.corporate-ir.net/ireye/ir_site.zhtml?ticker=CMLS&script=2100.

Elstein, A. (2002, March 20). XM Satellite Radio dismisses concerns of its auditor about long-term viability. *The Wall Street Journal*, B13.

Federal Communications Commission. (n.d.-a). *Low-power FM broadcast radio stations*. Retrieved March 23, 2004 from http://www.fcc.gov/mb/audio/lpfm.

Federal Communications Commission. (n.d.-b). *Broadcasting and localism*. Retrieved March 23, 2004 from http://www.fcc.gov/localism.

Federal Communications Commission. (n.d.-c). *AM stereo broadcasting*. Retrieved March 8, 2002 from http://www.fcc.gov/mmb/asd/bickel/amstereo.html.

Federal Communications Commission. (2004a). *Report to the Congress on the low power FM interference testing program*. Pub. L. No. 106-553. Retrieved April 20, 2004 from http://hraunfoss.fcc.gov/edocs_public/attachmatch/DOC-244128A1.pdf.

Federal Communications Commission. (2004b). *Broadcast station totals as of December 31, 2003*. Retrieved April 26, 2004 from http://www.fcc.gov/mb/audio/totals/bt031231.html.

Federal Communications Commission. (2004c). In the matter of digital audio broadcasting systems and their impact on the terrestrial radio broadcast services. *Notice of proposed rulemaking*. MM Docket No. 99-325. Retrieved April 20, 2004 from http://hraunfoss.fcc.gov/edocs_public/attachmatch/FCC-04-99A4.pdf

Garrity, B. (2004, March 6). A big price or a small item: Cost of new iPod raises questions. *Billboard*. Retrieved April 11, 2004 from Factiva database.

Goldman, M. (2004, March 8). CHUM bid challenges U.S. firms' pay-radio plans. *The Toronto Star*. Retrieved April 18, 2004 from Factiva database.

Harmon, A. (2002, February 21). Panel's ruling on royalties is setback for Web radio services. *The New York Times*, C11.

Huff, K. (1992). AM stereo in the marketplace: The solution still eludes. *Journal of Radio Studies*, *1*, 15-30.

iBiquity Digital Corporation. (n.d.-a). *HD Radio: The technology behind HD Radio*. Retrieved April 19, 2004 from http://www.ibiquity.com/hdradio/hdradio_tech.htm.

iBiquity Digital Corporation. (n.d.-b). *iBiquity Digital identifies initial markets for conversion to digital AM and FM broadcast technology*. Retrieved March 10, 2002 from http://www.ibiquity.com/news_broadcaster%20support.html.

iBiquity Digital Corporation. (n.d.-c). *What is iBiquity Digital?* Retrieved March 10, 2002 from http://www.ibiquity.com/01content.html.

iBiquity Digital Corporation. (2004a). Public radio moves toward digital future. *Company press release*. Retrieved April 22, 2004 from http://www.ibiquity.com/press/pr/-41504.htm.

iBiquity Digital Corporation. (2004b). HD Radio going live coast-to-coast ... and beyond. *Company press release.* Retrieved April 22, 2004 from http://www.ibiquity.com/press/pr/041904Coast2Coast.htm.
Lewis, T. (1991). *Empire of the air: The men who made radio.* New York: Harper Collins.
Mathews, A. (2002, February 25). A giant radio chain is perfecting the art of seeming local. *The Wall Street Journal*, A1, A10.
McConnell, B. (2001, January 1). Congress reins in LPFM. *Broadcasting & Cable*, 47.
McMullen, T. (2004, January 30). Personal finance & spending: Fitness fans have digital-music options. *The Wall Street Journal Europe.* Retrieved April 25, 2004 from Factiva database.
Morgan, C. (2002, February 1). IBOC: What engineers should know. *Radio World.* Retrieved February 25, 2002 from http://www.radioworld.com/reference-room/iboc/guestmorgan.shtml.
National Association of Broadcasters. (n.d.-a). *NRSC revises U.S. RBDS standard.* Retrieved March 3, 2002 from http://www.nab.org/SciTech/Nrscgeneral/rds.asp.
National Association of Broadcasters. (n.d.-b). *High-speed FM Subcarrier (HSSC) Committee.* Retrieved March 3, 2002 from http://www.nab.org/SciTech/Nrscgeneral/hsscsub.asp.
National Radio Systems Committee. (2001, November 29). *DAB Subcommittee evaluation of the iBiquity Digital Corporation IBOC system, part 1—FM IBOC.* Retrieved March 9, 2002 from http://www.nab.org/SciTech/Fmevaluationreport.asp.
O'Dell, J. (2004, March 24). Satellite radio eager to receive Howard Stern fans. *The Los Angeles Times.* Retrieved April 21, 2004 from LexisNexis Academic database.
Radio Advertising Bureau. (2003). *Radio marketing guide & fact book for advertisers 2003-2004 edition.* New York: Radio Advertising Bureau.
Sirius Satellite Radio Homepage. (n.d.). *Company information.* Retrieved April 10, 2004 from http://www.siriusradio.com/servlet/ContentServer?pagename=Sirius/CachedPage&c=Page&cid=1065475754271
Sirius Satellite Radio. (2004a, February 12). Sirius Satellite Radio raises year-end subscriber projection. *Company press release.* Retrieved March 22, 2004 from http://www.siriusradio.com/servlet/ContentServer?pagename=Sirius/CachedPage&c=PreReleAsset&cid=1074263718345.
Sirius Satellite Radio. (2004b). Sirius Satellite Radio announces fourth quarter and year-end 2003 financial and operating results. *Company press release.* Retrieved March 22, 2004 from http://www.siriusradio.com/servlet/ContentServer?pagename=Sirius/CachedPage&c=PreReleAsset&cid=1074263713102.
Sirius Satellite Radio. (2004c). Sirius Satellite Radio available at nation's largest retailer. *Company press release.* Retrieved April 22, 2004 from http://www.siriusradio.com/servlet/ContentServer?pagename=Sirius/CachedPage&c=PreReleAsset&cid=1082553117765.
Stavitsky, A., Avery, R., & Vanhala, H. (2001). From class D to LPFM: The high-powered politics of low-power radio. *Journalism and Mass Communication Quarterly, 78*, 340-354.
Sterling, C., & Kittross, J. (1990). *Stay tuned: A concise history of American broadcasting.* Belmont, CA: Wadsworth Publishing.
Stimson, L. (2002, February 1). Digital radio makes news at CES. *Radio World.* Retrieved March 22, 2002 from http://www.radioworld.com/reference-room/special-report/ces.shtml.
Tedeschi, B. (2004, March 22). Proponents say that the time has come for online radio, and now they hope mainstream advertisers come along. *The New York Times.* Retrieved April 25, 2004 from Factiva database.
U.S. Bureau of the Census. (2003). *Statistical abstract of the United States.* Washington, DC: U.S. Government Printing Office.
World DAB: The World Forum for Digital Audio Broadcasting. (n.d.-a). *The benefits.* Retrieved April 21, 2004 from http://www.worlddab.org/benefits.aspx.
World DAB: The World Forum for Digital Audio Broadcasting. (n.d.-b). *Country updates.* Retrieved April 14, 2004 from http://www.worlddab.org/cstatus.aspx.
XM Satellite Radio Homepage. (n.d.). *Corporate information.* Retrieved April 12, 2004 from http://www.xmradio.com/corporate_info/corporate_information_main.html.

XM Satellite Radio. (2000). Sirius Radio and XM Radio form alliance to develop unified standards for satellite radio. *Company press release*. Retrieved March 2, 2000 from http://www.xmradio.com/js/news/pressreleases.asp#.

XM Satellite Radio. (2004a). XM Satellite Radio Holdings Inc. announces fourth quarter and full-year 2003 results. *Company press release*. Retrieved April 10, 2004 from http://www.xmradio.com/newsroom/print/pr_2004_02_12.html.

XM Satellite Radio. (2004b, April 1). XM Satellite Radio tops more than 1,680,000 subscribers. *Company press release*. Retrieved April 10, 2004 from http://www.xmradio.com/newsroom/print/pr_2004_04_01a.html.

Zeidler, S. (2004, February 10). U.S. Copyright Office sets Webcaster royalty rates. *Reuters*. Retrieved April 25, 2004 from Factiva database.

III

Computers & Consumer Electronics

Developments in computer technology are at the core of every digital technology discussed in this book. It is therefore not surprising that personal computers themselves are experiencing change as profound as for any other communication technology. This year's computer technology will unquestionably be replaced in less than two years by a technology that has up to twice the performance at almost half the cost (a phenomenon known as "Moore's law"). These advances in computer technology, in turn, lead to advances in almost every other technology, especially those consumer products incorporating microprocessors or other computer components. The next nine chapters illustrate the speed, direction, and impact of continuous innovation in computer technologies across a wide range of computing and consumer electronics technologies.

The next chapter explores the latest developments in personal computers, providing a perspective on all of the digital technologies explored in this book. Chapter 11 addresses the fastest growing area of computer technology: mobile computing. The following chapter then explores one of the most pervasive applications of a "dedicated computer": video game consoles. Chapter 13 addresses the most significant emerging application of personal computers, the Internet and the World Wide Web. The development of Internet-based e-commerce has made such a strong impact on individual businesses, and the economy in general, to warrant detailed discussion in Chapter 14.

The most forward-looking chapter in this section, Chapter 15, discusses a set of technologies—virtual reality—that may have the greatest long-term potential to revolutionize the way we live and work. The production and distribution of video and audio programming is the subject of the next two technologies in this section. The home video chapter reports on the incredible popularity of existing analog video formats, and on new, digital technology that is providing a formidable challenge to the analog incumbents. The digital audio chapter reports on the continuing battle among competing analog and digital audio technologies, with digital casualties as well as victors. Finally, Chapter 18, new to this edition, provides a snapshot of the latest developments in digital photography.

In reading these chapters, the most common theme is the systematic obsolescence of the technologies discussed. The manufacturers of computers, video games, etc. continue to develop newer and more powerful hardware with new applications that prompt consumers to continually discard two- and three-year-old devices that work as well as they did the day they were purchased—but not as well as this year's model. Most software distribution, from movies and music to television and video games, has been based on the continual introduction of new "messages." The adoption of this marketing technique by hardware manufacturers assures these companies of a continuing outlet for their products, even when the number of users remains nearly static.

An important consideration in comparing these technologies is how long the cycle of planned obsolescence can continue. Is there a computer or piece of consumer electronics so good that a "better" one will never replace it? Will technology continue to advance at the same rate it has over the past two decades? How important is the equipment (hardware) versus the message communicated over that equipment (software)?

Finally, each of these chapters provides some important statistics, including penetration and market size, that can be used to compare the technologies to each other. For example, there is far more attention paid today to the Internet than to almost any other technology, but just over half of all U.S. households have access to the Internet (with even smaller penetration levels in other countries). On the other hand, the VCR is found in about 9 out of 10 U.S. homes, yet is considered a dying technology. In making these comparisons, it is also important to distinguish between *projections* of sales and penetration and *actual* sales and penetration. There is no shortage of hyperbole for any new technology, as each new product fights for its share of consumer attention.

10

Personal Computers

Chris Roberts, M.A.*

It has been more than two decades since *Time* declared the computer its "Machine of the Year" (Friedrich, 1983). Computers of that era are silicon dinosaurs compared with the faster, more powerful, and productive machines of today. While personal computers remained a novelty at the start of the 1980s, the machines have become ubiquitous commodities at the start of the 21st century. Consider that:

- In 1983, companies sold 2.8 million personal computers worldwide. In 2003, sales reached 152.7 million units worldwide (IDC, 2004).
- In 1983, the fastest processing chip made by industry giant Intel Corporation ran at 12 million cycles per second (Old-computers.com, 2004). In early 2004, the newest desktop computer based on an Intel chip ran at 3.4 *billion* cycles per second (Karagiannis, 2004)—a speed improvement of more than 28,000%.

Some uses for computers—processing words, building spreadsheets, and playing games—remain as important today as they were two decades ago. Today's computers, however, have untold new uses as prices have fallen, power has increased, Internet use has risen, and computers themselves have become easier to use and to program.

Personal computers are now in close to four-fifths of American households (Seitz, 2004). By comparison, telephones did not reach 80% of the nation's homes until 1954, nearly 85 years after the

* Doctoral Student, The University of South Carolina. Assistant Business Editor, *The State* (Columbia, South Carolina).

first home telephones debuted (FCC, 2000). Personal computers are in nearly every school in America, meaning the current generation of students cannot recall a time when computers were not part of their lives (Cattagni & Farris, 2001). Here is a look at the history of computers, the current state of technology, and what may be coming next.

Background

A Brief History of Computers

The U.S. Census Bureau had a problem in the late 1800s. As the United States became more populous and expansive, bureaucrats simply could not count everyone fast enough. Agency clerks needed seven years to complete work on the 1880 census, rendering the exercise nearly useless. So the agency turned to employee Herman Hollerith, whose mechanical invention ran through the 1890 data in about three months (Campbell-Kelly & Aspray, 1996). This exercise marked the first practical use of a computer.

Hollerith, who laid the foundation for International Business Machines Corporation, based his idea on punch tickets used by railroad conductors, but he and nearly all computer designers owe a debt of gratitude to 1800s British inventor Charles Babbage. While his "difference engine" was never built, Babbage's idea for a computer remains a constant regardless of the technology: Data is input into a computer's "memory," processed in a central unit, and the results output in a specified format.

Early computers, first called calculators because "computers" were people who solved math equations, were designed to tackle a specific task—counting people, calculating artillery firing coordinates, or forecasting the weather. The first general-purpose computers emerged at the end of World War II in the form of ENIAC, the "electronic numerical integrator and computer" (Brown & Maxfield, 1997). Sperry-Rand's UNIVAC reached the market first in 1950, but Rand and six other computer companies soon became known as the "Seven Dwarfs" compared with IBM. The combination of IBM's powerful marketing efforts and understanding of business needs grew its market share to more than two-thirds in 1976. By 1985, however, IBM controlled less than one-fourth of computer market share, as personal computers surpassed mainframes in sales and use.

The introduction of MITIS Altair 8800 in 1975 brought the first real "personal computer," but Apple Corporation's successful Apple II machine gave that company half of the personal computer market share by 1980. The early 1980s were marked by a Babel of personal computing formats, but a standard emerged within a few years after the August 1981 arrival of the IBM PC, powered by an Intel chip and the MS-DOS operating system from Microsoft (Campbell-Kelly & Aspray, 1996).

IBM's influence as a maker of PCs faded as competitors bested its products with machines with better price and performance. Microsoft, however, built a still-growing market share for its operating systems, the programs that manage all other programs in a computer. The company eventually controlled the market for text-based operating systems, and it managed to hold off competitors while creating its "Windows" operating systems. Windows employs a graphical user interface (GUI), which harnesses the computer's graphics capability to make operation of the machine simpler and more intuitive. The Macintosh (and its 1983 predecessor, the Lisa) were built upon a GUI, giving them an ease-of-use advantage over IBM-based systems. Microsoft's first version of Windows shipped in late

1985, but Windows did not reach widespread use until its third version of Windows shipped in mid-1990. The mass acceptance of Windows gave Microsoft further dominance in the business of selling operating systems, and the company leveraged that power by selling applications based on the Windows platform. By 2002, nearly 94% of personal computers shipped worldwide were delivered with Windows or another Microsoft operating system (IDC, 2003).

How Computers Work

The elements that make up a computer can be divided into two parts—hardware and software. Hardware describes the physical parts of a computer, such as the central processing unit, power controllers, memory, storage devices, input devices such as keyboards, and output devices such as printers and video monitors. Computer software is the term used to describe the instructions regarding how hardware manipulates the information (data) (Spencer, 1992). This definition of software differs from the umbrella perspective on communication technology, which defines software as the "messages communicated through a technology system." Under the umbrella perspective, the content of a word-processing document would be considered "software," while the word-processing program would be defined as part of the hardware.

Central to understanding hardware is the central processing unit (CPU), also known as the microprocessor. The CPU is the brain of the computer, performing math and logic according to the information given it. To do its work, the CPU requires memory to hold the data, and the instructions needed to process that information. That memory is based upon a series of switches that, like a light switch in a house, are flipped on or off.

The original memory devices required vacuum tubes, which were expensive, prone to breakage, generated a great deal of heat, and required a great deal of space. The miniaturization of computers began in earnest after December 23, 1947, when scientists perfected the first "transfer resistor," better known as a "transistor." Nearly a decade later, in September 1958, Texas Instruments engineers built the first "integrated circuit"—a collection of transistors and electrical circuits built on a single "crystal" or "chip." These semiconductors sharply reduced CPU sizes from buildings to wafers. Today, circuit boards hold the CPU and the electronic equipment needed to connect the CPU with other computer components. The "motherboard" is the main circuit board that holds the CPU, sockets for random access memory, expansion slots, and other devices. "Daughterboards" attach to the motherboard to provide additional components, such as extra memory or cards to accelerate graphics.

The CPU needs two types of memory to do its job: random access memory (RAM) and storage memory. RAM is the silicon chip (or collection of chips) that holds the data and instruction set to be dealt with by the CPU. Before a CPU can do its job, the data are quickly loaded into RAM—and that data are eventually wiped away when the work is done. RAM is measured in "megabytes" (the equivalent of typing a single letter of the alphabet one million times) or, increasingly, in "gigabytes" (roughly one billion letters). Microsoft's Windows XP operating system for home users functions with as little as 64 megabytes of RAM, but the company recommends at least 128 megabytes of RAM (Microsoft, 2004). Only the most basic personal computers today ship with less than 128 megabytes of RAM.

Think of RAM as "brain" memory, a quick but volatile memory that clears when a computer's power goes off or when it crashes. Think of storage memory as "book" memory—information that takes longer to access, but stays in place even after a computer's power is turned off. Storage memory devices use magnetic or optical media to hold information.

Major types of storage memory devices include:

- *Hard drives* are rigid platters that hold vast amounts of information. The platters spin at speeds of 5,400 to 15,000 revolutions per minute, and "read/write" heads scurry across the platters to move information into RAM or to put new data on the hard drive. The drives, which can move about 30 megabytes of information each second, are permanently sealed in metal cases to protect the sensitive platters. A drive's capacity is measured in gigabytes, and only the most basic computers today ship with less than 20 gigabytes of hard-drive capacity. The drives almost always hold the operating system for a computer, as well as key software programs (Wikipedia, 2003).

- *Floppy diskettes* hold 1.4 megabytes of data on a small magnetic oval protected in a 3.5-inch plastic case. Their limited capacity makes them less attractive as a storage device, and some computer makers sell floppy drives only as optional accessories.

- *Zip and "super floppy" diskettes* hold at least 100 megabytes of data. Iomega's Zip and Imation's SuperDisk devices are also less fashionable today, since newer storage devices are smaller, cheaper, and hold much more data.

- *Keydrives* are tiny storage devices that earned the name because they are small enough to be attached to a keychain. They use flash memory—a solid-state storage device with no moving parts—and plug into a computer using the Universal Serial Bus (USB) port, which also provides power to the device. Some of the larger-capacity keydrives currently can hold at least 2 gigabytes of data.

- *Other flash memory devices* can be connected to a computer. Some personal computers ship with devices that can read small memory cards used in digital cameras, music players, and other devices.

- *Compact discs* were introduced more than two decades ago. These 12-centimeter wide, one-millimeter thick discs hold nearly 700 megabytes of data or more than an hour of music. They ship in three formats: CD-ROM (read-only memory) discs that come filled with data and can be read from but not copied to; CD-R discs that can be written to once; and CD-RW discs that, like a floppy diskette, can be written to multiple times. Most computers ship with CD drives capable of recording ("burning") CDs.

- *DVDs*, known as "digital versatile" or "digital video" discs, are increasingly replacing CDs as the storage medium of choice. They look like CDs, but hold much more information—typically 4.7 gigabytes of computer data, which is more than six times the capacity of a conventional CD. DVD players and burners are becoming standard equipment with new computers, since DVD video has reached critical mass acceptance and because DVD players and burners are backward-compatible with CDs. As illustrated in Table 10.1,

DVD technology includes multiple formats, not all of which are compatible with each other.

Table 10.1
DVD Formats

Format	Storage Size	Pros	Cons
DVD-ROM	4.7 GB	Works in set-top DVD players and computers	Read-only
DVD-R	4.7 and 9.4 GB	Works in most set-top DVD players and computers	Can be written to only once; may not work in DVD+R drives
DVD-RAM	2.6 to 9.4GB	Is rewritable many times	Works only in a DVD-RAM drive
DVD-RW	4.7 GB	Can be written to up to 1,000 times; can be used in many DVD players and computers	DVD-RW discs may not play back on some older or entry-level DVD systems
DVD+R	4.7 GB	Works in most set-top DVD players and computers equipped with DVD-ROM drives	Can be written to only once; may not work in DVD-R drives.
DVD+RW	4.7 GB	Works in most set-top DVD players and computers equipped with DVD-ROM drives	DVD+RW discs may not play back on some older or entry-level DVD systems

Source: C. Roberts

Another key category of hardware is known as input and output devices. Input devices—named because they deliver information to the computer—include keyboards, mice, microphones, scanners, and other devices. Output devices—which deliver information from the computer to users—include monitors and printers. Other devices that let computers communicate with the outside world are modems (modulator/demodulators that translate the digital data of a computer into analog sound that can travel over telephone lines) and network cards that allow the computer to send and receive high-speed, digital data signals through computer networks. Figure 10.1 illustrates the relative cost of each of the components discussed above in a typical computer.

Figure 10.1
PC Component Prices, February 2004

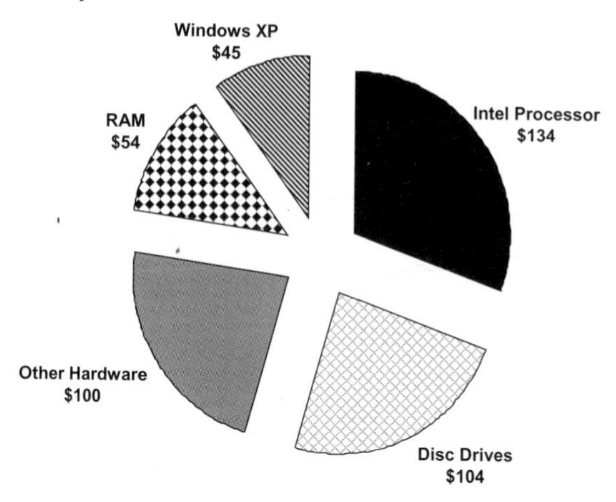

Source: C. Roberts

Software

A computer is little more than a silicon doorstop until software breathes life into the machine. The most important software is the operating system, which coordinates with hardware and manages other software applications. The operating system controls a computer's "look-and-feel"; stores and finds files, takes input data and formats output data; and interacts with RAM, the CPU, peripherals, and networks. Microsoft's XP operating systems reign supreme in sales against competing operating systems such as Apple's OS X, UNIX, and various versions of GNU/Linux. For smaller computer devices known as personal digital assistants, Palm OS and Microsoft's Windows CE are the two main competitors.

Operating systems provide the platform for applications—programs designed for a specific purpose for users. Programmers have created literally tens of thousands of applications such as word processing, Web browsing, e-mail, Web design, games, video editing, and programming.

Programs designed to improve computer performance are known as utilities. The best-known utility programs improve how data is stored on hard drives and stop malicious computer code (such as viruses, worms, or Trojan horses) designed to destroy files, slow computer performance, or let outsiders surreptitiously take control of a computer. Software that identifies and sorts unsolicited commercial "spam" e-mail messages is also popular, as is software that removes "pop-up" advertising from Web sites.

Recent Developments

Hardware

It is no wonder why *Forbes* ranked Gordon Earle Moore among the world's richest people, with a net worth of $5.5 billion in 2003 (The world's, 2004). In 1965, the co-founder of computer chipmaker Intel predicted that the number of transistors on a computer chip could double every couple of years, meaning computing power roughly doubles along with it (Intel, 2004). This phenomenon became known as Moore's Law, and, as illustrated in Figure 10.2, remains prescient after nearly 40 years. Intel engineers, however, say the law may reach its end by 2018 "as we approach the energy barrier due to limits of heat removal capacity" (Kanellos, 2003). (Heat is the chief enemy of a CPU; most chips come with "heat sinks" and small fans designed to draw heat away from the chip.)

Figure 10.2
Moore's Law

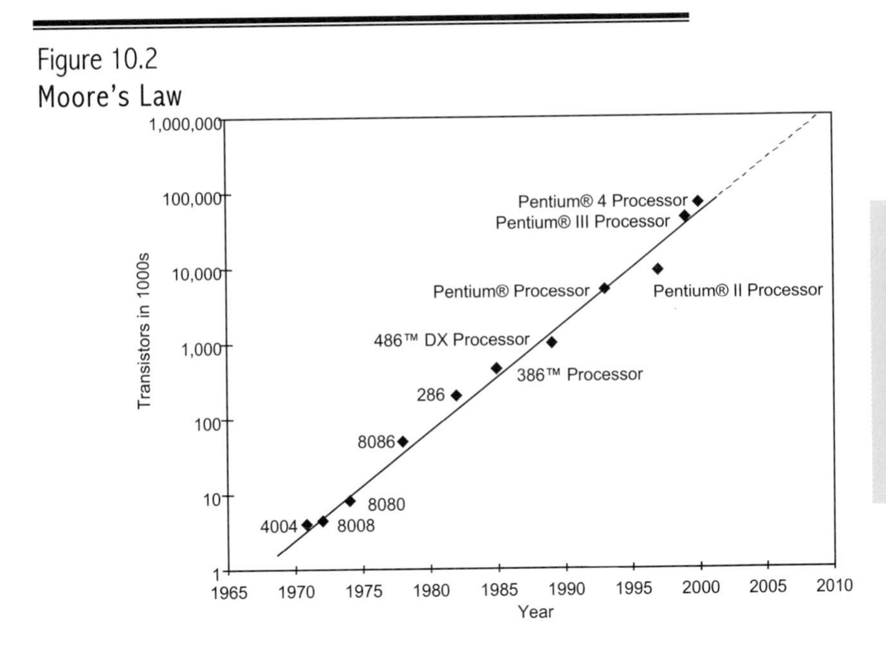

As predicted by Gordon Moore in the 1960s, the power of computers has been increasing geometrically over the past four decades. Note that the scale for the number of transistors is a logarithmic scale.

Source: Technology Futures, Inc.

The biggest news in CPU development is the arrival of 64-bit processors, which manipulate binary numbers that are 64 zeroes and ones long. This is double the power of 32-bit processors, which, in the late 1980s, replaced 16-bit processors as the standard for CPUs. In August 2003, Apple introduced its PowerMac G5 computer with a 64-bit processor; AMD and Intel followed with their own chips. With 64-bit machines, processors can work with up to 16 quintillion (that is, 16 billion billion) bytes of RAM, an exponential leap from the maximum 4 billion bytes of RAM that 32-bit machines could handle. Computers may never have that much RAM, even as memory prices plunge, but, as a practical matter, the 64-bit processor will mean better graphics on faster computers (Markoff, 2003).

Storage devices also are growing in speed and capacity. Video and audio files are becoming easier to bring into a computer, store, manipulate, and output, but these files demand large chunks of hard-drive space. Drive capacities have continued growing to meet the need, with the current biggest-capacity drive able to hold 400 gigabytes of data, a number that surely will be surpassed before this book goes to press (Fraunheim, 2004).

Video files are also helping push mass acceptance of DVD burners on personal computers. The competing standards of DVD-R and DVD+R remain unsolved, but newer burners have bridged the gap by being able to create discs for either standard. At least as important, prices on DVD burners fell below $100 during 2003, and newer stand-alone DVD players that connect to television sets are compatible with PC-made DVDs (DVDhelp.com, 2004).

Memory is also becoming more portable and less expensive. The past two years has seen USB "keydrive" and other flash-media storage devices reach critical mass, with prices falling while capacity increases. Newer keydrives can hold two gigabytes of data (some computers have difficulty with larger-capacity keydrives) and use the USB 2.0 standard. Sales are also on the rise for external hard drives, conventional platter-like drives that connect to a computer through a USB or "firewire" port and move data nearly as quickly as an internal hard drive.

As hardware becomes more powerful, faster, and smaller, it is no wonder that laptops make up a larger share of computer sales. Laptops still command a higher price than desktop models, but the gap is closing in performance between state-of-the-art laptops and state-of-the-art desktops. Nearly 35 million laptops were sold during 2003, and experts predict a 20% growth during 2004. Portable PC sales accounted for 40% of the market in 2003 (Kovar, 2004). By 2007, nearly half of all computers sold in the United States will be laptops (Spooner, 2004).

Another change in the past few years has been the wider adoption of flat-paneled screens, thanks to price drops. Sales of flat-panel displays in the United States topped $21 billion during 2003, the first time these monitors have outsold heavier, larger cathode-ray tube monitors. About half of the nation's computers now sell with flat-panel monitors, a percentage expected to grow as prices continue to drop (Kessler, 2004).

What also has been falling—beside (and because of) falling prices—is the number of companies making computers. Hewlett-Packard's 2002 merger with Compaq marked the largest computer-related merger in history (Lohr &Gaither, 2002). In March 2004, struggling Gateway Inc. completed its $236 million purchase of eMachines Inc., a maker of low-cost computers. The deal made it the third-largest U.S. maker of personal computers, behind Dell and Hewlett-Packard (Marshall, 2004).

Apple, meanwhile, holds less than 5% of the market share—but an important single-digit market share. Its Macintosh computers remain dominant in publishing sectors, multimedia creation, and among people interested in the company's trend-setting design and engineering. The company also has been successful in the audio industry, with its best-selling iPod audio devices and its iTunes music store that provides music for sale online.

Computer Software

Major developments in the software industry almost always revolve around Microsoft, the dominant company in the industry. A collection of competitors and government say this dominance has allowed Microsoft to stray into improper monopolistic behavior.

One key event for the company was the November 2002 settlement of the monopoly lawsuit filed against it by the U.S. government. The agreement required the company to share some computer code with third parties, but the deal fell short of requiring Microsoft to split itself into separate companies (one for its operating systems, another for applications such as Office). Nine states and the District of Columbia disagreed with the settlement and continued litigation against Microsoft (Wikipedia, 2004). Meanwhile, in March 2004, the European Union fined Microsoft more than $600 million, saying its "near-monopoly" status trampled competitors. The company plans to appeal (Meller, 2004).

Microsoft also finds itself fighting against open-source software—programs in which the code is freely available to the public. Microsoft has cut prices of its operating system and Office products in the face of competition from open-source competitors, particularly in foreign countries. The company also sued the maker of the "Lindows" operating system, leading the company to change its name to Linspire in April 2004 (Sharma, 2004). Microsoft also played a behind-the-scenes role in SCO Group's copyright infringement lawsuits against companies using the open-source Linux operating system (Lohr, 2004). Microsoft quelled unrest from investors in 2003 by introducing a 16-cents-per-share quarterly stock dividend, signaling that it is now a "mature" company that does not need all its money for research and capital expenses.

Further, Microsoft said it is behind schedule on "Longhorn," the next edition of an operating system for personal computers. A test, or "beta," version of the software may become available in late 2004, with a commercial version ready as early as late 2005. A version of the Windows operating system that makes fuller use of 64-bit processors is expected to be available this year to join a Linux operating system, but few applications currently are designed to use the boost in computing power (Kharif, 2004).

Microsoft competitors continue to make inroads against the company's dominance in operating systems and productivity suites, particularly outside the United States. The various versions of open-source Linux operating systems hold more market share outside of Microsoft's home, and their market share is expected to continue growing as PC markets offer Linux as a pre-installed option on machines they sell. As more software companies develop products for Linux, more buyers may be persuaded to replace Windows with Linux (or at least experiment with Linux). In the productivity sector, Open Office and Star Office offer a free or less-expensive option to Microsoft's Office suite. In September 2003 Massachusetts, the lone state still suing Microsoft for monopolistic activities, said it plans to migrate its state computers from Microsoft products to open-source products (Associated Press, 2003). Other national governments, such as Brazil and Germany, are also looking for alternatives to Microsoft products.

Current Status

Although computers are in nearly four-fifths of U.S. households, computer sales continue to climb as consumers replace aging computers and buy second or third computers for their homes. Sales growth reached double digits in 2003, the first time since 2000 (IDC, 2004).

The computer industry sold $41 billion in equipment during 2002, with desktop and laptop personal computers accounting for nearly two-thirds of sales value (U.S. Census Bureau, 2004).

Dell shipped nearly 17% of the 152.7 million computers made worldwide in 2003, giving it a slight lead over Hewlett-Packard's 16.4%. IBM was a distant third at 6% (IDC, 2004).

Most of those computers ship with central processing units made by Intel, which held 83% of the market share in early 2004 for the "x86" processors that run Windows-based computers. Advanced Micro Devices (AMD) is a distant second, with about 16% of the market (Krazit, 2004). Motorola and IBM continue to make CPUs for Apple computers, which shipped about 3% of computers in the United States during 2003 (Apple finishes, 2004).

Prepackaged software sales in 2003 were estimated to have contributed nearly $80 billion to the U.S. gross domestic product, up nearly 50% since 1996 after adjusting for inflation (Economic and Statistics Administration, 2004).

Factors to Watch

Pending Activity

More Microsoft antitrust actions. While Microsoft won its round with the federal government, more lawsuits continue at home and abroad. The company faces action from RealNetworks over media formats and access to Windows desktops and also faces action from states that dropped out of the federal antitrust action because the settlement was too lenient. As of mid-2004, other legal proceedings continued in connection with the European Union's antitrust fine (Reuters, 2004).

More pushback against Microsoft software. After Microsoft changed how it charges businesses for its software, more companies are beginning to look at alternatives to the company's operating systems and applications. Rivals Oracle, IBM, and others are at the forefront of those efforts against Microsoft, but Microsoft's dominance and marketing savvy continues to give it dominant market share in most of its ventures.

More fights with pirates. Nearly two-fifths of business software used in the world is pirated (Business Software Alliance, 2004), and companies are struggling to find ways to protect themselves without alienating customers. Intuit, for example, in 2004 backed away from an anti-piracy approach in 2003 that made its popular TurboTax software more difficult to use and install (Moran, 2003). Meanwhile, fights against companies that make it easy to copy protected DVDs and Internet sites that make it easy to move copyright-protected movies and music can expect to see more days in court.

More worries about viruses. Most computers have some sort of connection to the Internet and e-mail, which means those computers can be infected by a virus. New strains of viruses seem to pop up daily (or at least weekly), requiring users to remain vigilant.

Likely Developments

Will TVs and PCs merge? The intersection of TVs, PCs, DVDs, flat-paneled monitors, and digital video recorders are leading some hardware makers to offer machines that blur the lines between televisions and computers. Consumers were not impressed by efforts in past decades by Apple, Sony, and Gateway, all of which sold computers designed to be "entertainment centers." But companies are trying again, this time with operating systems aimed at easy-to-use "media centers" (Nothing says, 2004). Target markets include consumers living in small confines and technologically savvy people who want to use multiple file platforms through a single device. Researchers note, however, that the fundamental differences between computers and televisions make it difficult for one technology to be used as the other (one example: computer users sit much closer to the screen than television users) (Morrison & Krugman, 2001).

How small can PCs become? Laptop computers continue to gain market share as they become as powerful (for roughly the same price) as desktop computers. At the same time, personal digital assistants (PDAs) are becoming even more useful as they gain telephone and Internet capabilities. PalmOne and Microsoft make competing operating systems for the PDA market. In the next few years, "screenless" monitors and "projected" keyboards could make computers even smaller and more useful (Ho, 2004). Chapter 11 contains a detailed discussion of these mobile computers.

Will tablets take off? Several companies now make "tablet computers," which are about the size of a legal pad and rely on wireless communications and a stylus. About one million were sold in 2003. They are used primarily by insurance adjustors or healthcare workers, but the computer industry hopes improvements from earlier generations of tablet computers will lead to greater adoption (Brown, et al., 2003)

Will instant-on computers ever arrive? A chief complaint about computers is that they take too long to "boot up." They are different from television or radio receivers, which deliver programming within seconds of receiving power. Hardware and software companies continue to find ways to speed up the "booting" process, but no instant-on computer seems on the horizon.

In the past quarter-century, personal computers have grown from underpowered novelties into one of the world's most common, and most useful, technological devices. Their connectivity gives users access to the world, and they are primary aids to personal creativity and productivity. As the power of computers continues to rise exponentially, the machines have become more affordable. The key to the future of computers, then as now, is to make computers more powerful, more reliable and easier to use.

Bibliography

Apple finishes '03 as no. 5 PC vendor with 3.2% share. (2004). *Mac News Network*. Retrieved March 14, 2004 from http://www.macnn.com/news/22955.

Associated Press. (2003). Massachusetts wants to use Linux-type systems. *USA Today*. Retrieved April 16, 2004 from http://www.usatoday.com/tech/news/techpolicy/2003-09-26-mass-wants-linux_x.htm.

Brown, B., Brown, M., & Karaginnis, K. (2003). Tablet PC secrets revealed. *PC Magazine*. Retrieved March 15, 2004 from http://www.pcmag.com/article2/0,1759,1271392,00.asp.

Brown, C., & Maxfield, C. (1997). *Bebop bytes back: An unconventional guide to computers*. Madison, AL: Doone Publications.

Business Software Alliance. (2004). *Fact sheet*. Retrieved March 15, 2004 from http://www.bsa.org/usa/press/Fact-Sheets.cfm.

Campbell-Kelly, M., & Aspray, W. (1996). *Computer: A history of the information machine*. New York: Basic Books.

Cattagni, A., & Ferris, E. (2001, May). *Internet access in U.S. public schools and classrooms: 1994-2000*. National Center for Educational Statistics. Retrieved April 17, 2004 from http://nces.ed.gov/pubsearch/pubsinfo.asp?pubid=2001071.

DVDhelp.com. (2004). Retrieved March 14, 2004 from http://www.dvdrhelp.com/dvd.

Economic and Statistics Administration. (2004). *Digital economy 2003*. Retrieved March 14, 2004 from http://www.esa.doc.gov/DigitalEconomy2003.cfm.

Federal Communications Commission. (2000). *Penetration rates—Inception*. Retrieved January 21, 2004 from http://www.fcc.gov/Bureaus/Common_Carrier/Notices/2000/fc00057a.xls.

Fraunheim, E. (2004). Hitachi to unveil 400GB drive. *C/Net News*. Retrieved March 14, 2004 from http://news.com.com/2100-1015-5171944.html.

Friedrich, O. (1983, January 3). Machine of the year: The computer moves in. *Time, 142,* 14.

Ho, D. (2004, March 14). Thin air about to replace viewing screens. *The Atlanta Journal-Constitution,* D1.

IDC. (2003). *IDC says Microsoft is moving into dominant role in server operating environments, even as Linux grows*. Retrieved February 29, 2004 from http://www.idc.com/getdoc.jhtml;jsessionid=WISKN0NDABVUGCTFA4FCFGAKMUDYWIWD?containerId=pr2003_09_29_140158.

IDC. (2004, January 14). *PC shipments reach record levels as battle for market leadership continues, according to IDC*. Retrieved March 15, 2004 from http://www.idc.com/getdoc.jhtml?containerId=pr2004_01_13_185937.

Intel Corporation. (2004). Silicon: Moore's Law. *Intel.com*. Retrieved March 14, 2004 from http://www.intel.com/research/silicon/mooreslaw.htm.

Kanellos, M. (2003). Intel scientists find wall for Moore's Law. *C/Net News*. Retrieved March 14, 2004 from http://news.com.com/2100-7337-5112061.html?tag=nefd_lede.

Karagiannis, K. (2004, March 2). Prescott brings more cache to Intel's future. *PC Magazine, 23,* 26-27.

Kessler, M. (2004, February 24). Flat-panel monitors flatten boxy rivals in market share. *USA Today,* D1.

Kharif, O. (2004). The 64-bit question. *Business Week*. Retrieved March 14, 2004 from http://www.businessweek.com/technology/content/feb2004/tc20040226_9664_tc119.htm.

Kovar, J. (2004, March 1). Mobile PCs on move to outsell desktops. *Computer Reseller News,* 63.

Krazit, T. (2004). Intel, AMD see steady Q4 market share as prices rise. *Infoworld*. Retrieved March 14, 2004 from http://www.infoworld.com/article/04/02/03/HNintelamd_1.html.

Lohr, S. (2004, March 12). Microsoft said to encourage big investment in SCO Group. *The New York Times,* 5.

Lohr, S., & Gaither, C. (2002, March 20). Hewlett-Packard claims a victory. *The New York Times,* A1.

Markoff, J. (2003, August 18). How an extra 32 bits can make all the difference for computer users. *The New York Times,* 4.

Marshall, C. (2004, January 31). Gateway makes deal to acquire eMachines. *The New York Times,* C2.

Meller, P. (2004, March 24). Europeans rule against Microsoft; Appeal is promised. *The New York Times,* A1.

Microsoft. (2004). *Windows XP Home Edition system requirements*. Retrieved March 14, 2004 from http://www.microsoft.com/windowsxp/home/evaluation/sysreqs.asp.

Moran, J. (2003, October 16). Intuit apologizes for its copy protection zeal. *The Hartford Courant,* 3.

Morrison, M. &. Krugman, D. (2001). A look at mass and computer mediated technologies: Understanding the roles of television and computers in the home. *Journal of Broadcasting & Electronic Media, 45* (1), 135-161.

Nothing says home more than the living room. (2004, April 6). *PC Magazine,* 91.

Old-computers.com. (2004). *PC XT - Model 5160*. Retrieved February 21, 1994 from http://www.old-computers.com/ museum/computer.asp?st=1&c=286.

Reuters. (2004). EU Commission gets backing for Microsoft action. *Forbes*. Retrieved March 15, 2004 from http://www.forbes.com/technology/newswire/2004/03/15/rtr1298430.html.

Seitz, P. (2004, January 9). Personal computer still eludes a fifth of U.S. households; Cost remains biggest factor. *Investors Business Daily,* 4.

Sharma, D. (2004). Looking for Lindows? Try Linspire. *ZDNet News*. Retrieved April 14, 2004 from http://zdnet.com.com/2100-1104_2-5191333.html.

Spencer, D. (1992). *Webster's new world dictionary of computer terms* (Fourth ed.). New York: Prentice Hall.

Spooner, J. (2004). Consumers keep notebook sales on a roll. *C/Net News*. Retrieved March 14, 2004 from http://news.com.com/2100-1044-5138348.html.

The world's richest people 2003. (2004). *Forbes*. Retrieved March 14, 2004 from http://www.forbes.com/static_html/bill/2003/rank.html.

U.S. Bureau of the Census. (2004). *Statistical abstract of the United States* (123 ed.). Washington, DC: U.S. Census Bureau.

Wikipedia. (2003). *Hard disk*. Retrieved March 13, 2004 from http://en.wikipedia.org/wiki/Hard_drive.

Wikipedia. (2004). *Microsoft antitrust case*. Retrieved March 14, 2004 from http://en.wikipedia.org/wiki/Microsoft_antitrust_case.

11

Mobile Computing

Mark J. Banks, Ph.D. & Robert E. Fidoten, Ph.D.*

In the larger sense, anything that contains a microprocessor and is portable or battery-operated falls under the spectrum of "mobile computing." Examples of mobile computing devices include cell phones, portable video games, satellite global positioning devices, personal digital assistants (PDAs), laptop and tablet computers with or without wireless capability, advanced walkie-talkies, and the full range of wireless networks. Mobile computing devices also enable an endless array of specialized applications, ranging from hand-held inventory record-and-control devices to students on campus using wireless access to research journals to UPS (United Parcel Service) tracking of shipments to the portable computers used by car rental agencies to process a car return on the spot. This chapter focuses on the use of mobile computing through wireless technologies.

Mobile computing is fast becoming one of the largest links to the end-user in the vast panoply of data connectivity. While it takes enormous processing and storage capacity to perform the functions of wired networks (e.g., Internet, corporate mainframes, telephone systems), the *wireless* end of that linkage, though still too often sluggish and sporadic in its accessibility, is rapidly overcoming many of the obstacles. Wireless networks are expanding on a global scale. Bandwidth has grown larger and larger, even for the mobile links. End-user devices have enjoyed an increase in function, capacity, and speed, as well as an increase in their multifunctional features. The whole landscape of connectivity is becoming more and more seamless at a rapid pace.

Not only will connectivity become more seamless, it will also become more automated or invisible. "End users" will not only be people, but other machines as well, leading to what some call the

---

* Mark Banks is Professor and Chair and Robert Fidoten is Associate Professor. Both are faculty in the Communication Department at Slippery Rock University (Slippery Rock, Pennsylvania).

"Internet of things," where machines communicate with machines, much like a person's cell phone keeps in touch with cell networks to always convey its location and availability (Reinhardt, 2004).

This chapter gives a short description of the background and history of mobile computing, recent and upcoming developments with some data on the current status of mobile computing, and a discussion of factors and issues to watch as this phenomenon continues to evolve.

Background

As with most technology-laden areas, the mobile computing area has its share of acronyms and jargon: 3G, CDMA, TDMA, UWB, Wi-Fi, WiMAX, Bluetooth, RFID, WPAN, EV-DO, GSM, GPRS-EDGE, and so on. Unraveling the myriad of developments produces only a snapshot in time, since the technologies and their corresponding acronyms persistently continue to change.

Mobile wireless communication has existed for many years and, in its early stages, included pagers, clunky bandwidth-hogging pre-cellular mobile telephones, and some limited-use radio communications such as amateur radio and citizen's band. Practical business and professional applications were very limited, and service costs were high. Although the range, penetration, and application of mobile computing and communication seem to be rather extensive today, there is also a sense that the proliferation is just really beginning.

Today's cellular and personal communication services leave the mobile user with pervasively slow and limited access, particularly in wideband capability and penetration. Mobile services still largely preclude wideband automotive and other transportation uses. Security of information is also still rather primitive, especially as the number of wireless mobile users increases and the availability of "hot spots" proliferates, leaving wireless signals vulnerable to interception (Nasaw, 2004). It seems wireless services are on the verge of a growth spurt, as computer makers and software developers get on the high-speed wireless bandwagon (Flynn, 2004).

Recent Developments

Cell Phones

The technology behind cell phones is well known, and is discussed in greater detail in Chapter 24. Essentially, cell phone technology incorporates a large number of low-power telephones that share the same frequencies because their transmission area is limited to geographic cells. When a user moves from cell to cell, as in a moving vehicle, the transmission is handed off to an available frequency in the new cell, thus freeing the frequency used in the former cell for another user.

There are more than 1.317 billion cell phones in use throughout the world; they are used by more than 20% of the world's population, and usage is increasing at over 10% per year (Rerisi, 2003). Wireless data usage has had similar dramatic growth, and is anticipated to reach over 67 million workers by 2006, representing over half the U.S. workforce according to Access Markets International. Such workers will be using cellular, Wi-Fi hotspots, and other broadband mobile computing technologies (Schwartz, 2004)

Other Wireless Connectivity

In the wireless connectivity of portable computers, including laptops and PDAs, there are five major technologies: Wi-Fi, WiMAX, Mobile-Fi, ZigBee, Bluetooth, and Ultrawideband.

Wi-Fi uses the 802.11 formats and has a varying range of up to 350 feet, depending on which format is used, and a capacity of up to 54 megabytes per second. This initial entry in the area of wireless computing is supported by Intel technology. It is anticipated that competitive technologies will enter the market in 2004-2005. The anticipated proliferation of new wireless computing technologies will lead to an expansion of Internet reach from feet to miles, providing "connection anywhere, anytime" (Green, 2004).

ZigBee technology "coordinates communication among thousands of tiny sensors" such as thermostats, lighting controls, and smoke detectors. The sensors may be highly scattered, passing data over radio waves throughout a highly diffuse geographic area. The data can be loaded into computers en route or transferred to another wireless technology. It is anticipated that ZigBee will reach the market sometime in 2004 (Green, 2004).

WiMAX is similar to Wi-Fi technology. Wi-Fi can service computers within a range of several hundred feet, and it is designed to create "hot spots," or areas around a central antenna. WiMAX, by comparison, has a range of 25 to 30 miles. Equipped portable computers can then access the Internet over relatively long distances. Current WiMAX technology can only be used in a static work environment, and it does not permit access from vehicles in motion. It is anticipated that an offshoot of WiMAX called Mobile-Fi will be available in a few years that will provide speeds equal to or faster then today's broadband technology (Green, 2004).

Ultrawideband is designed to move huge files at high speed over short distances. The technology has a short range of only about 33 feet, but it can channel as much as 480 megabytes per second, almost 10 times that of Wi-Fi. There are currently competing systems for the development of an ultrawideband standard for wireless personal area networks (WPANs). Motorola is proposing one system based on a well-tested spread spectrum (DSSS) technology, while most other industry companies are proposing a different standard. At issue is which standard is best, and whether the standards are compatible with each other (Motorola versus, 2004).

Bluetooth has a much shorter range and a limited capacity of about 1 megabyte per second. The technology is often used to interconnect devices at ranges of 33 feet or less. Bluetooth often interconnects devices within an office, such as a mouse, keyboard, printer, etc, without use of wires.

A major issue to be resolved for these emerging computer communication technologies is arriving at a standard that will be adapted by the computer chip, computer system, and communications companies. The need for a standard is driven by competitive considerations as well as the technical problems to be resolved.

Parallel to these efforts is work underway at cellular telephone organizations designed to provide faster Internet connections for mobile telephones and mobile computers. The largest U.S. cellular telephone service provider, Verizon, has already installed 3G (third generation) facilities in two major geographic areas and has plans to reach 100 markets by 2005-2006. This technology is slower than the emerging WiMAX technology.

As with all communication technology product and service launches, the availability of spectrum is a major problem. Efforts are underway to have the Federal Communications Commission (FCC) release or transfer spectrum for use by the mobile communication/computing industries. Large spectrum resources are held by the major television broadcasters, and the commission is being pressured to reassign or find an alternate procedure to make some portion available for the mobile communications/computer market.

Current Status

In late 2004, dual-mode cellular/voice-over-Wi-Fi handsets will be available, and ABI Research predicts they will constitute about 7% of all handsets by 2007. The advantage of such handsets is their ability to use the cell phone function when out of Wi-Fi range, and to automatically switch to Wi-Fi when within range. A business, for example, can have its employees continue using cell phones in the office without incurring the expense of cell phone minutes (Enterprise to drive, 2004).

Although most of the technologies discussed in the previous section have not yet begun widespread diffusion, predictions of their market potential suggest a robust, emerging market for wireless connectivity. For example, Synergy Research and IMS Research predict that Wi-Fi will grow to more almost 75 million units shipped in 2005, up from slightly more than 25 million in 2003. Bluetooth will grow from about 50 million to almost 200 million units in the same time period (Finn, 2004).

Factors to Watch

If the goal of mobile computing is "anytime, anywhere communications," progress is moving toward that goal on many fronts. The primary technology of this movement was, of course, the cellular telephone. Soon, PCS technologies began to fill in following the cellular designs, using a larger portion of the spectrum than provided for the original cellular systems, although both cellular networks and PCS are now substantively the same (Mobile basics, 2004).

ABI Research identifies eight areas to watch for further developments:

- Location-based services, including carriers, subscribers, devices, and applications.
- Wireless subscribers and forecasts of subscriber growth for all the services.
- Wireless infrastructure, contracts, and business development.
- Broadband wireless, last-mile solutions, and outlook for end-user development.

- In-building wireless networks and active, passive, and hybrid developments.
- RF power devices, the development and progress of integrated circuits, and batteries.
- WiMAX: the development of 802.16 and 802.20 standards for broadband wireless.
- Smart antenna markets, trends, and standards for next-generation systems for wireless (Infrastructure program, 2004).

Location-based services are position locators, which may use global positioning satellites or geographic information systems, based on maps and certain geographic features such as streets or rivers. At the basic level, information can be given about one's location. Other applications include location-based billing for services, emergency location services, and tracking, which is what trucking firms do to locate vehicles or operators.

The number of wireless subscribers will continue to increase. As wireless devices converge, there will likely be diminishing use of PDAs and an increase in the use of handsets that provide cellular, Bluetooth, Wi-Fi, and PDA functions.

In the broadband wireless realm, look for significant increases in speed in the last link to the user. Ultrawideband capability will increase in both capacity and in distance. As handsets become smaller and more efficient, with larger capacity computer processors, the need for broadband functions will also grow to include all forms of multimedia including streaming.

For cell phones, a major development will be improvement of flash memory—those small cards that digital cameras use. The capacity of flash memory is increasing at a rate comparable to that of hard drives a few years ago, and flash has the advantage of being solid-state with no moving parts. Today, a flash memory card can hold as much as a gigabyte of information, and soon we may have high-resolution "camera phones, phone camcorders, phone videoconferencing, and more" that rely upon this technology. As this memory capacity increases and prices diminish, and as battery power and life increases (though not as fast), there will likely be an expansion of functions for cell phones themselves, but also an increased flexibility in the configurations of mobile devices, perhaps leading to such things as phones built into eyeglasses or more "wearable" computers (Gomes, 2004).

In-building wireless networks will likewise expand, using the various technologies described. As network capability increases, so also do many of the problems associated with that increase: increased costs for hardware and network systems, increased complexity and the need for associated training, and increased vulnerability to security threats.

Probably the most dramatic changes will come in increases in the capacity and capability of RF (radio frequency) devices, notably the handsets or mobile devices that will use these technologies. Along with the increased speed and capacity will come increased mobility. For example, the development of automotive wireless is just underway, and although automotive Bluetooth was introduced in 2003, more expansive developments are now occurring, including the role of Wi-Fi, satellites, and other technologies (Automotive wireless, 2004).

WiMAX system development will also dramatically increase the speed and capacity of wideband services. There is also enormous potential for many more radio frequencies available for PCS after over-the-air television stations relinquish the channels being replaced by digital channels.

More Convergence

As many desktop functions have converged—telephone + computer + television + multifunction office devices + interconnectivity— there will be a growing convergence of mobile computing functions. The cell phone already takes photos and can be combined with a PDA or a laptop computer to include not only work functions and interconnectivity but also movies, games, and music.

Along with this growing convenience will also be a growing aggravation. Cell phones connected to the Internet are plagued by pop-up ads, spam, unwanted e-mail, viruses, pornography, and a host of other things currently unwanted through the Internet—all at the cost of per-minute cell phone charges. Cell phone e-mail spam has become such a large problem in Japan that cellular companies have had to take major steps to curb it. Further, as cell phones become more versatile and able to download programs, they become more vulnerable to viruses as well, though such viruses are currently very rare (Pringle, 2004).

Security Concerns

The need for security increases with the proliferation of technologies that use airwaves, for obvious reasons. In Wi-Fi networks, anyone can hitchhike through a company's hot spots, which do not usually honor the boundaries of walls or property. Encryption and firewalls become all the more important as one's signals lie bare to intrusion, sabotage, and spying. So-called "virtual private networks (VPNs)" provide some security, although hackers can find ways around them. Managers of mobile systems must be more attentive to matters of network security, such as keeping administrative functions off the network and protecting the privacy of sensitive information, such as financial or health information, and, at the same time, enabling the mobile networks the greatest flexibility and usage (Nasaw, 2004).

Usability, Practicality, and Perceived Benefits

An important concern with any and all new technological developments is the balance between the benefits to be derived and the costs in both money and time to implement new technologies. For example, a recent conversation with a medical specialist revealed that many practicing physicians have an aversion to employing mobile computing as part of their routine daily practice. This physician, reflecting her views and those of colleagues, essentially feels that the computer technology is simply too complex and not worth the time to learn and master. Further, there appears to be a fundamental mistrust with respect to reliability, cost, privacy, and quickly outdated technologies. In the past year, privacy has also become a substantive legal issue, with the implementation of the Health Insurance Portability and Accountability Act to protect people from misuse and unwarranted access to information.

Consider the physician's typical patient-laboratory-hospital-professional colleague interactions. As a session with a patient commences, most physicians quickly review what is often a formidable paper-based patient medical history file. Prior visits, tests performed, surgeries, medications previously prescribed, allergies, and vital medical information is hastily checked since the total scheduled patient visit may only be 20 to 30 minutes long. Thus, perhaps 20% or more of the patient visit time is spent recapping medical history. "[D]espite pressure from large employers, unions, and healthcare advocacy groups—and aggressive marketing by vendors—only a few dozen medical centers across the country are making full use of the latest computerized patient safety systems" (Freudenheim, 2004, p. C6). At Cedars-Sinai in Los Angeles, one such system was withdrawn when physicians complained that "the computerized system was too great a distraction from their medical duties" (Freudenheim, 2004, p. C6).

Mobile computing could combine easy-to-use preformatted patient history files organized by the physician's prioritized access preference. A brief summary of the medical history, in abstract format, could be provided to help orient the physician to the patient. This could be linked to detailed medical history, if required. With a well-designed system, such input as preformatted prescription order forms, medical test orders, follow-up that may be required, and instructions to other related care providers could all be easily accessed and entered through minimal keyboarding and perhaps a touch system.

This form of mobile computing is being put into practice is at Montefiore Medical Center in Bronx (New York). "The doctor wanted to order blood thinner for the patient. The computer checked whether the patient was already taking the medication, whether he was allergic to it, whether there might be a negative interaction with another drug he was taking, whether any of his lab tests indicated a danger in taking the new one, whether the dosage was correct for his size and age, and whether there was cheaper alternative. In a blink, the computer warned of a potentially dangerous interaction with two antibiotics the patient was on, suggested a much lower dose of blood thinner, proposed a less expensive drug, and calculated the approximate dose" (Perez-Pena, 2004, p. C6). At this institution, there are "thousands of fixed computer terminals, tied into a central database … and there are 350 wireless terminals on small carts, a few in each ward, that doctors and nurses wheel up to patients bedsides rather than consulting paper charts." This application of wireless computing cost more than $100 million to develop and install, and took many years to develop, debug, and refine. Obviously, mobile computing for personal convenience and the same technology applied in an intense and critical work environment differ substantially in magnitude and complexity (Perez-Pena, 2004).

In another realm, the "virtual office" has emerged as a growing phenomenon for contemporary and future work environments. Many types of traditional office work that required a fixed physical setting can be relocated to a wide variety of alternative sites. The employee's home, automobile, client/customer locations, or even temporary hourly/daily space can substitute for traditional centralized office space. Among the advantages of a virtual office are "increased productivity, lower real estate and travel costs, reduced employee absenteeism and turnover, increased work/life balance and improved morale, and access to additional labor pools, including disabled workers, to ease skills shortage" (Hrisak, 1999).

Telecommuting technologies also permit freedom of location, instantaneous interaction, fast response, and spontaneity. From a business perspective, the virtual office provides substantial economic benefits. Enterprises are partially relieved of relatively high-cost real estate investment or rental. But the virtual office, or "dispersed collaboration," even in the wake of September 11 when there was a small surge in this development, is not as common or as easy as generally supposed. It requires a commitment across an entire hierarchy of an organization, a safeguard against too much blending of one's work life and personal life, an investment in adequate technology to support telecommuting, and attention to cultural and geographic factors such as work across time zones (Virtual office, 2002).

A major implication of this approach is the need for homes or telecommuting sites to be equipped with multiple high-speed communication lines, as well as wireless cellular technology. The speed and quality of residential communication lines shifts from primarily providing voice-oriented facilities to one that provides rapid data and image transport as well as videoconferencing capability.

Employers must often provide up-to-date computers with organizationally standardized and compatible software, fast modems, communications services, network access, and other related facilities. Further, it is essential that security be given additional emphasis, since there is markedly increased difficulty in maintaining control and limiting unauthorized access to proprietary information.

Since the office can also move into mobile virtual locations, employers may also provide laptop portable computers, modems, PDAs, and fax facilities so that office workers can have almost infinite flexibility in reaching clients, colleagues, and the home office.

Probably the largest growth in mobile computing will be the increased capability of audio and video "streaming" over the Internet through wireless systems. Wireless technologies have enjoyed substantial increase in accessibility and bandwidth, with 3G, Wi-Fi, and other improvements yet be developed.

All told, there is an incessant progress in both technological and system development toward the "anytime, anywhere" Utopia, and it affects not only workers and the workplace, but is pervasive in personal and social lives as well. While the "last mile" used to refer to the last link in the wired telephone system, the term is more appropriately applied now to that last, still a bit sluggish, link in the wireless connection to the end user. That last mile is fast becoming a fast and vast link as well.

Bibliography

Automotive wireless networks. (2004, April 24). *ABI Research*. Retrieved April 24, 2004 from www.abiresearch.com/reports.AWN.html.

Enterprise to drive dual mode cellular/voice over Wi-Fi handsets. (2004, April 20). Press release. *ABI Research*. Retrieved April 24, 2004 from www.abiresearch.com/abiprdisplay2.jsp?pressid=260.

Finn, E. (2004, May). Be careful when you cut the cord. *Popular Science*, 30.

Flynn, L. (2004, May 4). Silicon Valley gears up for another big thing. *The New York Times*, E3.

Freudenheim, M. (2004, April 6). Many hospitals resist computerized patient care. *The New York Times*, C1, C6.

Gomes, L. (2004, April 26). Cell phones get smarter as flash memory gets cheaper, better. *The Wall Street Journal*, B1.

Green, H. (2004, April 26). No wires, no rules. *Business Week*, 95-102.

Hrisak, D. (1999, December). Millions move to the home office. *Strategic Finance, 81* (6), 54-57.

Infrastructure program. (2004, April 24). *ABI Research*. Retrieved April 24, 2004 from www.abiresearch.com/program/CA04-P01.jsp.

Mobile basics. (n.d.). *MobileIn.com*. Retrieved April 25, 2004 from http://www.MobileIn.com/mobile_basics.htm.

Motorola versus the rest of the world: ABI Research observes the ultrawideband battle. (2004, March 16). Press release. *ABI Research*. Retrieved April 24, 2004 from www.abiresearch.com/abiprdisplay2.jsp?pressid=247.

Nasaw, D. (2004, April 26). Wary of wireless. *The Wall Street Journal*, R10.

Perez-Pena, R. (2004, April 4). Bronx hospital embraces online technology that others avoid. *The New York Times*, C6.

Pringle, D. (2004, April 26). Cell phones + Internet = … A host of unexpected problems. *The Wall Street Journal*, R11.

Rheinhardt, A. (2004, April 26). A machine-to-machine "Internet of things." *Business Week*, 102.

Rerisi, E. (2003, October 21). Where's the growth? Searching for profits in hyper-competitive markets for wireless technology. *ABI Research*. Retrieved April 24, 2004 from www.abiresearch.com/insights/39.html.

Schwartz, M. (2004, April 21). Q&A: Securing mobile workers. *Enterprise Systems*. Retrieved May 5, 2004 from http://esj.com/security/article.asp?EditorialsID=934.

"Virtual office" not yet common. (2002, March). *Financial Executive, 18* (2), 10.

12

Interactive Video Game Technologies

John Long, Ph.D.*

The terms video game and computer game are often used interchangeably, although their markets vary moderately. To use one or the other, there are some common components. Both have a display module consisting of a television, monitor, or liquid crystal display (LCD). The internal system configuration, detailed extensively throughout this chapter, is a necessary component, as is a user control interface that includes a controller, joystick, keyboard, or mouse, or more recently, voice recognition.

This chapter details the technological evolution of the video game console, the emergence of hand-held units, the industry dynamics of the manufacturers, and the status of the software publishing industry. It takes into account the impact of the consumer in terms of use factors and demographics and provides insight into its position in a consumer market that is intensively growth-oriented and highly volatile.

Background

The birth and emergence of video games into our culture is the result of systematic advances in technological sophistication and mutually interdependent relationships with other growth-oriented

* Chair & Professor, Department of Communication Design. California State University, Chico (Chico, California).

technologies. Since their inception, these developing video game technologies have been conditioned by corporate investment, contemporary economics, social interests, politics, and processor speeds. Entering their third generation of users worldwide, video games are now clearly established as a major industry with unimaginable growth potential.

Retrospectively, the video game industry is comprised of seven different stages, each with its own protocols, systems, and software. A preliminary phase began in 1966 when Ralph Baer, often referred to as the Thomas Edison of video games, supervised a team of 500 engineers that constructed the first video game console (Hart, 1996a). Funded by the Pentagon and kept top secret, this system evolved into a rudimentary hockey game. Although unimpressed, the military reasoned there may be future application and continued to fund the project.

Magnavox introduced the first home video game system, Odyssey, in 1972, after years of negotiation with Ralph Baer and manufacturers of television sets. This system used a few logic switches (not a microprocessor) and offered 12 games, separate controllers, and graphic overlays. Shortly thereafter, Nolan Bushnell founded the Sygzy Company, renamed it Atari (which means "prepare to be attacked" in the Japanese game *GO*), and marketed the very simple stand-alone game *Pong* for home use (Polsson, 2002). Odyssey, the first home video game for non-military personnel, had a lucrative first year, selling over 100,000 units at $100 per system. *Pong* won the consumer battle due to the development of low cost LSI (large-scale integrated) circuits (Hart, 1996c). Atari's sales in 1975 were about $40 million.

The emergence of second-generation systems (1977-1981) was fueled by the implementation of the microprocessor. The result of this advancement provided a dramatic quality increase in graphical and auditory effects. Several corporate developments propelled Atari into the forefront as it introduced the VCS/2600 system. First, Sears and Atari agreed to terms providing the marketing infrastructure. Second, Warner Communications invested heavily in the Atari system, knowing that most profits would be realized from game cartridges. Atari cornered the market through sole licensing for *Space Invaders* and for exclusive right to games based on the popular motion pictures, *E.T.* and *Raiders of the Lost Ark* (Polsson, 2002). Over $5 billion worth of VCS/2600 systems and peripheral products were sold over the next five years (Hart, 1996b).

In the 1981-1984 era (third-generation systems), Atari was still the industry trendsetter. However, it made some critical errors leading to what is known as the "dark ages" of the video game industry. Atari released the 5200 system in 1982 as a follow-up to the VCS/2600. The system included ROM (read-only memory) cartridges with minimal storage, as opposed to computer disks of much higher capacity. Arcade games gained consumer preference due to storage capacity 20 times that of the 5600. With no other viable competition, consumer reaction was disastrous for Atari and the industry. By 1983, Atari was losing $2 million daily, forcing Warner Communications to split Atari and sell off its money-losing home division (The history, 1999).

The computer electronics industry provided impetus for fourth-generation systems (1988-1989). Two events, the reduced cost of dynamic random access memory (DRAM) and the development of the 8-bit processor, allowed home system manufacturers to compete with the arcade industry in terms of graphics quality and operational speed (Hart, 1997). With Atari mired in debt and lacking direction, two Japanese companies, Sega and Nintendo, ventured into the arena armed with new perspectives. Sega was the first of these with its Sega Master System (SMS). It was unique in that it had two cartridge parts, one a standard cartridge, the other much smaller but much less expansive than the memory-intensive standard cartridge. In tandem, these parts could provide a rudimentary "virtual reality" experience for its users when used with 3-D glasses.

Six months after the release of the SMS, Nintendo rushed a basic home video system to the U.S. market. Wrought with defects because of the short design period, but coupled with a clever and effective advertising campaign, the Nintendo Entertainment System (NES) gained a huge market share over SMS. It became the number one selling toy in the United States in 1987 due largely to its game, *The Legend of Zelda*, in concert with the NES being packaged with *Super Mario Bros* and *Duck Hunt* (Smith, 1999). Nintendo's corporate tactics also assisted in market share and notoriety. The company was charged with antitrust violations in all 50 states, price fixing, and intimidation of retailers and competing companies. In spite of these tactics—perhaps because of them—Nintendo attained near monopoly control of the gaming business over this time period.

On the cusp of the transition from fourth- to fifth-generation systems (1989-1995), Nintendo furthered its dominance by releasing its handheld console, Game Boy, which included *Tetris*. A version of *Super Mario* and other popular games would soon follow. However, Nintendo failed to acknowledge the axioms of progress in the industry. Another Japanese company, NEC, entered the marketplace with an arcade-like system known as the NEC Turbographix-16 that relied on a 16-bit graphics platform. Initially, consumers reacted by purchasing the new system over NES by a three to one margin. At this point, however, NEC was doomed to failure because most game developers were under contract to Nintendo, and its CPU (central processing unit) was still only 8-bit (only the graphics processor offered16-bit power) (Stahl, 2003). In contrast, when Sega entered the 16-bit marketplace with the Sega Genesis console six months later, it had a star game attraction: *Altered Beast*. Some companies left Nintendo to work with the technologically-superior Sega system. Nintendo countered by joining forces with NEC to resurrect the Turbographix-16 by providing more games. With the slow processor, however, the NEC system was no match for Genesis' 7.6 MHz CPU (History timeline, 2000). Nintendo soon abandoned the system and NEC.

Realizing that Genesis remained an elite system, Nintendo reacted by developing the Super Nintendo Entertainment System (SNES), and released it in late 1990. SNES did have better graphics, audio, and controllers than Sega's Genesis. Consumers, however, regarded the system as inferior due to its slower processor. With rapid improvements and the help of *Super Mario*, the sales of SNES and Genesis were roughly equal by the end of 1991. Throughout the rest of the fifth generation, these two corporations remained roughly as equals, sharing the marketplace in systems and games sales.

The sixth generation (1995-2001) of systems brought with it 64-bit and 128-bit processor architectures that provided a more realistic and immersive experience for users. Memory became more intensive through use of compact discs (CDs) and later DVDs (digital videodiscs). Sony entered the marketplace in 1994 with the original PlayStation. The Sega Saturn was launched, and Atari re-entered the scene in 1993 with the Jaguar. Nintendo retained its market share with a rollout of its Nintendo 64 in 1996 (Stahl, 2003).

In 1999, Sega released the 128-bit, Internet-ready Dreamcast console; it launched SegaNet, the first online gaming network one year later. About the same time, Sony marketed the PlayStation 2 (PS2). In 2001, the system marketplace received a major, but not unexpected, shock with the advent of Microsoft's Xbox. Realizing the futility of its efforts, Sega stopped producing Dreamcast and focused on the software game market. Finally in 2001, Nintendo released its 128-bit console, the GameCube.

The high-end home video console market now consists of Sony, Nintendo, and Microsoft. Each has is proponents due to game availability, ease-of-use, compatibility, or speed of operation. The major technological differences are compared in Table 12.1.

Table 12.1
Power Comparisons for the Major Consoles

	Sony Playstation 2	Nintendo GameCube	Microsoft Xbox
Processor Type	"Emotion Engine"	"Gekko" IBM Power PC	Intel Pentium III
Speed	300 MHz	485 MHz	733 MHz
Graphics	"Graphics Synthesizer"	"Flipper" ATI Graphics Chip	nVidia 3-D Graphics Chip
Speed	750 MHz	162 MHz	250 MHz
RAM	32 MB RDRAM	40 MB	64 MB
Medium	4.7 GB DVD	1.5 GB Optical disc	4.7 GB DVD
Controller	2 controller ports	4 game controller ports	4 game controller ports
Notable Features	2 USB ports, Firewire port	2 high speed serial ports	8 GB hard drive, expansion port

Source: J. Long

Recent Developments

The events of the early part of this century (2001 to present) have shaped the seventh generation of the video gaming business. Sega had been driven into the software business, Nintendo held a strong share in the hand-held marketplace, Microsoft made its grand entrance, and Sony emerged as the market leader with the release of PS2. In a $10 billion industry, the stakes remain high, and the modes of competition continue to be fierce.

Since its release in 2002, the PS2 has sold over 60 million units in contrast to about 15 million combined for the Gamecube and Xbox (Changing dynamics, 2004). Sony has been able to maintain its dominance through a marketing strategy known as "backward-compatibility." Instead of creating new consoles with new software, Sony has created a more sophisticated system that enables its new system to use the game software developed for previous systems. By allowing consumers to retain games purchased at an average of $50, Sony built in an allegiance to the console without additional cost. Sony's 2003 launch of the PSX in Japan extended this trend. Although pricey ($910), the PSX is a home entertainment device that combines a PS2 with a DVD recorder, a hard-disk based video recorder, satellite and analog TV tuners, and photo album and MP3 features. It is Sony's boldest attempt to create a media hub, a single gateway that aggregates all the digital media in the home (GNETAsia, 2003). This move by Sony has upped the technological stakes for its competition.

Microsoft will likely consider the issues of backward-compatibility when it introduces Xbox2 in 2005, an iteration that will confront the next generation of games, Sony's PS3 and the Nintendo "N5." The Xbox2 will be powered by a trio of IBM-designed PowerPC processors and carry 256 MB of RAM. The precise architecture of the system, especially the DVD technology, remains a trade secret until game publishers provide adequate feedback. However, industry insiders have indicated that Xbox2 will not be shipped with a hard drive installed, but as an optional extra. Further, some claim that the system will employ not three, but four or more IBM PowerPC processors. This structure requires a rethinking of game development to fully take advantage of the increased sophistication (Fahey, 2004). Elimination of the hard drive would make backward-compatibility even more difficult, potentially decreasing its appeal and marketability.

All three of the next generation console makers have announced they will use multiple power processors. The PS3 will likely use up to eight of its cell microprocessors, and in its arrangement with IBM, Nintendo will likely use chips similar to Microsoft in the N5 system. Only Sony will use an internal hard drive in its PS3 similar to its PSX model, but with substantially more record time (Fahey, 2004).

In the hand-held, portable market, Nintendo has held a dominant market share with its Game Boy Advance, which is compatible with the Game Cube and is the top selling video game system in the world. In early 2004, Sony indicated that it was developing a disc-based hand-held known as the Play Station Pocket or the PSP (Nintendo unveils, 2004). Like the Game Boy Advance, it would be compatible with the manufacturer's home console, the PS2, and is expected to have superior features to the Nintendo line. Nintendo has its own long-range development effort known as Project: Nitro, which reportedly will result in a dual-screen hand-held known as NDS (Hueber, 2004). The system provides separate processors for each screen and memory up to one Gigabit. With the dual screens, games for other handhelds will not be compatible with the NDS. Recent reports on the PSP have suggested that Sony will delay its release until 2005 because it wanted more time for Sega to prepare game software. The PSP is designed to attract a different niche in the hand-held market, as it will provide the ability to play games, movies, and MP3 music files (Sony delays, 2004).

Advances in mobile phone technology are providing another outlet for the video gaming industry. Already, Nokia, Motorola, Ericsson, and Siemens have formed the Mobile Games Interoperability (MGI) Forum (Phone play, 2004). Third-generation (3G) networks are likely to support wireless services to mobile phones, pagers, and pocket computers. Sega has been engaged in developing mobile games for Motorola's IDEN phone, and UI Evolution, another publisher, uses a version of Java to create such phone games as *Slider* and *Sokoban Challenge*.

The technology that enables enhanced mobile games is emerging quickly. For example, in 2004, SanDisk introduced a new type of memory known as T-Flash that is designed for mobile phones with storage-intensive multimedia applications, such as interactive multiplayer video games (SanDisk, 2004). ATI Technologies, closely allied with Nintendo, has announced a new family of graphics co-processors for mobile phones. The company claims these chips will enable 3D gaming on mobile phone platforms (ATI introduces, 2004). Similarly, Texas Instruments has unveiled OMAP 2410 and 2420 processors that allow users to play advanced video games that take up to 4-megapixel digital pictures (Krazit, 2004). Nokia, the world's largest maker of cell phones, has recently acquired Symbian, the company that makes the "smartphone" operating system Nokia phones use. Symbian is closely allied with Cellsoft, a Pleasanton (California) company that makes a multi-player game (Preimesberger, 2004).

The thinking in the 3G industry is that video gaming and other services are sufficiently in demand to further support the funding and growth of a next generation of mobile networks designed to facilitate interactive, networked games. Although in a growth cycle with a proven technological infrastructure, the software is considerably less developed in relation to the Sony, Microsoft, and Nintendo platforms. However, the demand is growing, and marketplace trends are predictably bright.

Current Status

Figures for 2003 indicate that sales of video game hardware, software, and peripherals in the United States fell 3% from the previous year to about $10 billion (Takahashi, 2004b). (In contrast, Hollywood box office revenue was $9.3 billion.) There are reasons for this decline. Many astute console users are waiting for the big three to rollout their next-generation systems in 2005. Also, in reviewing the data, hardware revenues diminished 16% largely due to price cutting. Revenues for software were up 5% and U.S. industry's software sales were up 15%. The market share of the PS2 slid from 66% to 50% in 2003. At the same time, the Game Cube market share rose from 9% to 26%, and the Xbox remained steady at 24% (Takahashi, 2004a).

The video game publishing industry also has an intense impact on console sales. As games increase in sophistication and variety, the industry itself is becoming much more concentrated. Electronic Arts, Activision, and Take-Two Interactive Software have surfaced as the major publishers and are quickly driving out the competition (Berkowitz, 2004). Other major publishers, Acclaim Entertainment and Bauer Entertainment, both carry heavy debt. Recently, Acclaim had to sell its rights to *Burnout*, its top seller, to Electronic Arts to avoid bankruptcy (Hwang, 2004).

The publishing industry is keeping a close watch on the consumer marketplace. A recent study determined that 21% of gamers were children or teens, 47% were in their 20s, and 32% were over 30, half of which were over 35 (Emling, 2004). The industry reacted by producing games of WWII simulations and pocket billiards. Recent reports indicate that, in 2003, $1.2 billion of the $5.8 billion of total video game sales were sports titles, such as *Tiger Woods PGA Tour* (Knight, 2004). These games are directly marketed to the over-35 audience.

The Entertainment Software Association (ESA) has reported a drop in the popularity of mature-rated games. Only one title rated mature, *Grand Theft Auto: Vice City*, became a top 10 seller in 2003 (Boyle, 2004). In terms of top-selling games genres in 2003, Figure 12.1 illustrates the relative categories leading into 2004. Games such as *Grand Theft Auto: Vice City* have sparked considerable controversy on the legislative front. Much like the Motion Picture Association of America, the video game industry has developed its own ratings system to help serve as a basis for purchasing decisions. The Entertainment Software Rating Board (ESRB) oversees the rating of video games by assigning each a category and descriptor regarding the content of the age specific categories. The ratings include Early Childhood (EC), Everyone (E), Teen (T), Mature (M), and Adult Only (AO). "M"-rated video games are directed toward those 17 and older, whereas the "AO" category designates material suitable for those over 18. In addition, 31 descriptors aid the consumer regarding the overall theme of the game (use of drugs, tobacco, blood and gore, nudity, and the like) (Atwood, 2004).

Figure 12.1
Top Sellers for 2003

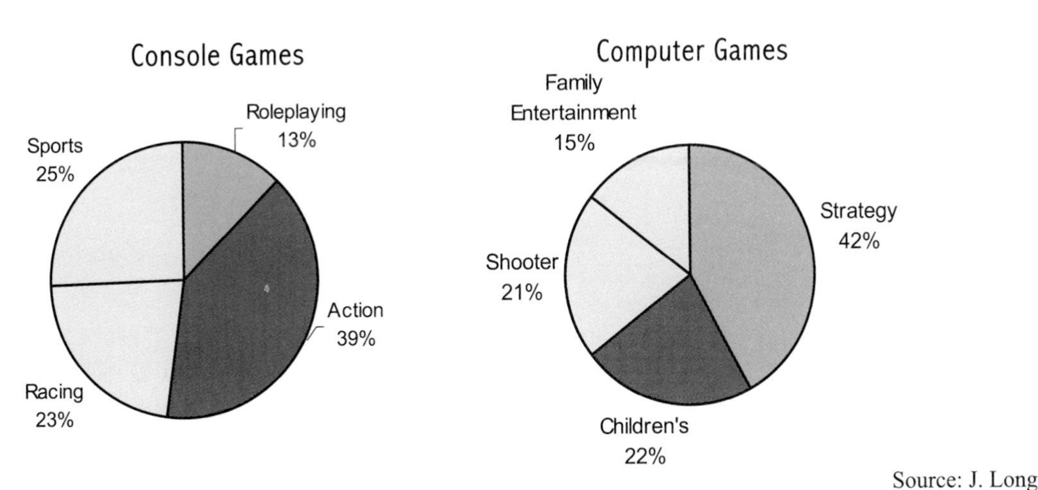

Source: J. Long

In spite of the genuine attempt at self regulation by ERSB, video game software suppliers have been assailed regarding their products' deleterious effects on child behaviors. As with television, some researchers report that violent video games are linked to aggressive, violent acts in some children (Video game, 2003). Others assert that the interactive nature of video games affect aggression by priming aggressive thoughts (Violent video, 2004).

These studies, in concert with a growing body of supporting literature, have brought video games into direct confrontation with the legal community. In response to a 2001 Federal Trade Commission (FTC) report showing that children age 13 to 16 could purchase M-rated games, Congressman Baca (D-CA) introduced a bill (Protect Children from Video Game Sex and Violence Act of 2003) that would impose penalties on those who sell or rent to minors games that depict harmful and sexual activity (Morris, 2003a). That bill died in subcommittee, but Baca has reworked the document for debate later in 2004. His bill was based on successful legislation in St. Louis making it illegal to sell or rent violent video games to minors without parental consent. The ESRB challenged this ruling on First Amendment grounds, but the court initially ruled that video games are not entitled to full constitutional protection. An appeals court reversed the restriction, finding that video games are a protected form of expression (Federal court, 2004).

At the state level, Washington passed a law levying a $500 fine to anyone selling violent video games to those under age 17 (Morris, 2003b). In California, State Assembly member Yee (D-San Francisco) is proposing a bill that prohibits minors from buying M-rated video games involving acts that would be criminal in real life (Nelson, 2004). Even the New York City Council has introduced bills fining merchants $1,500, with the threat of incarceration, for selling violent games to minors (NYC may, 2004).

Further criticism has been leveled at the video game industry regarding its link to childhood obesity. One study suggests a relationship between video game usage and obesity, without finding a similar link for extensive TV usage (Levin, 2004). Others speculate that overweight children tend to play more video games because they are socially ostracized (DeNoon, 2004). Clearly, there are a myriad of interrelated factors that require further examination before any conclusions can be drawn.

The Interactive Digital Software Association reports that males are still the most frequent users of video games in the United States, and that at least one-half are below the age of 34 (Interactive Digital Software Association, 2003). Nielsen has taken note of the behavior of this demographic and claims that members of this group are viewing decreasing amounts of prime-time television. Nielsen also reports that the 18 to 34 male demographic used 9% more DVDs, VCRs, and video games in 2003 than two years earlier (Nielsen Media Research, 2003).

Losing this demographic at an alarming rate is particularly problematic for the advertising industry. The solution appears to be a blending of publishers and advertisers. For example, players of race car games have routinely sped by billboards advertising Ford and Chrysler products. The popular *Tony Hawk* skateboarding game and *The Sims* online game (Electronic Arts) include McDonald's restaurants in its virtual world (Borland, 2004). Massive Incorporated has taken the trend toward advertising in video games one step farther. It has developed an ad server that enables advertisements to be instantaneously downloaded into the game environment. It proposes that saleable advertising for such things as clothing, stores, or cars can be provided anywhere in the game (Massive, Inc., 2004). As online or cell phone gaming becomes omnipresent, this technology will likely be an integral part of the advertising industry.

Factors to Watch

The big three console manufacturers, Nintendo, Sony, and Microsoft, are each poised to release competing versions of digital hub multifunction systems. All will likely have wired and wireless broadband connections for Internet gaming, enhanced stand-alone graphics, and proprietary software appealing to specific audience niches. With so much revenue at stake—some predictions top $15 billion in 2005—other companies will be showing interest. Scientific-Atlanta has developed a version of its Explorer set-top box that offers well-known game titles that can be played over a digital cable network. This gaming device would not compete in the retail segment, but would be distributed directly by cable operators (Baumgartner, 2004).

The human interface is beginning to be addressed. The traditional controller, joystick, or mouse may be replaced by voice recognition. Electronic Arts (EA) has entered into an agreement with Fonix to license its speech recognition software and integrate this technology into its game development strategy. Voice commands are a likely next step in expanding the versatility of the game interface (Fonix signs, 2004). Other systems are being developed that watch gamers' movements and even monitor their pulse and heart rates. In *Wild Divine*, players attach three plastic biofeedback rings, called Magic Rings, that connect to a computer's USB port and then perform tasks, such as surfing. Tasks are better performed by those able to control their emotional feedback state (Slagle, 2004).

There is a synergistic relationship between console manufacturers and the game publishing industry. If firms such as EA and Activision continue to develop games that use nontraditional interfaces such as voice recognition, manufacturers will have to respond. In an industry with a predictably high growth curve, the consumer will likely be the beneficiary of these technological advances.

Bibliography

ATI introduces graphics chip for mobile phone. (2004). *Game Marketwatch*. Retrieved on February 6, 2004 from http://www.gamemarketwatch.com/news/item.asp?nd=2841.

Atwood, B. (2004). ESRB: Understanding video game ratings. *Amazon.com*. Retrieved on April 20, 2004 from http://www.amazon.com/exec/obidos/tg/feature/-/71576/002-5427759-1515260.

Baumgartner, J. (2004). Report: S-A mulling high-end set-top/gaming combo. *CED*. Retrieved on March 4, 2004 from http://www.cedmagazine.com/cedailydirect/2004/0304/cedaily040304.htm,

Berkowitz, B. (2004). Smaller U.S. video game publishers under pressure. *USA Today*. Retrieved on March 1, 2004 from: http://www.usatoday.com/tech/techinvester/2004-02-22-game-squeeze_x.htm.

Borland, J. (2004). Prime time for games. *Globe*. Retrieved on February 26, 2004 from: http://www.globetechnology.com/servlet/story/RTGAM.20031224.htgamefeatdec29/BNSt.

Boyle, E. (2004). Sales of violent video games drop. *Connected Home*. Retrieved on March 1, 2004 from http://www.connectedhomemag.com/HomeTheater/Articles/Index.cfm?ArticleID=41621.

Changing dynamics of video-game industry. (2004, January 16). *India Times: Economic Times*. Retrieved February 13, 2004 from: http://economictimes.indiatimes.com/articleshow/msid-425564.curpg-1.cms.

DeNoon, D. (2004). Video games—Not TV—Linked to obesity. *WebMD*. Retrieved April 20, 2004 from: hppt://my.webmd.com/content/article/84/98017.htm.

Emling, S. (2004). Video game companies tapping older consumers. *Salt Lake Tribune*. Retrieved February 26, 2004 from http://www.sltrib.com/2003/Dec/12222003/Monday/122181.asp.

Fahey, R. (2004, February 2). Xbox2 set to go multiprocessor: Hard drive may not be built-in. *Games Industry*. Retrieved February 13, 2004 from http://www.gamesindustry.biz/content_page. php?section_name-dev&aid=2897.

Federal court rules regulation of video games unconstitutional. (2004). *Xbox solution*. Retrieved April 20, 2004 from: http://www.xboxsolution.com/article820.html.

Fonix signs EA to global licensing agreement for video command solutions. (2004). *Business Wire*. Retrieved March 1, 2004 from http://home.businesswire.com/portal/site/google/ index.jsp?ndmViewld—news_view&newsl.

GNETAsia. (2003, December 16). Sony bets on hybrid PSX games-DVD recorder. *C/Net News*. Retrieved February 28, 2004 from: http://asia.cnet.com/newstech/personaltech/ o,39001147,39161743,00.

Hart, S. (1996a). Guns, games, and glory: The birth of home video games. *Geek Comix*. Retrieved February 6, 2004 from: http://www.geekcomix.com/vgh/g gg2.shtml.

Hart, S. (1996b). Second generation systems, 1977-1981. *Geek Comix*. Retrieved February 6, 2004 from: http://www.geekcomix.com/vgh/second/.

Hart, S. (1996c). Video game infrastructure phenomenon, part 2. *Geek Comix*. Retrieved February 6, 2004 from: http://www.geekcomix.com/vgh/first/inf2.shtml.

Hart, S. (1997). Atari VCS/2600. *Geek Comix*. Retrieved February 6, 2004 from: http://www.geekcomix.com /vgh/second /at2600.shtml.

History timeline—The 1990s (2000). Retrieved on February 28, 2004 from: http://www.mintfresh.Ocatch.com/ time90s.html.

Hueber, J. (2004, February 25). War of hand-held systems brewing in 2004. *Indiana Statesman*. Retrieved March 1, 2004 from: http://www.indianastatesman.com/vnews/display.v/ART/2004/02/25/ 403c338c0a9c7.

Hwang, J. (2004). Video game kings. *The Motley Fool*. Retrieved on March 1, 2004 from http://www.fool.com/ News/mft/2004/mfi04022418.htm.

Interactive Digital Software Association. (2003). Essential facts about the computer and video game industry. *The Entertainment Software Association*. Retrieved on February 6, 2004 from: http://www.theesa.com/ pressroom html.

Knight, D. W. (2004). Sales of sports video games on the rise. *Indianapolis Star*. Retrieved on March 1, 2004 from http://www.indystar.com/articles/9/124264-1069-107.html.

Krazit, T. (2004). TI unveils multimedia phone chips. *PC World*. Retrieved on March 1, 2004 from http://www.pcworld.com/news/article/0,aid,114889.00asp.

Levin, A. (2004). Video games, not TV linked to obesity in kids. *Center for the Advancement of Health*. Retrieved on April 20, 2004 from http://www.cfah.org/hbns/news/video03-1704.cfm.

Massive, Inc. (2004). Massive ad server. *Press Release*. Retrieved March 1, 2004 from http://www.massiveincorporated.com/textbox.html.

Morris, C. (2003a). Mr. Nukem goes to Washington. *Money*. Retrieved on April 20, 2004 from http://money.cnn.com/2003/01/15/commentary/game_over/column_gaming/.

Morris, C. (2003b) Washington to ban "violent" game sales. *Money*. Retrieved on April 20, 2004 from http://money.cnn.com/2003/04/18/commentary/game_over/column_gaming/.

Nelson, N. (2004). Cooperation not legislation to control violent video game sales. *San Francisco Chronicle*. Retrieved April 12, 2004 from http://www.sfgate.com/cgi-bin/article.cgi?f=/g/archive/2004/12/jnelson.DTL.

Nielsen Media Research. (2003). *Research paper*. Retrieved on March 1, 2004 from http://212.26.200.226/VNU/uploads/executive%20summary %20conclusion.pdf.

Nintendo unveils new portable game system. (2004, January 21). *Game Marketwatch*. Retrieved February 6, 2004 from http://www.gamemarketwatch.com/news/item.asp?nid-2848.

NYC may restrict violent video game sales. (2004, April 6). *Join Together Online*. Retrieved April 12, 2004 from: http://www.jointogether.org/gv/news/summaries/reader/0,2061,570271,00.html.

Phone play. (2004). *About, Inc*. Retrieved on February 13, 2004 from: http://portables.about.com/library/weekly/aa080101.htm.

Polsson, K. (2002). Chronology of video game systems. *Islandnet*. Retrieved February 14, 2004 from http://www.islandnet.com/~kpolsson/vidgame/.

Preimesberger, C. (2004). Is Nokia trying to take over Symbian? *IT Manager's Journal*. Retrieved March 1, 2004 from http://mobile.itmanagersjournal.com/mobility/04/02/09/2019258.shtml?tid-46&nd=75&tid.

SanDisk. (2004). SanDisk introduces T-flash—World's smallest removable flash storage module for mobile phones: Motorola first to support T-flash with its new line of mobile phones. *Business Wire*. Retrieved on March 1, 2004 from http://home.businesswire.com/portal/site/google/indes.Jsp?ndmViewld=news_view&newsl.

Slagle, M. (2004, February 26). Bio gaming. *The Sacramento Bee*, D4.

Smith, B. (1999). Read about the following companies: Nintendo, Sega, Sony. *University of Florida Interactive Media Lab*. Retrieved February 26, 2004 from http://iml.jou.ufl.edu/proects/Fall99/SmithBrian/aboutcompany.html.

Sony delays U.S., Europe launch of PSP to 2005. (2004, February 27). *ABS-CBN News*. Retrieved March 1, 2004 from http://www.abs-cbnnews.com/NewsStory.aspa?section=INFOTECH&old=45750.

Stahl, T. (2003, December). Chronology of the history of videogames. *The History of Computing Project*. Retrieved February 29, 2004 from http://www.thocp.net/software/games/games.htm.

Takashashi, D. (2004a). Some gamers scoring amid sales slump. *Silicon Valley.com*. Retrieved on February 26, 2004 from http://www.siliconvalley.com/mld/mercurynews/business/7769102.

Takashashi, D. (2004b). Video game hardware revenues down in 2003. *Mercury News*. Retrieved on February 26, 2004 from http://www.mercurynews.com/mld/mercurynews/news/local/7763478.htm.

The history of Atari. (1999). *Helsinki University of Technology*. Retrieved February 29, 2004 from http://www.hut.fi/~eye/videogames/ atari_history.html.

Video game playing leading factor in kids public health issues. (2003). *National Institute on Media and the Family*. Retrieved on April 20, 2004 from http://www.mediafamily.org/press/20031208.shtml.

Violent video games can increase aggression. (2004). *American Psychological Association*. Retrieved on April 20, 2004 from http://www.apa.org/releases/videogames.html.

13

The Internet & the World Wide Web

Jim Foust, Ph.D.*

In about a decade, the Internet has evolved from a technical curiosity to a major influence on nearly every aspect of life in developed countries. The Internet has become a social force, influencing how, when, and why people communicate; it has become an economic force, changing the way corporations operate and the way they interact with their customers; and it has become a legal force, compelling re-examination and reinterpretation of the law.

No longer merely the domain of technically advanced "geeks," the Internet has, for many, become an indispensable tool for commerce, research, communication, and leisure. It is estimated that more than half of all Americans use the Internet, and that number is growing at an impressive rate. At the same time, there are concerns about a "digital divide" between those who have access to the Internet and those who do not. Despite the fact that, in developed countries, an increasing number of people are now able to access the Internet with lightning-fast broadband connections (see Chapter 20), significant populations of those countries do not have *any* reliable access to the Internet.

Although the terms "Internet" and "World Wide Web" are often used interchangeably, they have distinct—and different—meanings. The Internet refers to the worldwide connection of computer networks that allows a user to access information located anywhere else on the network. The World Wide Web refers to the set of technologies that places a graphical interface on the Internet, allowing users to interact with their computers using a mouse, icons, and other intuitive elements rather than

---

* Associate Professor, Department of Journalism, Bowling Green State University (Bowling Green, Ohio).

typing obscure computer commands. The two technologies can be combined to make possible a variety of types of communications, discussed in more detail in the next section.

This chapter addresses the basic structure and operation of the Internet and World Wide Web, including legal, economic, and social implications. The World Wide Web has become the foundation for a host of other technological developments that bring their own legal, economic, and social effects. Technologies such as Internet commerce (Chapter 14), online games (Chapter 12), and broadband access (Chapter 20) are discussed separately. In this chapter, the focus is on the Internet itself.

Background

In the 1950s, the U.S. Department of Defense started researching ways to create a decentralized communications system that would allow researchers and government officials to communicate with one another in the aftermath of a nuclear attack. A computer network seemed to be the most logical way to accomplish this, so the military formed the Advanced Projects Research Agency (ARPA) to study ways to connect networks. At the time, there was no reliable way to combine local area networks (LANs), which connected computers in a single location, and wide area networks (WANs), which connected computers across wide geographic areas. ARPA sought to create a combination of LANs and WANs that would be called an "internetwork"; ARPA engineers later shortened the term to Internet (Comer, 1995).

By 1969, ARPA had successfully interconnected four computers in California and Utah, creating what came to be called ARPANET. A key innovation in the development of ARPANET was the use of TCP/IP (transmission control protocol/Internet protocol), a method of data transmission in which information is broken into "packets" that are "addressed" to reach a given destination. Once the data reaches its destination, the packets are reassembled to recreate the original message. TCP/IP allows many different messages to flow through a given network connection at the same time, and also facilitates standardization of data transfer among networks. Interest in ARPANET from academia, government agencies, and research organizations fueled rapid growth of the network during the 1970s. By 1975, there were about 100 computers connected to ARPANET, and the number grew to 1,000 by 1984 (Clemente, 1998). In 1983, ARPANET became formally known as the Internet, and the number of computers connected to it continued to grow at a phenomenal rate (see Table 13.1).

The Domain Name System

Each computer on the Internet has a unique Internet protocol (IP) address that allows other computers on the Internet to locate it. The IP address is a series of numbers separated by periods, such as 129.1.2.169 for the computer at Bowling Green State University that contains faculty and student Web pages. However, since these number strings are difficult to remember and have no relation to the kind of information contained on the computers they identify, an alternate addressing method called the domain name system (DNS) is used, which assigns text-based names to the numerical IP addresses. For example, personal.bgsu.edu is the domain name assigned to the computer at Bowling Green State University referred to above. Domain names are organized in a hierarchical fashion from right to left, with the rightmost portion of the address called the top-level domain (TLD). For computers in the United States, the TLD identifies the type of information that the computer contains.

Thus, personal.bgsu.edu is said to be part of the ".edu" domain, which includes other universities and education-related entities. To the immediate left of the TLD is the organizational identifier; in the example above, this is bgsu. The organizational identifier can be a domain as well; the computer called "personal" is thus part of the "bgsu" domain. Table 13.2 lists TLDs in use as of mid-2004.

Table 13.1
Number of Host Computers Connected to the Internet by Year

Year	# of Host Computers	Year	# of Host Computers
1981	213	1993	1,313,000
1982	235	1994	2,217,000
1983	562	1995	4,852,000
1984	1,024	1996	9,472,000
1985	1,961	1997	16,146,000
1986	2,308	1998	29,670,000
1987	5,089	1999	43,230,000
1988	28,174	2000	72,398,092
1989	80,000	2001	109,574,429
1990	313,000	2002	162,128,493
1991	535,000	2003	171,638,297
1992	727,000		

Source: Internet Software Consortium

Table 13.2
Top Level Domain Names

Extension	Definition	Extension	Definition
.aero	Air-transport industry sites	.int	International sites
.arpa	Internet infrastructure sites	.mil	Military sites
.biz	Business sites	.museum	Museum sites
.com	Commercial sites	.name	Individuals' sites
.coop	Cooperative organization sites	.net	Networking and internet-related sites
.edu	Educational institution sites	.org	Sites for organizations
.gov	Government sites	.pro	Sites for professions
.info	General usage sites		

Source: J. Foust

Domain names are administered by a global, nonprofit corporation called the Internet Corporation for Assigned Names and Numbers (ICANN), and the only officially authorized TLDs are those administered by ICANN (http://www.icann.org/). A series of computers called root servers, also known as DNS servers, contain the cross-referencing information between the textual domain names and the numerical IP addresses. The information on these root servers is also copied to many other computers. Thus, when a user types in personal.bgsu.edu, he or she is connected to the computer at 129.1.2.169.

Outside the United States, computers are identified by country code top-level domains (ccTLD). In these cases, the last part of the domain name identifies the country, not the type of information. For example, computers in Japan use the .jp ccTLD, while those in Canada use .ca.

Text-Based Internet Applications

Several communications applications developed before the rise of the graphical-based World Wide Web. Some of these applications have been largely replaced by graphical-based applications, while others merely have been enhanced by the availability of graphical components. All of these applications, however, rely chiefly upon text to communicate among computers.

Electronic mail, or e-mail, allows a user to send a text-based "letter" to another person or computer. E-mail uses the domain name system in conjunction with user names to route mail to the proper location. The convention for doing so is attaching the user's name (which is often shortened to eight or fewer characters) to the domain name with an "at" (@) character. For example, the author's e-mail address is jfoust@bgnet.bgsu.edu. E-mail can be sent to one or many recipients at the same time, either by entering multiple addresses or by using a list processor (listproc), which is an automated list of multiple e-mail addresses. E-mail can also contain computer files, which are called attachments. The rise of graphical-based e-mail programs has also made possible sophisticated text formatting, such as the use of different font styles, colors, and sizes in e-mail communication.

Newsgroups are an outgrowth of early computer-based bulletin board systems (BBSs) that allow users to "post" e-mail messages where others can read them. Thousands of newsgroups are available on the Internet, organized according to subject matter. For example, the "alt.video.dvd" newsgroup caters to DVD (digital videodisc) enthusiasts, while "alt.sports.hockey" caters to hockey fans. One of the advantages of newsgroups is that they allow users to look back through archives of previous postings.

Chat allows real-time text communication between two or more users. One user types information on his or her keyboard, and other users can read it in real time. Chat can be used either in private situations or in open forums where many people can participate at the same time. To use chat, users normally enter a virtual "chat room," where they are then able to send and receive messages. A related technology, instant messaging (IM), allows users to exchange real-time text-based messages without having to be logged in to a chat room.

Telnet allows a user to log onto and control a remote computer, while file transfer protocol (FTP) allows a user to exchange files with remote computers. However, since both telnet and FTP are exclusively text-based, both have been, in most cases, supplanted by Web-based applications.

The World Wide Web

By the early 1990s, the physical and virtual structure of the Internet was in place. However, it was still rather difficult to use, requiring knowledge of arcane technical commands and programs such as telnet and FTP. All of that changed with the advent of the World Wide Web, which brought an easy way to link from place to place on the Internet and an easier-to-use graphic interface.

The WWW was the brainchild of Tim Berners-Lee, a researcher at the European Organization for Nuclear Research. He devised a computer language, HTML (hypertext markup language), that allows users with little or no computer skills to explore information on the Internet. The primary innovations of HTML are its graphical-based interface and seamless linking capability. The graphical interface allows text to intermingle with graphics, video, sound clips, and other multimedia elements, while the seamless linking capability allows users to jump from computer to computer on the Internet by simply clicking their mouse on the screen (Conner-Sax & Krol, 1999). WWW documents are accessed using a browser, a computer program that interprets the HTML coding, and displays the appropriate information on the user's computer. To use the Internet, a user simply tells the browser the address of the computer he or she wants to access using a uniform resource locater (URL). URLs are based on domain names; for example, the author's webpage URL is personal.bgsu.edu/~jfoust.

The advent of the World Wide Web was nothing less than a revolution. As illustrated in Table 13.1, the impressive growth rate of the 1970s and 1980s paled in comparison with what has happened since, as users discovered they did not have to have a degree in computer science to use the Internet. Internet service providers such as America Online (AOL) brought telephone line-based Internet access into homes, and businesses increasingly connected employees to the Internet as well. Since HTML is a text-based language, it is also relatively easy to create HTML documents using a word processing program (Figure 13.1). However, more complex HTML documents are usually created using WYSIWYG (what you see is what you get) programs such as Microsoft Frontpage or Macromedia Dreamweaver. These programs allow users to create Web pages by placing various elements on the screen; the program then creates the HTML coding to display the page on a browser (see Figure 13.2).

Figure 13.1
Simple HTML Coding

Hello World!

```
<html>
<head>
<title>Hello Test Page</title>
<meta http-equiv="Content-Type"
content="text/html; charset=iso-8859-1">
</head>

<body bgcolor="#FFFFFF" text="#000000">
<h1>Hello world!</h1>
</body>
</html>
```

Source: J. Foust

Figure 13.2
Complex HTML Coding

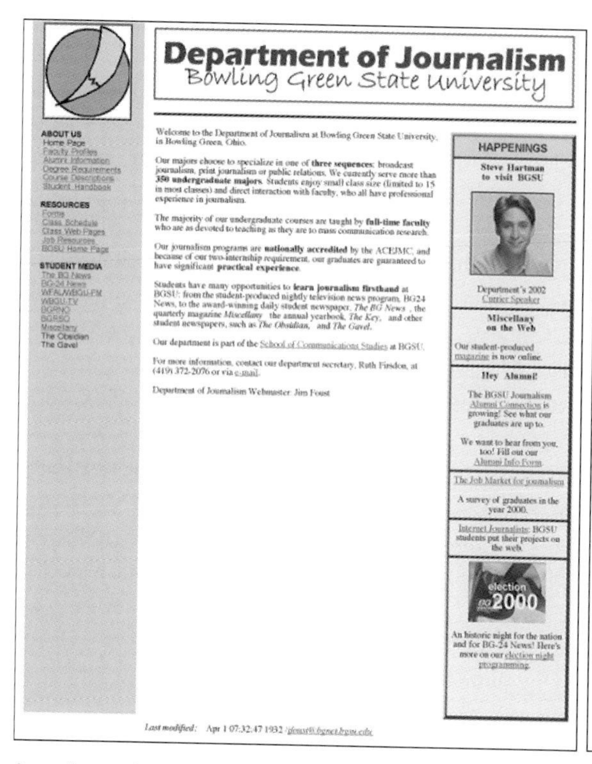


```
<html><!-- #BeginTemplate "/Templates/HomePageTemplate.dwt" -->
<head>
<!-- #BeginEditable "doctitle" -->
<title>BGSU Department of Journalism</title>
<!-- #EndEditable -->
<meta http-equiv="Content-Type" content="text/html; charset=iso-
8859-1">
</head>

<body bgcolor="#FFFFCC" leftmargin="0" topmargin="0"
marginwidth="0" marginheight="0">
<table width="704" border="0" cellspacing="0" cellpadding="0"
align="center" bgcolor="#FFFFFF">
  <tr>
    <td width="150" height="135" bgcolor="#FF9900">
      <div align="center"><img src="images/logo.gif" width="115"
height="120" vspace="0" hspace="0" name="Logo" alt="Journalism
Logo"></div>
    </td>
    <td width="549" height="135">
      <p><img src="images/headertext.gif" width="549" height="89"
alt="Department of Journalism Bowling Green State University"></p>
      <hr>
    </td>
  </tr>
</table>
<table width="704" border="0" cellspacing="0" cellpadding="0"
align="center" bgcolor="#FFFFFF">
  <tr>
    <td width="150" bgcolor="#FF9900" valign="top">
      <table border="0" cellspacing="0" cellpadding="5"
width="97%">
        <tr valign="top">
          <td nowrap>
            <p><b><font face="Arial, Helvetica, sans-serif"
size="2">ABOUT US<br>
               </font></b><font face="Arial, Helvetica, sans-
serif" size="2">Home
              Page </font><b><font face="Arial, Helvetica, sans-
serif" size="2"><br>
               </font></b><font face="Arial, Helvetica, sans-
serif" size="2"><a href="faculty.html">Faculty
              Profiles</a><br>
               <a href="alumni.html">Alumni
Information</a><br>
```

(Continued)

A small part of the HTML code used to create the Web page on the left is illustrated on the right. Because the Web page is so complex, the actual code is quite extensive; printing it would take many pages.

Source: J. Foust

Recent Developments

As noted at the beginning of this chapter, the Internet has become an integral part of existing economic, legal, and social structures. In business, what started out as merely an alternate marketing channel has now become an integral part of the overall economy. Similar conclusions could be reached when considering the Internet's effects on legal and social issues. Still, its relative newness and uniqueness as a communications medium often raise issues that have not been dealt with before. Principles, precedents, and even ethical guidelines often need to be reconsidered when applied to an inherently unique medium that has become so ubiquitous so quickly. That very ubiquity has itself changed the nature of the Internet, making it much less a novelty and much more a tool.

For example, when terrorists attacked the United States on September 11, 2001, the Internet both reflected and affected the human experience and aftermath. On the day of the attacks, Internet-based instant messaging and e-mail allowed people in the vicinity of the attacks to send brief "I'm okay" messages to loved ones, while other types of communications were disabled (Beacham, 2001).

More recently, the Internet has continued to grow in importance for e-commerce, information, and entertainment. Although former Vermont governor Howard Dean was ultimately unsuccessful in his bid to become the Democratic presidential candidate in 2004, he rose from relative obscurity to frontrunner status based largely on his skillful use of the Internet as a grassroots fundraising tool. This phenomenon was certainly not lost on other political candidates, and it is clear that Internet communications will be an important part of presidential—and even state and local—elections from now on (Wolf, 2004).

Economic Developments

During the late 1990s, Wall Street investors created what came to be known as the "dot.com bubble," over-enthusiastically investing in Internet-related companies, regardless of whether the companies were making money—or even seemed to have the potential for making money. In 2001, however, the so-called "dot.com crash" brought an end to such exuberance, leaving hundreds of Internet-related companies out of business, and thousands of their employees out of work (Regan, 2001).

However, in the years since, many companies based on Internet commerce have recovered and thrived, perhaps revealing the "dot.com crash" to be as much hype as the "dot.com bubble" was. In addition, traditional companies have made the Internet an integral part of their business models, and consumers increasingly look to the Internet for purchases large and small. The U.S. Bureau of the Census estimates that total e-commerce sales for 2003 were $54.9 billion, an increase of 26.3% over 2002 (U.S. Bureau of the Census, 2004). (Chapter 14 discusses e-commerce in detail.)

Legal Developments

The use of the Internet to distribute illegal copies of artistic and other works has become the most important legal issue facing the technology. Here, the most famous example involved Napster, an Internet-based service that allowed users to share music files in MP3 format for free. The recording industry eventually forced Napster to end its free service, as courts agreed with the record companies that Napster violated copyright laws. In the meantime, other lower-profile sites emerged that continued to allow users to download and share free copies of music, movies, and other materials. Beginning in late 2003, the Recording Industry Association of America (RIAA) began suing individual users for downloading copyrighted material (Boehlert, 2003). Napster has since reappeared as a subscription service that charges users for copyrighted material, then pays the copyright owner. Apple's iTunes music store operates in much the same way.

In other areas, the legal issues are less clear, thus presenting significant challenges to government regulators. For example, federal and state entities have attempted to regulate unsolicited mass e-mails, or so-called "spam." Such messages usually are advertising for online sites or other offers, often involving pornography, prescription medications, or sexual enhancements. In 2003, President George W. Bush signed into law the so-called "CAN-SPAM Act," placing restrictions on unsolicited e-mail messages. The act prohibits deceptive subject lines in e-mail messages, and requires that the e-mail provide a way for the user to opt out of the e-mail list. However, these and other regulations must walk a fine line between restricting unwanted "junk e-mail" and preserving constitutionally protected free speech (Electronic Frontier Foundation, 2001).

Online gambling involves a similarly tricky balancing of issues. The U.S. Department of Justice (DOJ) says that online gambling is illegal based on a 1961 law designed to stop organized crime entities from using telephones for gambling. The DOJ has been prosecuting operators of online gambling Web sites and pressuring credit card companies not to work with the sites, many of which are located outside the United States. However, the World Trade Organization (WTO) has ruled that such U.S. actions are themselves illegal violations of free trade agreements. Meanwhile, Congress is considering passing laws banning online gambling, but it is not clear how these would affect offshore operations (Coates, 2004).

Social Developments

Computer viruses are an increasing danger for Internet users. Sophisticated viruses are able to circumvent protection software and can not only damage files on individual users' computers, but bog the Internet itself down by infecting Web servers. It is estimated that one in every 300 e-mail messages contains a virus of some sort, and there is increasing fear that viruses may be unleashed by terrorists. While individual users can protect themselves by not opening unfamiliar e-mail attachments and making sure they have the latest updates to their operating system, application, and virus protection programs, the Internet itself remains vulnerable to several types of virus threats (Legard, 2001).

Privacy is also a growing concern, not only from the standpoint of information about users that may be available on the Internet, but also in terms of sites users visit. Today's browser programs allow Websites to store and read "cookies"—small bits of information about users and sites they have visited. This information can be used to track users and, potentially, to retrieve information users would like to stay private. It is also possible to track users according to their IP addresses.

So-called spyware software is another potential threat to privacy. Spyware is software that is installed onto a user's computer without his or her knowledge or consent, often as a result of visiting a certain Website or downloading another program or file. The software can then track the user's Web habits, even, in some cases, logging keystrokes used to type passwords or other data. This tracking information can then be sent back to the original source of the spyware program. The U.S. Congress is currently examining the issue of spyware, and Utah has already passed an anti-spyware law (Carlson, 2004).

There is increasing fear that tracking information will be used improperly by government or business. Internet service provider Comcast alarmed many customers in early 2002 when it was revealed that the company had been tracking the Web usage of nearly a million of its customers without telling them. After the news became public, the company stopped the practice, but the event only heightened fears of privacy advocates. Such information is potentially very valuable to companies that want to know the habits of customers and potential customers (No tracking, 2002). Privacy advocates are no less alarmed by the USAPATRIOT Act, passed by Congress in the wake of the September 11 attacks. The act gives law enforcement agencies greatly expanded powers to monitor Internet use of suspected criminals and demand information from Internet service providers (Stenger, 2001).

Current Status

It is estimated that about half a billion people worldwide have access to the Internet from their homes, and it is a safe assumption that many more who do not have home access have it at work or school. Forty percent of Internet users are in the United States, but the highest rates of growth are in Asia and Europe. Nielsen/NetRatings estimates that 75% of U.S. homes now have access to the Internet (Nielsen/NetRatings, 2004).

Meanwhile, the number of domains continues to grow, although not nearly at the rate of the late 1990s. It is estimated that there were more than 31 million domains as of October 2003 (Domain name, 2003).

A 2002 survey by the Department of Commerce revealed that there remains a "digital divide," especially among low-income people and the less educated:

- 75% of Internet non-users reside in households where income is less than $15,000 per year.
- 87.2% of people with less than a high school diploma do not use the Internet.
- 68.4% of Hispanics and 60.2% of blacks are not online.

The main reason given by those who did not have access was that it was too expensive. There was some good news, however. Users of the Internet in rural areas have nearly closed the gap with their urban counterparts, and access to computers in schools has helped ease the Internet access gap between low- and high-income children and teenagers (U.S. Department of Commerce, 2002).

In December 2002, President Bush signed into law the Dot Com Kids Implementation and Efficiency Act of 2002. The law authorizes the creation of the "kids.us" domain "as a haven for material that promotes positive experiences for children and families using the Internet, provides a safe online environment for children, and helps to prevent children from being exposed to harmful material on the Internet" (New top, 2004). The content of kids.us Web sites will be monitored by registering agencies, and sites that post material deemed inappropriate will be removed.

Factors to Watch

It is unlikely that the growth of Internet use will wane. "In just a handful of years," noted an analyst for Nielsen/NetRatings, "online access has managed to gain the type of traction that took other media decades to achieve" (Nielsen/NetRatings, 2004). This means that the Internet has already become an important part of most people's media landscape, and it is likely that it will continue to grow in importance.

The increasing availability of broadband Internet connections is likely to have a significant effect on how people use the Internet. One implication is that bandwidth-intensive media such as video will become more common, perhaps replacing broadcast, cable, or satellite television for some users. In

2003, President Bush called for "universal, affordable access to broadband technology" by 2007 (Bush wants, 2004).

Another area of potentially great growth is in virtual private networks (VPNs) that use shared public networks such as the Internet to create private connections between two or more computers. In a way, VPNs are the Internet come full circle: at first, the struggle was to interconnect and make accessible a series of discrete networks. Now, businesses are working to carve their own private networks out of the vast public network that is the Internet (Tyson, 2002).

Similarly, intranets and extranets are likely to become more important to businesses, government agencies, and other large organizations. An intranet is essentially a self-contained network that uses Internet protocols and a Web interface for communication and data transfer. Many large businesses have intranets that allow employees to access various types of data in much the same way that they use the Internet. Extranets are similar, but they allow authorized people from outside the organization to access the data as well.

Software advances will be an important part of facilitating this type of information sharing. Extensible markup language (XML) is rapidly gaining acceptance as a protocol for sharing information over computer networks. XML is unique because it defines parts of documents according to the type of data, not the appearance of the data. Thus, rather than defining a color or size of a font as you would in HTML, you simply tag the *function* of the data, such as "headline" or "unit cost." Then, data can be more seamlessly shared across different computers and networks. For example, the simple objects access protocol (SOAP), based on XML, allows various programs to share bits of data among one another, regardless of operating system or computer platform. XML, SOAP, and other protocols that facilitate sharing of information across programs and systems are collectively known as "Web services."

Meanwhile, a consortium of nearly 200 universities and 60 corporations is working on the so-called "Internet 2" project. Although it is not, as the name might imply, a replacement for the existing Internet, it seeks to fundamentally change online communication by increasing and then exploiting the speed of the existing Web. Participating universities are investing $80 million per year in the project, and corporations have pledged an additional $30 million over the life of the venture. The major goal of Internet 2 is "to ensure the transfer of new network technology and applications to the broader education and networking communities" (FAQs, 2004). The project is simultaneously developing technologies that will make the Internet faster, and applications such as digital libraries, virtual laboratories, and tele-immersion that will take advantage of that increased speed. Internet 2 seeks to do this by "recreat[ing] the partnership of academia, industry, and government that helped foster today's Internet in its infancy" (FAQs, 2004). Once again, the Internet may be coming full circle.

Bibliography

Beacham, F. (2001, October 17). 9/11: What worked, what didn't. *TVTechnology.com*. Retrieved April 17, 2004 from http://www.tvtechnology.com/features/Net-soup/f-fb-whatworked.shtml.

Boehlert, E. (2003, November 6). Send lawyers, guns and money. *Salon.com*. Retrieved April 14, 2004 from http://www.salon.com/ent/feature/2003/11/06/cd_sales/.

Bush wants cheap high-speed Internet access for all by 2007. (2004, March 26). *CNN.com*. Retrieved April 14, 2004 from http://www.cnn.com/2004/TECH/internet/03/26/bush.broadband.dc.reut/.

Carlson, C. (2004, April 26). House panel to probe spyware. *eWeek*. Retrieved April 29, 2004 from http://www.eweek.com/article2/0,1759,1573022,00.asp.

Clemente, P. (1998). *State of the net: The new frontier*. New York: McGraw-Hill.

Coates, R. (2004, March 25). U.S. online gambling laws ruled offside. *Silicon.com*. Retrieved April 14, 2004 from http://www.silicon.com/management/government/0,39024677,39119542,00.htm.

Comer, D. (1995). *The Internet book*. Englewood Cliffs, NJ: Prentice Hall.

Conner-Sax, K., & Krol, E. (1999). *The whole Internet: The next generation*. Sebastapol, CA: O'Reilly & Associates.

Domain name facts. (2003, October). *netfactual.com*. Retrieved April 14, 2004 from http:///www.netfactual.com.

Electronic Frontier Foundation. (2001, October 16). *Public interest position on junk e-mail: Protect innocent users*. Retrieved April 14, 2004 from http://www.eff.org/Spam_cybersquatting_abuse/Spam/position_on_junk_email.html.

FAQs about Internet 2. (2004). *Internet 2*. Retrieved April 14, 2004 from http://www.internet2.edu/html/faqs.html#.

Legard, D. (2001, September 20). Viruses are getting faster, tougher. *CNN.com*. Retrieved April 17, 2004 from http://www.cnn.com/2001/TECH/internet/09/20/faster.virus.idg/index.html.

New top level domains. (2004). *AllDomains.com*. Retrieved April 29, 2004 from http://www.alldomains.com/newtlds/kids.html.

Nielsen/NetRatings. (2004, March 25). *Three out of four Americans have access to the Internet, according to Nielsen/NetRatings*. Retrieved April 14, 2004 from http://www.nielsen-netratings.com/pr/pr_040318.pdf.

No tracking: Comcast to stop recording customer Web browsing. (2002, March 11). *ABCNews.com*. Retrieved April 16, 2004 from http://abcnews.go.com/sections/scitech/DailyNews/comcast020212.html.

Regan, T. (2001, December 27). After the dot.com crash. *Christian Science Monitor*, 13.

Stenger, R. (2001, September 13). Feds enlist ISPs in terrorist probe. *CNN.com*. Retrieved April 17, 2004 from http://www.cnn.com/2001/TECH/internet/09/13/fbi.isps/index.html.

Tyson, J. (2002). How virtual private networks work. *howstuffworks.com*. Accessed April 17, 2004 from http://www.howstuffworks.com/vpn.htm/printable.

U.S. Bureau of the Census. (2004, February 23). *Retail e-commerce sales in fourth quarter 2003 were 17.2 billion, up 25.1 percent from fourth quarter 2002, Census Bureau reports*. Retrieved April 14, 2004 from http://www.census.gov/mrts/www/current.html.

U.S. Department of Commerce, National Telecommunications and Information Administration. (2002, February). *A nation online: How Americans are expanding their use of the Internet*. Retrieved April 17, 2004 from http://www.ntia.doc.gov/ntiahome/dn/index.html.

Wolf, G. (2004, January). How the Internet invented Howard Dean. *Wired*. Retrieved April 14, 2004 from http://www.wired.com/wired/archive/12.01/dean.html.

14

Internet Commerce

Nicole Hacker & Jessica Sifford

Also referred to as electronic commerce (e-commerce), Internet commerce is a rapidly changing and expanding industry that has been significantly enhanced by the growth in overall Internet usage and computer access. More and more corporations are turning to the Internet to increase business and remain competitive. The terms Internet/e-commerce and electronic business (e-business) are often used interchangeably without the realization of similarities and/or differences. According to the U.S. Bureau of the Census, e-commerce is characterized as "any transaction completed over a computer mediated network that involves the transfer of ownership or rights to use goods or services" (Mesenbourg, 1999). Similarly, e-business involves any business organization that conducts at least one of the following processes over a computer-mediated network: production, customer, or management focused processes (Mesenbourg, 1999). It is important to recognize the relationship between the two in order to understand specific details within business-to-business (B2B), business-to-consumer (B2C), and consumer-to-consumer (C2C) transactions.

Because of the rise in Internet use and subsequent online business and shopping, it has become imperative that businesses in every sector employ e-commerce technology. The new breed of consumer is the multichannel shopper who utilizes more than one avenue to research, browse, and/or purchase goods or services. DoubleClick, an Internet advertising broker, reported that multichannel shoppers grew from 56% (of all consumers) in 2002 to 65% in 2003 (Rush, 2004). Therefore, it is now more important than ever that businesses take an integrated approach to providing and selling their respective goods and services.

Background

Since the late 1990s, the focus of the Internet has shifted from growth to profitability (Kessler, 2003). It is only natural that the technology that offered efficiency in communication, research, and entertainment would also provide efficiency in connecting businesses to businesses and businesses to consumers (Kilker, 2002).

Notably, the first online retail transaction took place on August 11, 1994, enabled by early encryption software was designed to guarantee privacy for financial information. Soon after, software companies began developing security browsers for use in e-business (Important dates, 1997).

Just as the Internet is an international medium, it is important to realize that e-commerce is conducted around the world as well. The Internet has allowed B2B and B2C transactions that otherwise would be costly or even impossible. For example, companies such as Amazon.com, Monster.com, eBay, and Yahoo!, have significant market share in Europe (EU e-commerce, 2002). This allows for efficient transactions between individuals and/or businesses across international boundaries.

E-commerce has provided an immeasurable amount of convenience to the average online consumer. Some examples include:

- Quick and easy comparative pricing.
- Confidentiality in shopping.
- In-home browsing and purchasing.
- Direct shipping and receiving of gifts.
- Opinion and experience sharing concerning goods and services.
- Availability of specialty goods (niche items).

Recent Developments

Legislation

According to the Congressional Research Service (CRS), state governments garner one-third of their total tax revenue from sales and use taxes. Because of the expansion of the Internet and its use in conducting business, many observers believe that the Internet and e-commerce "threatens to diminish the ability of state and local governments to collect sales and use taxes" (Maguire, 2003).

There are two types of Internet taxes: state sales tax on products purchased on the Internet and state sales tax on Internet access services. Although Congress consistently debates the definition of Internet access, the Internet Tax Freedom Act of 1998 (ITFA) defined Internet access as a service that allows users to access information, electronic mail, or other services associated with the Internet

(Maguire, 2003). Employing this definition, the ITFA "placed a three-year moratorium on the imposition of new taxes on Internet access services or any multiple or discriminatory taxes on electronic commerce by state or local governments" (Luckey, 2004). Simply stated, during the moratorium period, states could not impose a tax on Internet access fees nor on goods or services purchased over the Internet. However, those states imposing and enforcing Internet access taxes prior to October 8, 1998 maintained that ability through a grandfather clause (Luckey, 2004). The 107th Congress then passed the Internet Tax Nondiscrimination Act of 2001 to extend this moratorium through November 1, 2003. Presently, this moratorium has expired.

In the 108th Congress, several bills have been introduced that would affect the Internet tax moratorium. Identical legislation in both the House of Representatives and the Senate (S. 52 and H.R. 49) was introduced in January 2003. The bills sought to make the moratorium permanent, expand the definition of *Internet access* to include all forms of technology enabling such access, and repeal the grandfather clause. States protected by this clause would then lose that revenue source. Although H.R. 49 was passed in the House of Representatives, after consideration, the Senate refused the temporary or permanent extension of the moratorium (Luckey, 2004).

On February 12, 2004, S. 2084 was introduced in order to:

> (1) extend the Internet tax moratorium [approximately] two years through October 31, 2005; (2) include in the moratorium taxes on Internet access delivered through DSL [digital subscriber line]; (3) grandfather all Internet access taxes that were imposed before November 1, 2003; (4) clarify the definition of Internet access services; and (5) implement an accounting rule that would allow the taxation of Internet access if access were offered as part of bundled package and the access provider did not separate Internet access charges from the other services (Maguire 2004).

Interestingly enough, there are at least two known misconceptions regarding state taxation of the Internet. First, States *do* have the right to place sales tax on in-state sales made through the Internet, even after the ITFA was enacted. Second, states *do* have the right to tax transactions made by their own residents even when the seller is located out-of-state. For example, a South Carolina resident purchasing an item from California may only be taxed by the state of South Carolina. Therefore, it can be inferred that the ITFA was not enacted to prevent taxation as a whole, but to prevent double taxation across state lines. The problem from the states' perspective is how to collect taxes from a transaction between a resident of that state and an out-of-state company. States rely upon individuals to report such purchases and pay taxes on them, but many individuals do not bother to do so.

In Europe, legislation is currently centered on e-commerce consumer protection and the development of set Internet standards. The Electronic Commerce Directive, adopted in 2000, establishes a legal framework for e-commerce covering B2B and B2C transactions. The directive attempts to guarantee legitimate online transactions for online consumers in participating countries, not yet including France, The Netherlands, and Portugal.

Technology

When it comes to developing a site to be used for e-commerce, there are two types of Web sites available: informational and dynamic. An informational Web site contains company information, location, products offered, and basic technology that allows users to place orders. Dynamic Web sites, however, utilize a more advanced technology that produces "suggested" special offers based on each user's search history (Simmons, 2003). For instance, Amazon.com uses this personalization software to produce the "gold box" and other personal recommendations.

Personalization software used in sites similar to the aforementioned has recently been criticized as an expensive technology that fails to produce significant increases in online sales. In a study done by Jupiter Research, 14% of consumers cited personalization as their reason to buy more often from an online store, while only 8% said personalization encouraged repeat visits. Moreover, the same study found that basic site improvements, such as faster-loading pages and easier navigation, were much more effective in the eyes of consumers (Gonsalves, 2003).

Another recent development in technology is "price-bot" software, which uses "computerized algorithms that automatically adjust prices to prevailing market conditions" (Deck & Wilson, 2003). Because price-beating policies are used in bricks-and-mortar businesses, it seemed only natural that Internet retailers practice these procedures as well (Murphy, 2003).

Three well-established automated pricing algorithms include undercutting, low-price matching, and trigger pricing. The undercut algorithm allows retailers "to beat any competitor's price by a percentage of the price difference, a fixed percentage, or a fixed dollar amount" (Deck & Wilson, 2003). A more common pricing policy involves the low-price matching algorithm, where retailers pledge to match any competitor's price. The trigger algorithm, on the other hand, is a bit more complicated. It attempts to achieve the lowest price based on previously set price thresholds of sellers. Although the purpose of the "price-bot" system was to provide competitive prices for consumers, very few price-shopping sites remain in existence. According to Nielson/NetRatings, only 15% of online consumers used "price-bot" services (Murphy, 2003). The lack in use of shopping-bot services can perhaps be contributed to consumer brand loyalty.

In the past, online credit card fraud has been a concern of merchants who are typically responsible for its associated costs (Kilker, 2002). Unfortunate incidents such as these have prompted the development of new and improved encryption and authentication software and services from companies such as NTA Monitor. This Internet security company introduced an e-commerce testing service in December 2002 to perform onsite testing of e-commerce systems. The service was slated to evaluate the level of security concerning authentication methods and design aspects of e-commerce systems for individual companies (NTA Monitor, 2002).

Users

After its first exposure to the public, there was a rapid increase in the use of the Internet for different purposes, particularly for electronic commerce. Likewise, usage in every demographic is growing rapidly. A recent study conducted by comScore Media Metrix found that nearly 27 million male Internet users (aged 18 through 34) spent an average of 32 hours per person online in September 2003, whereas the average Internet user only spent an average of 27 hours online that month. Furthermore, while men between the ages of 18 and 34 were browsing for gaming information, their female counterparts were shopping online for products such as jewelry, apparel, and fragrances. A study conducted simultaneously by Nielson/NetRatings reported that children between the ages of 2 and 17 represented 21% of active in-home Internet users (Young men, 2003).

Generally, online consumers are purchasing computer hardware and software, appliances, furniture, housewares, personal care products, CDs, books, and much more. Consumers not only purchase products online, they are also using and purchasing services online, such as financial (banking, credit cards), dating, travel, and self-help services.

A significant portion of children and teenagers do not have access to credit cards for use in online shopping. Therefore, many businesses, especially those targeting children and teens, are using their Web sites to encourage shopping in bricks-and-mortar stores (Net, shopping, 2002).

Kiplinger Business Forecasts predicted in August 2003 that the leading growth in online retail sales will occur in health and beauty, closely trailed by sales in apparel, gifts and flowers, garden products, and more. However, the newest trends tend to lean toward specialty items from specialty sites and away from everyday products such as books and computers (Ostroff & Moore, 2003).

Current Status

According to the U.S. Department of Commerce, online retail sales are rising more than 25% every year. In the fourth quarter of 2003, U.S. e-commerce revenues exceeded $17.2 billion, which accounted for 1.9% of total retail sales. Jupiter Research expects Internet retail revenues to account for 5% of total retail sales in the United States by 2008. Although this amount seems small, it is important to note that large retail categories, such as grocery items, are not successfully sold online (Greenspan, 2004).

Supporting the rise of online sales are C2C business transactions, proliferated by the ever-popular eBay. eBay started as an online auction site that specialized in antiques and collectibles, and later grew to encompass anything that people were willing to both buy and sell online. Now, not only is eBay a successful C2C company, it is has recently developed into a B2C company, as well. Companies such as Motorola, Dell, and BMW North America are taking advantage of eBay's success by selling their products on the site, thus drawing on additional consumer markets (Legg, 2003). As of March 2004, eBay's auto market has become the "largest unit, selling vehicles, parts and accessories worth $7.5 billion last year, up from just $1 billion in 2001" (Sawyers, 2004). eBay's total revenue in 2003, including all B2B, B2C, and C2C transactions, was $2.17 billion, up 78% from 2002.

Even though not all items are purchased online, the Internet remains a valuable research tool for consumers. DoubleClick's multichannel shopping study of 2003 shows that consumers used a combination of store, catalog, and online channels to do their holiday shopping and, more important, the number of multichannel shoppers grew significantly since 2002. While Internet and catalog sales each increased by 2%, bricks-and-mortar retail sales decreased by 2%. All in all, the number of multichannel shoppers increased by 9% in December 2003 (DoubleClick, 2004).

In terms of demographics, triple-channel shoppers using store, online, and catalogs were female. Men were most often dual-channel shoppers and purchased through bricks-and-mortar stores and the Internet. Coincidentally, triple-channel shoppers spent 17% more than dual-channel shoppers.

DoubleClick also reported that 57% of the study's respondents said that they often researched products in one medium and purchased those products in another during the 2003 holiday season:

> Reasons for channel hopping varied depending on the relevant channels. While catalogs appeal to shoppers for their selection, price, and convenience, the majority of multichannel shoppers prefer to use the Internet or a store for purchasing because they are more convenient and offer the ability to see or sample the products (Rush, 2004).

Overall, a University of Michigan study found that Internet retail sales ranked higher in customer satisfaction than other channels (Greenspan & Ecommerce Guide Staff, 2004).

Currently, up-and-coming e-commerce technology involves analytical software that helps businesses quickly understand and utilize information received regarding consumer buying patterns. In February 2004, WebSideStory, an on-demand Web analytics provider, introduced a new version of the HitBox analytics service, now called HBX, that not only counts the number of hits to a Web site, but also analyzes an individual's browse sessions (Newcomb, 2004b). WebSideStory has paired their improved analytics service with Atomz, an on-demand Web site search provider. HBX, when combined with Atomz, "is designed to deliver a new level of reporting on internal searches, including search terms across categories, relevant searches, failed searches, and search term conversions" (Newcomb, 2004a). These new technologies will enable online retailers to adjust and improve their sites to suit the needs of their consumers, resulting in increased traffic and revenues.

Factors to Watch

E-commerce will continue to grow exponentially over the next several years. As revenues continue to increase, the debate concerning the U.S. position on Internet taxes will go on. With the introduction of new technologies that allow for Internet access, Congress will be forced to come to an agreement regarding the definition of *Internet access*, along with a decision finalizing the appropriate implementation of Internet sales and use taxes.

The growth of e-commerce will encourage growth in Web site technology. Internet retailers will discover "must-have" software that will help increase their revenues. Examples will include software such as RichFX, which provides zoom capabilities in order to better view merchandise, and iPIX, that allows for a 360-degree view of products. Equally important will be the choice of payment options and secure purchasing technology. However, retailers should recognize that older computers will not have the capabilities for advanced software such as Flash animation. It will be imperative for retailers to carefully consider consumer preferences (Ostroff & Moore, 2003). In the future, consumers can expect real-time feedback from both retailers and financial institutions.

Industry watchers expect the number of devices attached to the Internet to grow significantly in the coming years. Cell phones, personal computers, automobiles, and home appliances are among the technologies that will connect consumers to the Internet. Because these new technologies will make Internet connection available almost anywhere, the number of online buyers will continue to increase, thus boosting overall e-commerce revenues. Although online sales are projected to increase dramatically by 2008, it is more significant to look at past and current sales and consumer trends. Figure 14.1 demonstrates the steady increase in e-commerce sales between 2000 and 2003.

Figure 14.1
E-Commerce Sales, 2000-2003

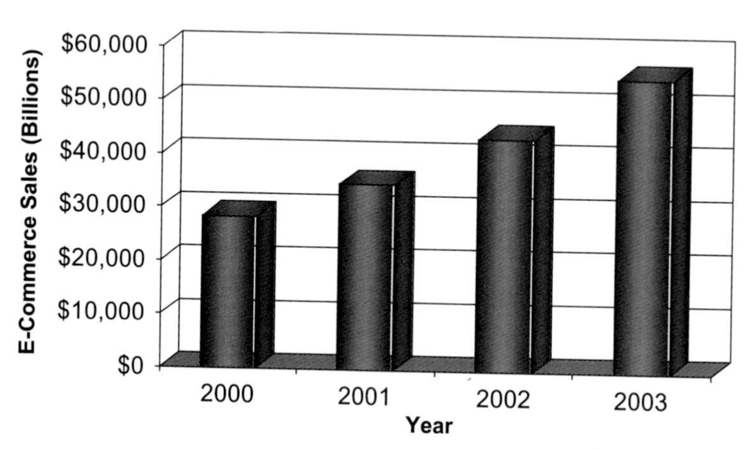

Source: U.S. Department of Commerce

Bibliography

Deck, C. & Wilson, B.. (2003, April). Automated pricing rules in electronic posted offer markets. *Economic Inquiry,* 208-225.

DoubleClick. (2004, January). *Multichannel shopping study-holiday 2003*. Retrieved on February 25, 2004 from http://www.doubleclick.com/us/knowledge_central/documents/research/dc_multichannel_holiday_0401.pdf.

EU e-commerce growth to outpace United States through 2003. (2002, November 1). *Market Europe*. Retrieved February 23, 2004 from http://web7.infotrac.galegroup.com/itw/infomark/274/694/44594109w7/purl=rc1_ITOF_0_A93899356&dyn=5!xrn_2_0_A93899356?sw_aep=usclibs.

Greenspan, R. (2004, February 25). Ecommerce penetration on the rise. *Clickz Network,* Retrieved on March 9, 2004 from http://www.clickz.com/stats/markets/retailing/print.php/3317811.

Greenspan, R. & Ecommerce Guide Staff. (2004, February 20). E-tailers earn high marks for customer satisfaction. *Ecommerce News*, Retrieved February 25, 2004 from http://ecommerce.internet.com/news/news//print/0,,10375_3315991,00.html.

Gonsalves, A. (2003, October 14). Personalization is ineffective on online retail sites; the majority of shoppers visiting a retail site already know what they're looking for, so they want to be able to find the product and make the purchase quickly, says Jupiter Research. *TechWeb News*. Retrieved March 1, 2004 from http://www.internetweek.com/e-business/showArticle.jhtml?articleID=15300234.

Important dates in the history of commerce on the Internet. (1997, December 1). *Internet History*. Retrieved March 10, 2004 from http://www.unc.edu/depts/jomc/academics/dri/pioneers4.html.

Kessler, S, (2003, March 17). Natural selection net-style; on the three year anniversary of the Nasdaq's peak, we take a look at why some dot-coms survived and other disappeared. *Business Week Online.* Retrieved on February 25, 2004 from http://www.businessweek.com/investor/content/mar2003/pi20030314_9616_pi044.htm.

Kilker, J. (2002). Internet commerce, In Grant, A. E. & Meadows, J. H., *Communication Technology Update,* (8th ed.). Oxford: Focal Press.

Legg, G. (2003, January 13). Techies discover eBay. *Design News,* 44-46.

Luckey, J. (2004, March 1). State sales taxation of internet transactions. *CRS Report for Congress.*

Maguire, S. (2003, January 1). Internet commerce and state sales and use taxes. *CRS Report for Congress.*

Maguire, S. (2004, February 24). Taxing internet transactions. *CRS Report for Congress.*

Mesenbourg, T. (1999). *Measuring electronic business: Definitions, underlying concepts, and measurement plans.* U. S. Bureau of the Census. Retrieved March 1, 2004 from http://www.census.gov.epcd/www/ebusines.htm.

Murphy, V. (2003, October 27). The revolution that wasn't. *Forbes Global,* 86.

Net, shopping go together. (2002, August 26). *MMR,* 72.

Newcomb, K. (2004a, February 25). Search meets analytic. *Ecommerce News.* Retrieved February 25, 2004 from http://ecommerce.internet.com/news/news/print/0,,10375_3317821,00.html.

Newcomb, K. (2004b, February 23). WebSideStory aims for new level of analytics sophistication. *Ecommerce News.* Retrieved February 25, 2004 from http://ecommerce.Internet.com/news/news/print/0,,10375_3317821,00.html.

NTA Monitor launches ecommerce testing service. (2002, December 10). *Corporate IT Update.* Retrieved February 23, 2004 from http://web7.infotrac.galegroup.com/itw/infomark/274/694/44594109w7/purl=rc1_ITOF_0_A95138481&dyn=14!xrn_4_0_A95138481?sw_aep=usclibs.

Ostroff, J., & Moore, G. (2003, August 25). Online sales will grow briskly. *Kiplinger Business Forecasts.* Retrieved March 1, 2004 from http://web7.infotrac.galegroup.com/itw/infomark/274/694/44594109w7/purl=rc1_ITOF_0_A107171062&dyn=22!xrn_1_0_A107171062?sw_aep=usclibs.

Rush, L. (2004, February 4). A holistic approach to ecommerce success. *Ecommerce News.* Retrieved February 25, 2004 from http://ecommerce.internet.com/news/news/print/0,,10375_3308671,00.html.

Sawyers, A. (2004, March 1). eBay auto unit shows it has clout. *Automotive News,* 1.

Simmons, K. (2003, March). Catch the wave: two years after the dot-com bust, the surf's still up—What does that mean for your shop's Website? *Impressions,* 30-33.

Young men surf, but young women shop. (2003, December). *Chain Store Age,* 110.

15

Virtual & Augmented Reality

John J. Lombardi, Ph.D.*

> After more than a century of electric technology, we have extended our central nervous system itself in a global embrace, abolishing both space and time as far as our planet is concerned. Rapidly, we approach the final phase of the extensions of man—the technological simulation of consciousness, when the creative process of knowing will be collectively and corporately extended to the whole of human society, much as we have already extended our senses and our nerves by the various media.
>
> —Marshall McLuhan (1994)

Virtual reality (VR) is a complex combination of computer technology, virtual reality hardware, and artistic vision. In its most basic form, VR represents "the forward edge of multimedia" computing (Biocca & Meyer, 1994, p. 185). Imagine being able to walk on, see, and touch the moon; fly an F-14 fighter plane; enter into a human aorta; or perform a delicate surgical procedure without fear of error. All of these experiences can be achieved through the realm of virtual reality.

Virtual reality systems are comprised of three basic components: high levels of user interactivity, high-quality three dimensional computer generated images, and varying levels of user immersion. The last of these is dependent upon the complexity of the VR interface (Pimental & Teixeira, 1995).

* Assistant Professor, Department of Mass Communication, Frostburg State University (Frostburg, Maryland).

Biocca and Meyer (1994) say that a prototypical VR system consists of:

1) A computer that generates and keeps track of virtual objects and renders new images as the user navigates through the virtual environment.

2) Output devices such as a head-mounted display with earphones (see Figure 15.1).

3) Input devices or sensors, such as datagloves, which detect the actions of the user.

Each piece of data that is sent or received by any one of the three components causes a change in the other components.

Figure 15.1
Head-Mounted Display

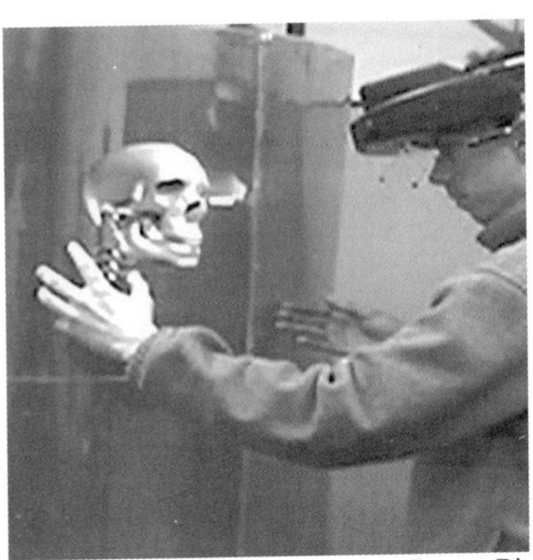

Source: Biocca, M.I.N.D.labs (www.mindlab.org)

The virtual reality experience begins with the creation of a virtual environment (VE). This is done by using computer technology to create realistic three-dimensional graphics. Someone wishing to enter a VE will utilize both output devices (to see and hear) and input devices (to touch and move). At this point, information has been sent only from the computer to the output device. Once the user sees the environment, he or she can make decisions as to how to navigate through the environment. As the user "moves" through the VE, information is sent from the input device back to the computer. Once the computer receives this information, it renders a new VE image, and the process begins again. Figure 15.2 illustrates this process.

Augmented reality (AR, also called mixed reality) is the popular term for using computers to overlay virtual information onto the real world. If you look at VR and true reality as two ends of a spectrum, AR would fall somewhere between the two. AR simply enhances the real environment, where VR replaces it (Tang, et al., 2003; Barfield & Caudell, 2001; Azuma, 1997). Although VR receives much media attention, AR may prove to be more useful, especially with the added range of information supplied from sources such as the Internet (Augmented reality, n.d.).

Figure 15.2
Virtual Reality Experience

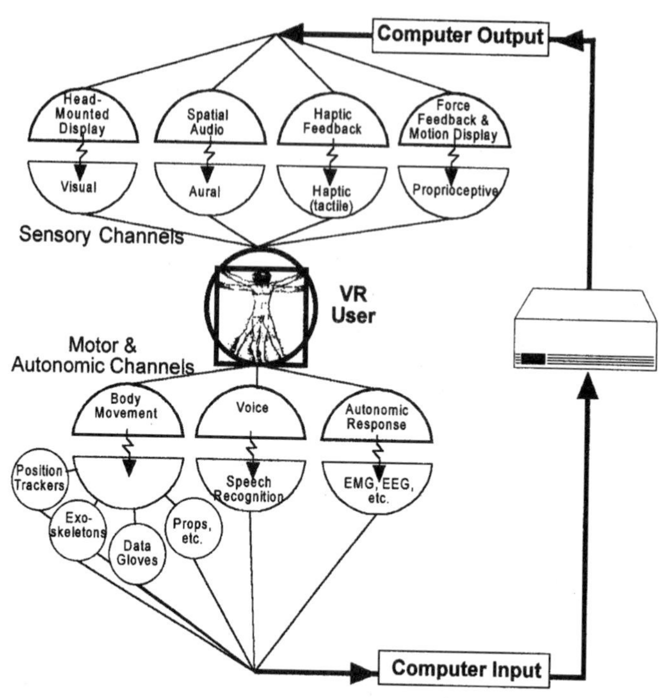

Source: Biocca, M.I.N.D.labs (www.mindlab.org)

Background

There is some argument as to when the concept of virtual reality began. Some believe the concept began with vehicle simulators in the 1920s (Hillis, 1999). Others believe that Morton Heilig's invention of the "Sensorama" in the mid-1950s was the starting point for VR (Welter, 1990). The Sensorama was an arcade-style attraction that allowed users to put their eyes to two stereo mounted lenses. Users would grasp motorcycle handlebars and watch a "movie" of Manhattan traffic. The seat and handlebars would vibrate while the movie gave users the illusion of riding through the city streets (Welter, 1990). Horn (1991) points to Tom Furness' work at Wright-Patterson Air Force Base in 1966 as the launching point of VR. Furness experimented with new methods of displaying visual information for the purposes of flight simulation.

Despite the argument surrounding the origins of VR, it is clear that sufficient expertise existed by the 1960s to move the concept of VR to the next level. Douglass Engelbart, a scientist exploring the idea of interfacing humans and computers, developed the idea of using computer screens as input and output devices (Pimental & Teixeira, 1995). Engelbart's work led to Ivan Sutherland's work on the "ultimate display." The ultimate display was called a "kinesthetic display" and allowed the user to interact with the computer (Sutherland, 1965). Such a display, as Sutherland envisioned, would be a room where objects could be completely controlled by a computer (Sutherland, 1965).

Sutherland's idea of the kinesthetic display led him to propose a system described as a "head-mounted three-dimensional display" (Sutherland, 1968). This display showed a slightly different, two-dimensional image to each eye. The brain would fuse the two images together to form a realistic three-dimensional image. As the user's head moved, the images changed. In addition to creating one of the first headmounted displays (HMD), Sutherland worked on VR developments in flight simulation (Hillis, 1999).

In the early 1970s, the entertainment industry began shaping this technology into what it is today. Moviemakers began using computers to generate thrilling special effects. The increased ability of computers to generate graphics led to various forms of data being displayed as dynamic images. For example, instead of looking at such things as DNA in the form of a pie chart, it was now possible to see a three-dimensional representation of an entire strand of DNA. Despite these advancements, one key component of virtual reality was still missing—interactivity (Mitchell, 1996).

By the late 1970s, the military had developed HMDs capable of real-time visual feedback, and computers were producing more and more sophisticated graphics (Pimental & Teixeira, 1995). When the two sides meshed in the early 1980s, the first primitive versions of virtual reality, complete with three-dimensional graphics and interactivity, emerged (Mitchell, 1996). However, it was not until 1989 when VR notable Jaron Lanier, founder of VPL Research, coined the term "virtual reality" (Heudin, 1999).

In the years that followed, there were no changes in the actual tenets of virtual reality technology. Instead, changes came in the level of sophistication available in virtual reality systems and with an increase in the number of applications for this technology.

Recent Developments

Discussions of the technology since the late 1990s have suggested that there are multiple uses for VR and AR technology, and technological improvements have continued to make this technology more powerful, more precise, and more flexible. However, despite decreasing prices, costs of VR and AR technology remain high. As a result, the main users of and investors in VR and AR technology continue to be large organizations such as the military and the aerospace industry. The most significant changes in VR and AR technology have come in the following areas:

Image Generation

At the heart of any VR or AR system is a high-speed computer capable of generating ultra-realistic graphics. These systems must be powerful enough to allow the computer to make changes to the virtual environment at a speed that appears to be realistic to the user. One such system is the Onyx4, which offers the latest in graphics generation capabilities. New rendering features can create highly realistic materials and self-animating objects, allowing for real-time cinematic-quality scenes (Silicon graphics, n.d.).

Although processing speed is very important in virtual and augmented reality projects, so is the graphics card that is used. NVIDIA Corporation is one of the world's leading manufacturers of graphics processing units (GPUs). In January 2004, NVIDIA announced that NASA will be using the company's photorealistic virtual reality technology to simulate Martian terrain from transmitted rover data, thus allowing scientists to explore Mars in 3D (NVIDIA helps, 2004).

Displays

Virtual reality displays vary depending upon their level of immersion. Displays such as head-mounted displays, virtual retinal displays (VRDs), cave automated virtual environments (CAVEs), and non-expensive automatic virtual environments (NAVEs) have a relatively high level of immersion. Others, such as workbenches and traditional computer monitors, have a relatively low level of immersion.

HMDs are still among the most popular form of high-immersion displays. These units are worn on the head and have small liquid crystal displays (LCDs) covering each eye. They also have stereo headphones covering each ear (NVIS, n.d.).

VRDs, a type of HMD, are smaller, lighter, and capable of generating higher-resolution images. VRD technology generates an image by optically scanning visible and infrared light and sending it directly to the user's eye (Chinthammit & Seibel, 2003). Researchers at the Human Interface Technology Lab (HIT Lab) developed this technology in the mid-1990s (Virtual retinal, n.d.).

CAVEs were developed by the Electronic Visualization Laboratory at the University of Illinois, Chicago in 1992. A CAVE is a type of stereovision system using multiple displays. Instead of wearing an HMD, the user is surrounded by images presented on three to six large screens (VR devices, n.d.). NAVEs are similar to CAVEs, but less sophisticated and less expensive. The current NAVE design is a three-screen environment. Each screen is eight feet wide and six feet tall. The two side screens are positioned at 120-degree angles to the main central screen, resulting in a three-sided display area that is 16 feet wide and approximately seven feet deep. One prototype was built at a total hardware cost of $60,000, including projectors, screens, audio, computers, and lumber (The NAVE, n.d.), compared with a possible cost of more than $1 million for a CAVE (Reaching out, n.d.).

Tracking

The principal component for virtual reality eye tracking research is an HMD-fitted binocular eye tracker, built jointly by Virtual Research and ISCAN. Unlike typical monocular devices, the binocular eye tracker allows the measurement of coordinated eye movements, which, in turn, provides the capability for calculating the three-dimensional world coordinates of the user's gaze (Virtual reality eye, n.d.). Coordinated eye movement is used to calibrate the action that an HMD wearer sees. Think about the contraption your ophthalmologist uses when checking your eyes. When he or she closes off one eye, the object you are looking at appears to be in a certain place. When he or she closes off the other eye, the same object appears to be in a different place. When both eyes are unblocked, the object appears to be in yet a third location. Measuring coordinated eye movement provides an accurate read on exactly where both eyes are looking. This measurement helps to more accurately track the eye movement and yields, at least in theory, a more accurate VR image.

Haptics And Tactile Devices

A haptic display enables users to experience artificially created tactile sensations in response to movements. A haptic display can recreate the experience caused by contact between a tool and an object. This capability is useful in a variety of applications, such as surgical simulators, because when a haptic simulation is combined with a graphic simulation, an enhanced sense of realism is created (Mahvash & Hayward, 2004). One such device is the Freedom 6S, which provides high-fidelity force feedback in six degrees of freedom, giving the user a realistic sense of touch and linking the virtual world to the real world.

CAD design. 3D touch capabilities can allow engineers to digitally design and manufacture engineering activities on the computer that are very similar to the physical world.

Medical simulation. Haptic technology allows surgeons and medical students to "feel" virtual patients while practicing operations. This technology results in reduced learning times and greater success rates.

Telemanipulators. This technology adds the sense of touch to the guidance of a robotic arm. It allows users more accuracy and fidelity in remote operations and manipulations (Freedom 6s, n.d.).

While VR is still being used to conduct flight simulations, as was done in the early years of VR research, this technology is being used in more and more settings. Here are some examples:

- Edietics Corporation has developed a method of using VR technology to conduct safe, accurate, fast, and cost-effective inspections of cargo, containers, and vehicles. This technology, combining state-of-the-art X-ray and high-definition imaging sensors, gives security inspectors the ability to spot items as small as a pinhead that might be in the cargo bay of a tractor-trailer truck. This method of inspection is obviously more important in the post-September 11 era (THSCAN™ inspection, n.d.).

- Researchers at Duke Medical Center are using VR to help reduce the adverse effects patients suffer while undergoing chemotherapy. According to a study from the Duke University School of Nursing and Case Western Reserve, patients have reported feeling less fatigued when using virtual reality as a distraction during treatments (Virtual reality helps, 2004).

- At the University of Buffalo, researchers are using virtual reality driving simulators to treat car accident survivors suffering from post-traumatic stress disorder (PTSD). PTSD researcher J. Gayle Beck says, "To be successful, you need a virtual reality system that taps into the fear structure of the patient. We've developed a very flexible software system that puts the patient in scenarios reminiscent of, or directly related to, their accident" (University of Buffalo, 2003).

- Virtual reality is also being used to help people overcome their addiction to cigarettes. Researchers at the University of Georgia have created a virtual environment based on a social gathering. As participants wander through the environment, they encounter certain settings in which they are tempted to smoke. The goal of this virtual environment is to identify situations that "trigger" the urge to light up. Once this is done, researchers believe that the addiction can be better treated (Johnston, 2003).

- Virtual reality is also finding its way into medical training. In this capacity, VR is used to create a "virtual cadaver" that allows students and established surgeons to practice various procedures, and it allows for new procedures to be developed (Cosman, et al., 2002). One example of such technology is the VRmagic Surgical Simulator used to train ophthalmic surgeons. The cost of training surgeons in this area is expensive, time consuming, and potentially dangerous to the patient. "Without the use of simulation, beginning surgeons can gain surgical experience only on actual patients, thus increasing rates of complications" (eMagin supplies, 2003).

- The advertising industry is also getting into the VR game. Brite Computers and X3D Technologies Corporation have entered into an exclusive agreement. The goal is to put upward of 15,000 50-inch X3D virtual reality displays in public venues such as movie theaters, shopping malls, airports, and hotels. These displays will present three-dimensional advertising video clips that will float inside the screen and up to three feet in the air outside the screen (Brite Computers, 2003).

Augmented Reality

Kooper and MacIntyre (2003) describe a prototype AR system that combines a three-dimensional display with the World Wide Web. The authors refer to this as the "Real-World Wide Web" (RWWW). This system merges the World Wide Web with the physical world to help users more fully comprehend the subject they are studying.

Another practical use of AR is the mobile augmented-reality system (MARS). MARS attempts to superimpose graphics over a real environment in real time and change those graphics to accommodate a user's head and eye movements. This way the graphics always fit user's perspective even while the user is moving. This is achieved using a transparent head-mounted display (with orientation tracker), a tracking system, a differential global positioning satellite system, and a mobile computer all incorporated into one wireless unit housed in a belt-worn device that relays information to an HMD. This allows the user to move freely (Augmented reality, n.d.).

Current Status

Machover Associates predicts the computer graphics industry, including virtual and augmented reality, will keep growing. Overall revenues are expected to grow by 13% compound annual growth rate (CAGR) between 1999 and 2004 (from $71.7 billion to $133.7 billion). The 3D segment is projected to grow faster at 20% CAGR from $24.9 billion in 1999 to $62 billion in 2004. Machover Associates report that revenues worldwide reached $81.7 billion in 2000 and will grow to $149.2 billion by 2005 (Netto & de Oliveira, 2002).

The visual simulation/virtual reality industry revenue has continually increased since the late 1990s. Revenues reached $13.6 billion in 1998, $17.7 billion in 1999, $27 billion in 2000, and $28 billion in 2001 (Delaney, 2001). In total, more than 308,000 VizSim/VR systems were sold in 2002 (VizSim/VR market, 2003).

Factors to Watch

VR and AR technology have numerous uses. However, this technology is certainly not without its drawbacks. The primary drawback of VR and AR technology is cost. A full virtual reality lab can cost more than $30 million. Therefore, it is unlikely that anyone other than large corporations or government agencies will utilize this technology. In the future, costs will continue to drop, and quality will continue to increase. Computer processors are at the heart of any VR/AR system. Because processing power is updated so rapidly, it is easy to understand how quickly improvements can be made to such systems.

Interestingly the ongoing war on terrorism is expected to be the impetus for major gains in the VizSim/VR industry. This is because two of the fastest growing applications for VR technology include military training and gas and oil exploration. A third area, not directly related to the war on terrorism, is virtual prototyping (Virtual reality industry, 2003).

The military uses VR and AR technology to train personnel in vehicle operations and maintenance. The military also uses VR/AR technology to train personnel for dangerous missions. The oil industry is using the advanced visualization components of VR/AR technology to help search for petroleum. VR/AR technology saves money on drilling and allows for more efficient management of refineries and pipelines. Virtual prototyping is used by all types of manufacturers. This type of VR/AR technology allows companies to create virtual products at a fraction of the cost of the real thing. Virtual prototypes can be sophisticated and detailed enough to allow for actual product testing without having to create the actual product (Virtual reality industry, 2003).

The National Research Council has made a set of recommendations to facilitate development of VR applications:

- A comprehensive national information service should be developed to provide information regarding research activities involving virtual environments.
- A number of national research and development teams should be created to study specific virtual and augmented reality applications.
- Federal agencies should begin experimenting with VR technologies in their workplaces.
- The federal government should explore ways to promote the acceptance of universal standards involving hardware, software, and networking technologies (Franchi, n.d.).

In the 1990s, VR was something seen only in movies. Today, it is used worldwide in a variety of settings. Tomorrow, it may be as common as television. Quoting Jaron Lanier (1989), the person who coined the term virtual reality, "VR is a medium whose only limiting factor is the imagination of the user" (p. 108).

Bibliography

Augmented reality explained. (n.d). *About.com.* Retrieved on February 16, 2004 from http://3dgraphics.about.com/library/ weekly/aa012303a.htm.

Azuma, R. (1997). A survey of augmented reality. *Presence: Teleoperators and Virtual Environment, 6* (4), 355-385.

Barfield, W., & Caudell, T. (2001). Basic concepts in wearable computers and augmented reality. In W. Barfield and T. Caudell (Eds.) *Fundamentals of wearable computers and augmented reality.* Mahwah, NJ: Lawrence Erlbaum Associates.

Biocca, F., & Meyer, K. (1994). Virtual reality: The forward edge of multimedia. In R. Aston and J. Schwarz (Eds.). *Multimedia: Gateway to the next millennium.* Boston: AP Professional.

Brite Computers to license. (2003, November 11). *PR Newswire.* Retrieved February 5, 2004 from the LexisNexis Academic database.

Chinthammit, W., & Seibel, E. (2003). A shared-aperture tracking display for augmented reality. *Presence: Teleoperators and Virtual Environments, 12* (1), 1-18.

Cosman, P., Cregan, P., Martin, C., & Cartmill, J. (2002). Virtual reality simulators: Current status in acquisition and assessment of surgical skills. *ANZ Journal of Surgery, 72,* 30-34.

Delaney, B. (2001, October). Moving to the mainstream. *Computer Graphics World, 24* (10), 18.

eMagin supplies OLED display. (2003, September 24). *Business Wire.* Retrieved February 5, 2004 from the LexisNexis Academic database.

Franchi, J. (n.d.). *Virtual reality: An overview.* Frostburg, MD: ERIC Processing and Reference Service. (ERIC Document Reproduction Service No. ED386178).

Freedom 6s force feedback hand controller. (n.d.). MPB Communications, Inc. Retrieved on March 10, 2004 from http://www.mpb-technologies.ca/space/freedom6_2000/f6s/freedom6s.html.

Heudin, J. (Ed.). (1999). *Virtual worlds: Synthetic universes, digital life, and complexity.* Reading, MA: Perseus Books.

Hillis, K. (1999). *Digital sensations.* Minneapolis: University of Minnesota Press.

Horn, M. (1991, January). Science and society: Seeing the invisible. *U.S. News & World Report, 28* (1), 56-58.

Johnston, L. (2003, November, 30). Want to kick the habit? UGA researcher develops virtual reality environment to treat the addiction. *The Associated Press, State and Regional.* Retrieved February 5, 2004 from the LexisNexis Academic database.

Kooper, R., & MacIntyre, B. (2003). Browsing the Real-World Wide Web: Maintaining awareness of virtual information in an AR information space. *International Journal of Human-Computer Interaction, 16* (3), 425-446.

Lanier, J. (1989). *Whole Earth Review. 64,* 108-119.

Mahvash, M., & Hayward, V. (2004, March/April). High-fidelity haptic synthesis of contact with deformable bodies. *IEEE Computer Graphics and Applications.*

McLuhan, M. (1994). *Understanding media: The extensions of man.* Cambridge: MIT Press.

Mitchell, K. (1996). *Virtual reality.* Retrieved February 15, 2004, from http://ei.cs.vt.edu/~history/Mitchell. VR.html.

Netto, A., & de Oliveira, M. (2002, September). *Industrial application trends and market perspectives for virtual reality and visual simulation.* Retrieved March 7, 2004 from http://www.icmc.sc.usp.br/~aneto /notas_rv.htm.

NVIDIA helps NASA reconstruct Mars rover data in virtual reality. (2004, January 19). *NVIDIA Corporation.* Retrieved on March 10, 2004 from http://nvidia.com/object/IO_10688.html.

NVIS, Inc. (n.d.). *New virtual imaging systems.* Retrieved on March 10, 2004 from http://www.nvisinc.com/index.htm.

Pimental, K., & Teixeira, K. (1995). *Virtual reality: Through the looking glass,* 2nd Edition. New York: McGraw-Hill.

Reaching out with technology. (n.d.). *Virginia Tech President's report.* Retrieved March 10, 2004 from http://www.president.vt.edu/presreports/pres9798/reachingout.html#reachingout.

Silicon Graphics Onyx4 UltimateVision. (n.d.). *Silicon Graphics U.S.A.* Retrieved March 10, 2004 from http://sgi.com/visualization/onyx4/overview.html.

Sutherland, I. (1965). The ultimate display. *Proceedings of the IFIP Congress, 2*, 506-508.

Sutherland, I. (1968). A head-mounted three dimensional display. *FJCC, 33*, 757-764.

Tang, A., Own, C., Biocca, F., & Mou, W. (2003). Comparative effectiveness of augmented reality in object assembly. *Proceedings of ACM CHI '2002.*

The NAVE virtual environment. (n.d.). *Georgia Tech Virtual Environments Group.* Retrieved March 10, 2004 from http://www.gvu.gatech.edu/virtual/nave/index.html.

THSCAN™ Inspection Systems. (n.d.). *Eidetics Corporation.* Retrieved March 6, 2004 from http://www.eideticscorp.com/THScan/summary.htm.

University at Buffalo. (2003, December 9). *AScribe Newswire.* Retrieved February 5, 2004 from the LexisNexis Academic database.

Virtual reality eye tracking. (n.d.). *Virtual Reality Eye Tracking Laboratory.* Retrieved March 10, 2004 from http://www.vr.clemson.edu/eyetracking/.

Virtual reality helps breast cancer patients cope with chemotherapy. (2004, January 30). *Science Daily.* Retrieved March 7, 2004 from http://www.sciencedaily.com/releases/2004/01/040130074412.htm.

Virtual reality industry value. (2003, October, 21). *Business Wire.* Retrieved February 5, 2004 from the LexisNexis Academic database.

Virtual Retinal Display (VRD) Group. (n.d.). *Human Interface Technology Lab.* Retrieved March 10, 2004 from http://www.hitl.washington.edu/research/vrd/.

VizSim/VR market to reach $36.2B in 2002. (2003, January 7). *Insight Media.* Retrieved March 3, 2004 from http://www.insightmedia.info/news/VizSimVRMarketoReach.htm.

VR devices. (n.d.). *Electronic Visualization Laboratory.* Retrieved March 10, 2004 from http://evlweb.eecs.uic.edu/research/vrdev.php3?cat=1.

Welter, T. (1990, October). The artificial tourist. *Industry Week, 1* (10), 66.

16

Home Video Technology

Steven J. Dick, Ph.D.*

In the early days, home video technology meant watching live over-the-air programs. The evolution of home video from simple reception devices to the media centers of today marks a tremendous investment for consumers and media companies alike. Each change in home video has given audiences more power, yet at a cost. Frequently, old equipment was abandoned in favor of new.

Background

As technology has improved, it is not enough to simply *receive* video. Increasingly, we have begun to *manipulate* video through storage and editing systems. Finally, *display* technology has grown dramatically in quality and picture size. The most visible part of the home video industry has been reception. Media companies have made billions in either direct costs of reception (subscriptions or sales) or indirect costs (advertising). The way the consumer receives the media has a great deal to do with business models and media options. There are three basic ways for a home to receive an electronic signal: by air, by wire, and by hand.

Reception by Air

U.S. commercial television began in 1941 when the Federal Communications Commission (FCC) established a broadcasting standard. Over-the-air television stations were assigned 6 megahertz of bandwidth in the very-high-frequency (VHF) band of the electromagnetic spectrum. Video is encoded using amplitude modulation (same as AM radio), but sound is transmitted the same as FM (frequency

---

* Assistant Professor, Department of Radio-Television, Southern Illinois University (Carbondale, Illinois).

modulation) radio. Initially, television was broadcast in black-and-white. In 1953, color information was added to the existing luminance signal. This meant that color television transmissions were still compatible with black-and-white televisions.

Television got off to a slow start for reasons that were more political than technical or audience driven. Television development virtually stopped during World War II with only six stations continuing to broadcast. Post-war confusion led to more delays, culminating in 1948 when an inundated FCC stopped processing license applications (Whitehouse, 1986), starting the infamous FCC television freeze. Four years later, the FCC formally accepted a plan to add ultra-high-frequency (UHF) television. UHF television transmissions were encoded the same as VHF, but on a higher frequency. Existing television sets needed a second tuner to receive the new band of frequencies, and new antennas were often needed. This put UHF stations in a second-class status that was almost impossible to overcome. It was not until 1965 that the FCC issued a *final* all-channel receiver law, forcing set manufacturers to include the UHF tuner.

Reception by Conduit

A combination of factors including the public's interest in television, the FCC freeze on new stations, and the introduction of UHF television created a market for a new method of television delivery. As discussed in Chapter 3, cable television's introduction in 1949 brought video into homes by wire. At first, cable simply relayed over-the-air broadcast stations. In the 1970s, however, cable introduced a variety of new channels and expanded home video capability.

Cable television companies are not locked to the same channel allocations as broadcasting since the coaxial cable is a closed communication system. For example, there is a large gap between VHF channels six and seven. Over-the-air, the gap is used for FM radio stations, aircraft communication, and other purposes. Cable television companies use the same frequencies for channels 14 through 22 (Baldwin & McVoy, 1986). Other cable channels are generally placed immediately above the over-the-air VHF channels. Cable companies did not have to use the VHF band, but it made the transition easier. The cable box (at first) simply supplied an outside tuner to receive the extra channels. The industry then promoted the creation of so-called "cable-ready" television sets that allowed reception of cable channels.

Reception by Hand

After Ampex developed videotape for the broadcast industry in 1956, the next logical step was to develop a version of the same device for the home. Sony introduced an open videotape recorder in 1966. Open reel tape players were difficult to use and expensive, so they did not have much effect on the market. Sony went on to develop the videocassette recorder (VCR). VCRs eliminated the need for tape handling since the machine threaded the tape, creating a consumer-friendly, easy-to-use device.

After demonstrating a VCR in 1969, Sony achieved their most powerful machine in 1977—the Betamax—that had a two-hour recording time. Two hours of recording time meant that most movies could be played back on a single tape. However, Sony's Betamax was already falling behind, as a group of companies were developing a competing standard called VHS. In 1977, JVC introduced a VCR with four hours of recording time, enough to record an evening's television programming. By 1982, a full-blown price and technology war existed between the two formats, and, by 1986, 40% of

U.S. homes had VCRs. Eventually, the VHS format became the standard, and Betamax owners were left with incompatible machines.

VCRs combined two functions into one use. Rented and purchased videotapes were essentially a video distribution technology. At the same time, VCRs, as the name implies, allowed home viewers to record (or manipulate) video. This record capability became the subject of a lawsuit (*Sony v. Universal Studios*, 1984, commonly known as the "Betamax" case), as a wary industry attempted to protect their rights to control content. However, the U.S. Supreme Court determined that the record and playback capability (time-shifting) was a justifiable use of the VCR. This decision legalized the VCR and helped force the eventual legitimization of the video sales and rental industry. VCR penetration quickly grew, from 10% in 1984 to 79% 10 years later.

In the 1980s, VCRs received a major challenge from two, incompatible videodisc formats. RCA's "Selectavision" videodisc player used vinyl records with much smaller grooves than audio recording to store the massive amount of information in a video signal. After selling a record-breaking half-million players in its first 18 months on the market, RCA ceded the market to the VCR and stopped making the player and discs (Klopfenstein, 1989).

MCA and Philips introduced a more sophisticated "Discovision" format in 1984 that used a laser to read an optical disc. Lack of marketplace interest led MCA and Philips to sell the format to Pioneer, which renamed it "Laserdisc" and marketed the format as a high-end video solution. Although laserdiscs never enjoyed massive popularity, well over a million players were sold before the format was abandoned in the late 1990s in the face of the emerging DVD (digital videodisc) format.

Video Manipulation

The first manipulation technology was the home video camera. If you track home video back to film, home cameras are much older than television itself. The practical home camera was introduced in 1923 in the 16-millimeter "Cine Kodak" camera and the "Kodascope Projector" (Eastman Kodak, n.d.). Video manipulation became practical with VCRs. The growth of home videotape cameras meant that users did not have to set up a separate system to view home videos. The big impediment to home video cameras was the image sensor. Professional quality cameras used an expensive vacuum pick-up tube as the image sensor. This meant the camera had to have at least one glass bulb a few inches long in it. Replacing the pickup tube with a photosensitive chip reduced camera size and fragility, while increasing reliability. JVC introduced two such cameras in 1976. The cameras weighed three pounds, but were attached to an outside VCR (at least 16 pounds). In 1982, JVC and Sony introduced the combination "camcorder," and the true home video industry was born (CEA, n.d.).

Display

As a fixture in American homes, the history of the television set itself deserves some discussion. The television "set" is appropriately named because it includes *tuner(s)* to interpret the incoming signals and a *monitor* to display the picture. Tuners have changed over the years to accommodate the needs of consumers (e.g., UHF, VHF, cable-ready). Subprocessors were added to the tuners to interpret signals for closed captioning, automatic color correction, and the V-chip.

In 1953, the National Television Standards Committee (NTSC) established the standard for analog color TV reception. The NTSC standard called for 525 lines of video resolution with interlaced scanning. Interlaced means that the odd numbered video lines are transmitted first, and then the display transmits the even numbered lines. The whole process takes one-thirtieth of a second (30 frames of video per second). Interlaced lines ensured even brightness of the screen and a better feeling of motion (Hartwig, 2000).

The first, and still popular monitor, is a cathode ray tube (CRT). The rectangular screen area is covered with lines of phosphors that correspond to the picture elements (pixels) in the image. The phosphors glow when struck by a stream of electrons sent from the back of the set. The greater the stream, the brighter the phosphor glows. Color monitors use three streams of electrons, one for each color channel (red, blue, and green).

Recent Developments

The modern age of home video began in the 1990s, as digital replaced analog technology throughout the media. Digital media is more efficient and generally offers higher quality. However, the transition caused problems for consumers as new services were introduced and home users faced a blizzard of new equipment options.

In the late 1990s, a new by-hand distribution system was introduced. The DVD was developed to be the new mass storage device for all digital content. It is based on the same technology as the compact disc—bits are recorded in optical format within the plastic disc. Unlike earlier attempts to record video on CDs (called VCDs), the DVD had more than enough capacity to store an entire motion picture in broadcast quality. The first DVDs were disadvantaged by the lack of a record capability. However, they were lighter and more durable than VHS tapes. The nonlinear capacity allowed motion picture companies to include bonus content on the discs such as better quality video, multiple language tracks, and even computer programs. As illustrated in Figure 16.1, DVD players and discs were introduced in 1997, and, by 2000, nearly 14 million players had been sold (CEA, 2002).

Although it has not yet had much impact on home video, the 1990s saw the introduction of one more by-wire distribution system: the Internet. With almost no users in 1990, Internet surfing had become a leisure activity for 17.9% of the population by 1999 (U.S. Bureau of the Census, 2001). Computer and media companies spent much of the 1990s jockeying for position. Software such as Real Player, QuickTime, and Windows Media Player threatened existing media companies with a possible competitor no one could ignore. Successful Internet media would make the telephone company a viable competitor. These players and the streaming media formats they use are discussed in detail in Chapter 8.

Internet distribution also enabled home video manipulation with even more advanced digital technologies. Home quality video cameras were introduced that used digital videocassettes (DVCs or DVs). Panasonic and Sony introduced the Mini DV camcorder in 1995. The combination of smaller tapes, flat LCD (liquid crystal display) screens, and chips rather than tubes for optical pickup made for smaller, sturdier cameras. These new cameras motivated users to take cameras where they would not have bothered before.

Figure 16.1
DVD Player Sales

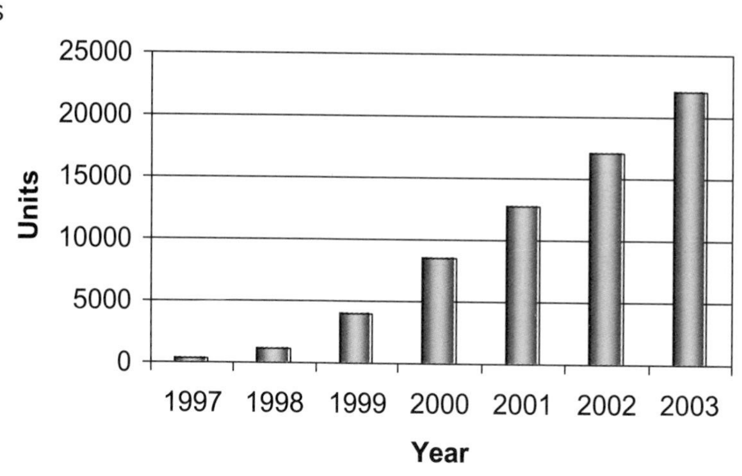

Source: Hunt (n.d.)

Current Status

The biggest factor in the home video industry as of mid-2004 is the decision that consumers have to make in choosing between continuing with VHS or changing over to DVD. By the end of 2004, DVD players are expected to be in two-thirds of U.S. households, with revenue of $16 billion from renting or buying DVDs. At the same time, VCRs are present in 92% of U.S. households, with close to $6 billion of annual video sales (Kipnis, 2004). The dichotomy between ownership and sales is forcing studios to decide if they wish to risk elimination of VHS distribution. The one exception is the children's video category that has strong VHS buy rates.

DVDs are ahead of the curve in one respect. While current DVDs cannot deliver true high-definition quality, they can deliver an enhanced widescreen format designed for newly available screens. These enhanced or anamorphic DVDs can store both a standard aspect ratio picture and widescreen format. People with DVD will get an improved image that will look even better when they eventually buy a new high-definition monitor (Hunt, 2000).

One new entrant has generated more excitement than sales. Digital video recorders (DVRs, also called personal video recorders) have generated a fairly small but extremely loyal following. Initial units were marketed under the ReplayTV and TiVo brands. DVRs are effectively small computers dedicated to managing video recording. The heart of the system is a high-capacity hard drive capable of recording 40 or more hours of video. However, the real power of the DVR is the computer brain that controls it. The DVR is able to search out and automatically record favorite programs. Since it is a nonlinear medium, the DVR is able to record and playback at the same time. This ability gives the system the apparent ability to pause—even rewind—live television.

The DVR's built-in computer power allows the system to do some more controversial things. First, the system could be designed to identify and skip commercials. In a concession to the industry, this feature is largely not implemented. Second, since the system must download program schedules, it can also be used to upload consumer use data. TiVo, for example, has already collected use data such as which programs have been watched and where users have manipulated the programs (e.g., paused or rewound a program).

As of mid-2004, DVRs have not been very successful in the marketplace. The basic hardware cost is at least the price of a high-quality DVD or VCR. In addition, some DVRs require a continuing subscription to an electronic program guide so the machine can search out desired programs. DVR companies have shifted their marketing efforts to license their technology to be incorporated into new cable and satellite boxes.

If consumers are not happy with video entertainment from the outside, a growing number are choosing to produce content at home. Gains made in home video cameras and multimedia computers have been matched by software. Both Microsoft (MovieMaker) and Apple (iMovie) have joined a growing field of companies distributing software designed for home video production. Most computer manufacturers will sell a system that has the ability to record the production on a DVD.

Home video continues to be a popular activity. In 2003, consumers spent more than $24 billion on home videos, a 10% increase from 2002 (Bull, 2004). The Video Software Dealers Association sales figures show a more challenging industry (Gerke & Associates, 2003). The most difficult time came in 1999 and 2000 as consumers switched from tapes to DVDs. Video store operating expenses increased, and the number of titles in the typical store increased dramatically. Figure 16.2 displays the typical video store revenue going from nearly $290,000 in 1998 to just over $335,000 in 2002. Stores have continued to be profitable by increasing revenue per customer and lowering operating costs. Most of the increased revenue comes from extended viewing fees and selling used products (Gerke & Associates, 2003).

Figure 16.2
Home Video Store Revenue Growth

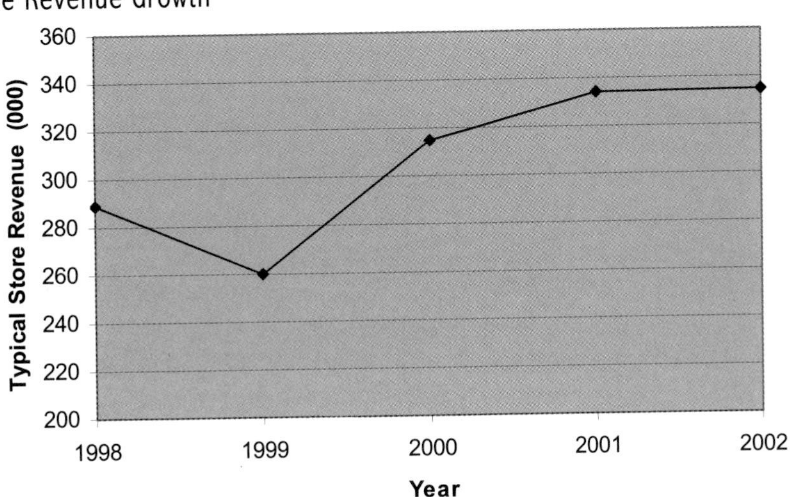

Source: Gerke & Associates (2003)

Factors to Watch

The home video landscape should be expected to continue to change over the next few years, with new delivery options, legal and regulatory issues and constraints, and the introduction of high definition home video playback equipment.

New delivery options include new paths to the home and new purposes for old program models. As broadband Internet penetrates the U.S. market, there is the growing ability to offer full motion movies via online delivery. Two services, Movielink and Cinemanow, have entered this marketplace. They charge video store rental prices and let the user download movies to their own computers with a 24-hour viewing period. Security certificates built into the viewing software limit access to enforce time limits. At the same time, both services promote methods to connect the computer directly to the television. Internet speeds are likely to increase, making distribution of video content even easier.

Another recent entrant has been the DVD-by-mail service. These companies take advantage of the fact that DVDs are light and easy to ship. Subscribers agree to a flat rate per month and are mailed three to seven DVDs at a time. After the subscriber returns a DVD (by mail), they are sent another to replace it. Through an aggressive advertising campaign, Netflix took the early lead and, at the end of the first quarter of 2004, boasted 1.9 million subscribers at $19.95 per month (Netflix, 2004). Traditional video stores are considering similar plans.

New media services have caused some legal concerns. The development of interactive media services (such as the DVR and its associated program guide) has created concerns of privacy, as some DVRs, most notably those made by TiVo, send data on all recording and playback to the company every night. The capability is related to the fact that a DVR is essentially a small computer in a set-top box with all the power to store and transmit information. For example, after the infamous Janet Jackson "wardrobe malfunction" during the 2004 Super Bowl, TiVo reported that 400,000 viewers rewound live television to view the incident. The fact that viewers had the ability is not as troubling to privacy advocates as the fact that TiVo knew this information (Stone, 2004). The balance between anticipating user needs and invasion of privacy is the subject of ongoing legal and policy battles.

Digital media, in general, have generated renewed concerns over copyright. With digital media, a virtually unlimited number of copies can be created from a single original. The motion picture industry lives in fear of a Napster-like, peer-to-peer file sharing service with the capacity to distribute movies. While current broadband transmission speeds preclude practical redistribution of video, video producers do not feel safe. Broadband capacity will continue to increase, making video redistribution easier. In addition, it is already possible to burn video to DVDs or CDs. The industry has responded by working with the FCC to create a "broadcast flag"—a signal encoded into the video that will help control content redistribution (FCC, 2003). At the same time, the DVD industry continues to try to control copying under the Digital Millennium Copyright Act (DVD Copy Control Association, n.d.).

Video Manipulation

A highly anticipated technology is true interactive media. Interactive media provides anything from video on demand to multiple versions of a program. A typical media application may allow the viewer to play along with a popular game show such as *Wheel of Fortune* or predict the next play in a football game. The challenges facing interactive media companies include finding a revenue model and establishing a technical standard. The key question is where to put the interactive content. "First screen" content would place the interactive material on the main television screen—the same as the program itself. "Second screen" content places the interactive material off the main television screen on a palmtop computer or other hand-held device. For more on interactive television, see Chapter 6.

The high-definition television DVD or HD DVR can deliver real interactive content to the consumer. At the very least, either device should allow the ability to record some video and play it back as desired. The prerecorded DVD could have high levels of interactivity, and the DVR could store content sent from an asynchronous digital stream. The challenge for the home user is obtaining systems that are truly able to handle HD content. There are two issues. The first is storage. Does the device have the capacity to store enough HD content to make it usable? The second is the transfer rate. Can the device send the data fast enough to create a quality HD signal? Given the limitations of broadcast spectrum, the data flow rate may initially be as high as 20 Mb/s (Megabits per second).

There are currently two standards for HD DVDs. The first, called Blu-Ray, is supported by Sony, Hitachi, Pioneer, and six others. It is compatible with current DVDs, but requires a more expensive production process. The second, called Blue-Laser, is supported by only NEC and Toshiba. However, it has greater storage capacity and lower production costs (HDDVD.org, 2003).

Display

Television screens will have to become more adaptable to digital media. There are likely to be multiple formats, and the monitor will have to correctly interpret each. Right now, the television industry is exploring two broad levels of quality. The first would be similar to current 525-line analog or standard definition television (SDTV). These pictures will probably have 480 lines. While there are fewer lines in the digital format, the quality of the digital picture will be equal or better. The second broad level of quality is HDTV. The HDTV standard is much broader. Clearly, there will be more lines, but how many? The most popular standards are 720 and 1080. The screen will be wider, 1.85:1 (wide by high), compared with our current analog pictures of 1.33:1. However, the images may have different frame rates (frames per second), audio quality, compression levels, and progressive or interlaced scanning. Each of these changes will dramatically affect the quality of the picture.

Buying an HDTV set is going to be more difficult than people imagine. The first main choice is whether the tuner is built into the monitor or separate. Those depending on cable or satellite to deliver HDTV content may save this cost and not have a built in tuner. The next major decision is the type of television screen. There is no reason why a CRT cannot be used for HDTV; pixels can be made smaller and packed more densely to produce a HDTV image. However, people want larger monitors. The CRT depends on an electron gun to excite the phosphor. That gun must be approximately equidistant from all phosphors. The bigger the screen, the farther away the gun must be. New flat plasma screens use invisible ultraviolet light to excite the phosphors. The flow of electricity causes a small envelope of gas (helium, neon, and xenon) behind each phosphor to give off the ultraviolet light, and that light energizes the phosphor to glow. Since each dot on the screen has it own energy source, the screen can be flat.

Due to advanced production techniques, plasma screens are still very expensive. They remain more of a goal than a reality to the mass market. The same can be said for many new entrants into home video. Too often, companies and individuals must justify product cost with real functionality. Until this is done, many products are going to languish. New display technologies, high-definition displays, DVD players, DVRs, and other interactive technologies are together causing a minor revolution in how we enjoy all forms of video programming in our homes. The irony is that the older technologies are not going away; rather they are moved from one room in the house to another, increasing the number of televisions and playback devices in the home. Any attempt to understand or analyze the impact of these new technologies should also include a bit of attention to what happens to the older technologies, as these remain an important part of the home video landscape.

Bibliography

Baldwin, T., & McVoy, D. (1986). *Cable communications,* 2nd edition. Englewood Cliffs, NJ: Prentice-Hall.

Bull, R. (2004, January 17). A funny thing happened on the way to the video store: Video-store alternatives give new meaning to staying home and watching a movie. *Florida Times-Union*, E-1. Retrieved May 4, 2004 from Lexis/Nexis.

Consumer Electronics Association. (n.d.). *Camcorders*. Retrieved April 1, 2004 from http://www.ce.org/publications/books_references/digital_america/history/camcorder.asp.

Consumer Electronics Association. (2002). *Digital video*. Retrieved April 1, 2004 from http://www.ce.org/publications/books_references/digital_america/history/digital_video.asp.

DVD Copy Control Association. (n.d.). *Announcements*. Retrieved May 4, 2004 from http://www.dvdcca.org/.

Eastman Kodak Company. (n.d.). *Super 8mm film products: History*. Retrieved April 4, 2004 from http://www.kodak.com/US/en/motion/super8/history.shtml.

Federal Communications Commission. (2003). In the matter of: digital broadcast content protection. *Report and order and further notice of proposed rulemaking*. MB Docket 02-230. Retrieved May 4, 2004 from http://hraunfoss.fcc.gov/edocs_public/attachmatch/FCC-03-273A1.pdf?date=031104.

Gerke & Associates. (2003). *2003 benchmarking report for the Video Software Dealers Association*. Retrieved May 4, 2004 from http://www.vsda.org.

Hartwig, R. (2000). *Basic TV technology: Digital and analog,* 3rd edition. Boston: Focal Press.

HDDVD.org. (2003). *The different formats*. Retrieved April 4, 2004 from http://www.hddvd.org/hddvd/difformatsblueray.php.

Hunt, B. (n.d.). CEA DVD player sales. *The Digital Bits*. Retrieved April 4, 2004 from http:://www.thedigitalbits.com/articles/cemadvdsales.html.

Hunt, B. (2000). The ultimate guide to anamorphic widescreen DVD. *The Digital Bits*. Retrieved April 4, 2004 from http:://www.thedigitalbits.com/articles/anamorphic/.

Kipnis, J. (2004). VCR still holds its own in a DVD world. *Billboard, 16* (12), 58. Retrieved April 4, 2004 from http://search.epnet.com/direct.asp?an=12536125&db=aph.

Klopfenstein, B. (1989). Forecasting consumer adoption of information technology and services—Lessons from home video forecasting. *Journal of the American Society for Information Science, 40* (1), 17-26.

National Cable & Telecommunications Association. (2004). *2003 Year-End Industry Overview*. Retrieved March 13, 2004 from http://www.NCTA.com.

Netflix. (2004). *Netflix announces first quarter 2004*. Retrieved April 4, 2004 from http://www.netflix.com/PressRoom?id=5251.

Sony Corporation v. Universal City Studios. (1984). 464 U.S. 417, 104 S. Ct. 774, 78 L. Ed. 2d 574.

Stone, B. (2004, February 16). TiVo's big moment. *Newsweek, 143* (7), 43.

U.S. Bureau of the Census. (2001). *Statistical abstract of the United States: 2001*, 121st edition. Washington: U.S. Government Printing Office.

Whitehouse, G. (1986). *Understanding the new technologies of mass media*. Englewood Cliffs, NJ: Prentice-Hall.

Digital Cinema

Sue Gilgenbach and Steven Dick

Digital cinema is a developing method for motion picture delivery. Digital technology will replace expensive and heavy films with digital media. The transition should make it easier to copy, store, and send motion pictures to theaters.

Currently, the industry is working toward a distribution standard using either a physical media (DVD-like optical disks) or direct electronic transmission. Direct electronic transmission sends films as digital files from a "network operations center" via satellite to theaters. The distribution system should set standards for picture resolution, encryption, and compression. The standard will be similar to the image compression technique used in digital still and video cameras such as JPEG and MPEG. The problem is that these standard compression techniques were designed to handle only 8 bits per pixel for each color channel (a total of 24 bits for the red, green, and blue colors at each pixel location). To achieve the image quality necessary for large-screen projection, 10 bits per pixel are needed for each color channel (30 bits per pixel). The final product must have enough pixels for good image quality, plus enough bits to allow the greater contrast ratio expected in movie theaters. The encryption system placed on the motion picture will minimize piracy, yet facilitate format interoperability.

In the change to digital delivery, theater owners will initially have to invest thousands in new equipment. The cost of digital cinema conversion will vary greatly depending on construction/ remodeling costs and the type of system used. Costs for a single theater with a conventional 35mm projector range from $30,000 to $40,000. Presently, a digital projector costs around $200,000. Additional funding will be needed for other facilities such as wiring, video servers, networks, and satellite reception. All of these will increase the cost per screen to around $250,000. This figure does not include the cost for installation, maintenance, and personnel.

For distributors, digital cinema will dramatically reduce costs. Exhibitors will offer consistent visual/audio presentations—no more film quality deterioration as dust and scratches take a toll with each showing. In addition, exhibitors may generate new revenue through advertising because projectors will be more compatible with video produced for television. New offerings such as live sporting events, teleconferences, documentaries, and motion pictures produced by independent filmmakers may increase attendance. As the motion picture industry continues to develop, digital cinema will enable the transfer of compressed movies between different studios. The final link may be a direct connection to the home or smaller venues such as bars and conference centers.

17

Digital Audio

Ted Carlin, Ph.D.*

Digital diversity continues to drive today's audio industry, and it is giving consumers more ways than ever to enjoy digital audio. The industry has released a number of digital audio options designed to enhance the realism and impact of prerecorded music through surround-sound and home theater technology, deliver new digital audio download and streaming options via the Internet, and provide Internet-delivered music without users being tied to a computer.

Following a weak economy and resulting declines in audio purchases in 2001 and 2002, audio sales rebounded strongly in 2003. According to the Consumer Electronics Association (CEA), the audio category is benefiting from rapid consumer adoption of digital products, in spite of facing declines across traditional home audio components and CD media.

For example, MP3 players surpassed all previous CEA estimates for a record-setting year in 2003, despite continued debates over digital downloading. The renewed interest in and launch of several Internet pay services seems a likely cause for the increased sales. Dollar revenues of MP3 players for 2003 increased to $556 million, a 171% increase over 2002. CEA projects dollar revenues of MP3 players to increase another 27% to $706 million for 2004 (CEA, 2004b). Although traditional audio component sales are down, the continued growth of home-theater-in-a-box sales, which improved by 15% from 2002 to 2003 to more than $630 million, are also creating news in the marketplace. These all-in-one systems can be found in one out of every three U.S. homes today (CEA, 2003).

Advances in home theater and MP3 technology are not the only developments pointing to renewed audio industry vibrancy. The last few years have included technological developments that

* Associate Professor and Chair, Department of Communications/Journalism, Shippensburg University (Shippensburg, Pennsylvania).

have given consumers compelling reasons to buy products that will rekindle their passion for listening to music—the passion that launched the hi-fi industry in the 1950s. Key manufacturers have worked to develop DVD (digital videodisc) based multichannel replacements for the two-channel compact disc (CD). Combination audio and video players are being sold that incorporate one or both of the two standards, DVD-Audio and Super Audio CD.

Audio suppliers, however, do not appear to be banking solely on technological advances to make their profits: 2002 and 2003 saw continuing changes in tradition-defying product designs that simplify purchase and hookup, fit unobtrusively in a home's décor, or fill a need that reflects new music-listening habits at home. Whereas the core audio customer was once a serious music listener who assembled a complex system of standard-size components in a room set aside for serious music listening, audio consumers today are listening to music in more than one room in a house and, especially, outside of the home. In addition, hundreds of electronic games exist in surround- and high-quality sound (CEA, 2004b).

As a result, traditional suppliers of home audio components have developed personal multimedia players and recorders for people on the go, and a broader selection of stylish "micro-sized" stereo systems for use in offices and homes. Suppliers also have improved the price/performance ratio of custom-installed whole-house audio systems that distribute music throughout the house to in-wall speakers powered from a single rack of audio components (CEA, 2004).

Convenience, portability, and sound quality have, therefore, re-energized the audio industry for the 21st century. Competition, regulation, innovation, and marketing will continue to shape and define this exciting area of communication technology.

Background

Analog versus Digital

Analog means "similar" or "a copy." An analog audio signal is an electronic copy of an original audio signal as found in nature. Microphones turn audible sounds into electronic versions of those sounds through mechanical reproduction and transduction. For example, an original sound wave travels through a dynamic microphone's port and causes an internal wire to vibrate. This vibration is transduced (changed) into electronic pulses that are sent through the audio system to a recorder, loudspeaker, etc., where they are stored or reproduced as analog sound waves. As this analog copy is created, some unwanted system distortion (noise) is also being recorded or broadcast. This is due primarily to the amount of electrical impedance present in the system components and cables. In an analog audio system, this distortion can never be totally separated from the original signal. Subsequent copies of the original sound suffer further signal degradation, called generational loss, as signal strength lessens and noise increases for each successive copy. However, in the digital domain, this noise and signal degradation can be eliminated (Watkinson, 1988).

Encoding. In a digital audio system, the original sound is encoded in binary form as a series of zero and one "words" called bits. The process of encoding different portions of the original sound wave by digital words of a given number of bits is called "pulse code modulation" (PCM). This means that the original sound wave (the modulating signal, i.e., the music) is represented by a set of discrete values. In the case of music CDs using 16-bit words, there are 2^{16} word possibilities (65,536). The PCM tracks in CDs are represented by 2^{16} values, and hence, are digital. First, 16 bits are read for one channel, and then 16 bits are read for the other channel. The rest are used for data management. The order of the bits in terms of whether each bit is on (1) or off (0) is a code for one tiny spot on the musical sound wave (Watkinson, 1988). For example, a word might be represented by the sequence 1001101000101001. In a way, it is like Morse code, where each unique series of dots and dashes is a code for a letter of the alphabet (see Figure 17.1).

Figure 17.1
Analog Versus Digital Recording

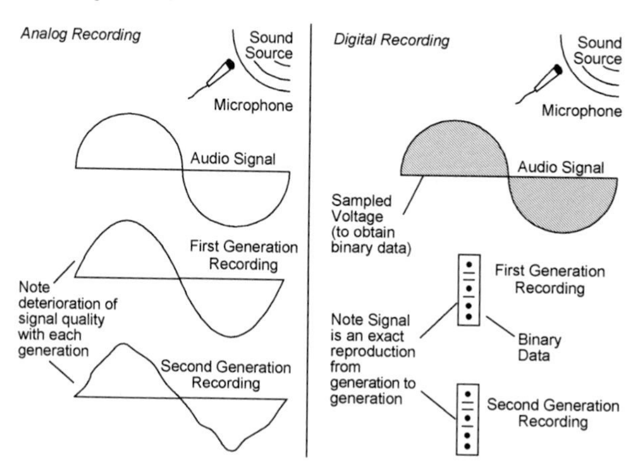

Source: Focal Press

Sampling. Digital audio systems do not create bit word copies for the entire original sound wave. Instead, various samples of the sound wave are taken at given intervals using a specified sampling rate. Three basic sampling rates have been established for digital audio: 32 kHz for broadcast digital audio, 44.1 kHz for CDs, and 48 kHz for professional digital audiotape (DAT) and videotape recording.

This digitalization process creates an important advantage for digital audio versus analog audio:

> A digital recording is no more than a series of numbers, and hence can be copied through an indefinite number of generations without degradation. This implies that the life of a digital recording can be truly indefinite, because even if the medium (CD, DAT, etc.) begins to decay physically, the sample values can be copied to a new medium with no loss of information (Watkinson, 1988, p. 4).

With this ability to make an indefinite number of exact copies of an original sound wave through digital reproduction comes the incumbent responsibility to prevent unauthorized copies of copy-righted audio productions in an effort to safeguard the earnings of performers and producers. Before taking a closer look at the various digital audio systems in use, a brief examination of the important legislative efforts and resulting industry initiatives involving this issue of digital audio reproduction is warranted.

Audio Home Recording Act of 1992

The Audio Home Recording Act (AHRA) of 1992 exempts consumers from lawsuits for copyright violations when they record music for private, noncommercial use and eases access to advanced digital audio recording technologies. The law also provides for the payment of modest royalties to music creators and copyright owners, and mandates the inclusion of the Serial Copying Management System (SCMS) in all consumer digital audio recorders to limit multigenerational audio copying (i.e., making copies of copies). This legislation also applies to all future digital recording technologies, so Congress is not forced to revisit the issue as each new product becomes available (HRRC, 2000).

Multipurpose devices, such as personal computers, CD-ROM drives, or other computer peripherals, are not covered by the AHRA. Exempt devices include digital download devices such as the Diamond Rio player. This means that manufacturers of these devices are not required to pay royalties or incorporate SCMS protections into the equipment. It also means, however, that neither the manufacturers of the devices, nor the consumers who use them, receive immunity from copyright infringement lawsuits (RIAA, 2002).

The Digital Performance Right in Sound Recordings Act of 1995

For more than 20 years, the Recording Industry Association of America (RIAA) has been fighting to give copyright owners of sound recordings the right to authorize digital transmissions of their work. Before the passage of the Digital Performance Right in Sound Recordings Act of 1995, sound recordings were the only U.S. copyrighted work denied the right of public performance.

This law allows copyright owners of sound recordings the right to authorize certain digital transmissions of their works, including interactive digital audio transmissions, and to be compensated for others. This right covers interactive services, digital cable audio services, satellite music services, commercial online music providers, and future forms of electronic delivery. Most non-interactive transmissions are subject to statutory licensing at rates to be negotiated or, if necessary, arbitrated.

Exempt from this law are traditional radio and television broadcasts and subscription transmissions to businesses. The bill also confirms that existing mechanical rights apply to digital transmissions that result in a specifically identifiable reproduction by or for the transmission recipient, much as they apply to record sales.

No Electronic Theft Law (NET Act) of 1997

The No Electronic Theft law (the NET Act) states that sound recording infringements (including by digital means) can be criminally prosecuted even where no monetary profit or commercial gain is derived from the infringing activity. Punishment in such instances includes up to three years in prison and/or $250,000 in fines. The NET Act also extends the criminal statute of limitations for copyright infringement from three to five years.

Additionally, the NET Act amended the definition of "commercial advantage or private financial gain" to include the receipt (or expectation of receipt) of anything of value, including receipt of other copyrighted works (as in MP3 trading). Punishment in such instances includes up to five years in prison and/or $250,000 in fines. Individuals may also be civilly liable, regardless of whether the activity is for profit, for actual damages or lost profits, or for statutory damages up to $150,000 per work infringed (U.S. Copyright Office, 2002a).

Digital Millennium Copyright Act of 1998

On October 28, 1998, the Digital Millennium Copyright Act (DMCA) became law. The main goal of the DMCA was to make the necessary changes in U.S. copyright law to allow the United States to join two new World Intellectual Property Organization (WIPO) treaties that update international copyright standards for the Internet era.

The DMCA amends copyright law to provide for the efficient licensing of sound recordings for Webcasters and digital subscription audio services via cable and satellite. In this regard, the DMCA:

- Makes it a crime to circumvent anti-piracy measures built into most software.
- Outlaws the manufacture, sale, or distribution of code-cracking devices used to illegally copy software.
- Permits the cracking of copyright protection devices, however, to conduct encryption research, assess product interoperability, and test computer security systems.
- Provides exemptions from anti-circumvention provisions for nonprofit libraries, archives, and educational institutions under certain circumstances.
- In general, limits Internet service providers from copyright infringement liability for simply transmitting information over the Internet. Service providers, however, are expected to remove material from users' Websites that appear to constitute copyright infringement.
- Limits liability of nonprofit institutions of higher education—when they serve as online service providers and under certain circumstances—for copyright infringement by faculty members or students.
- Calls for the U.S. Copyright Office to determine the appropriate performance royalty, retroactive to October 1998.
- Requires that the Register of Copyrights, after consultation with relevant parties, submit to Congress recommendations regarding how to promote distance education through digital technologies, while maintaining an appropriate balance between the rights of copyright owners and the needs of users (U.S. Copyright Office, 2002b).

The DMCA contains the key agreement reached between the RIAA and this coalition of non-interactive Webcasters (radio stations broadcasting on the Web), cablecasters (DMX, MusicChoice,

Muzak), and satellite radio services (XM, Sirius). It provides for a simplified licensing system for digital performances of sound recordings on the Internet, cable, and satellite. This part of the DMCA provides a compulsory license for non-interactive and subscription digital audio services with the primary purpose of entertainment. Such a compulsory licensing scheme guarantees these services access to music without obtaining permission from each and every sound recording copyright owner individually, and assures record companies an efficient means to receive compensation for sound recordings. This is similar to ASCAP and BMI compulsory licensing for music used on radio and television stations.

The U.S. Copyright Office has designated a nonprofit organization, SoundExchange, to administer the performance right royalties arising from digital distribution via subscription services. Before its spin-off in September of 2003 as an independent organization, SoundExchange was an unincorporated division of the RIAA originally created in 2000. Members of SoundExchange include record companies and artist representatives from major labels such as Sony Entertainment and Warner Music Group, as well as independents such as Alligator Records. An 18-member board of directors oversees SoundExchange operations, including the establishment of new royalty rates and compulsory license terms *every two years*. Once rates and terms are set, SoundExchange collects, administers, and distributes the performance right royalties due from the licensees to the record companies.

Of the royalties allocated to the record companies, the DMCA states that half of the royalties must be distributed to artists using the following method:

- 45% to the featured artists.
- 2.5% to non-featured musicians through a trust fund jointly administered by the record companies and the American Federation of Musicians (AFM).
- 2.5% to non-featured vocalists through a trust fund jointly administered by the record companies and the American Federation of Television and Radio Artists (AFTRA) (RIAA, 2004a).

SoundExchange only covers performance rights, *not* music downloads or interactive, on-demand Internet services. These are governed by the reproduction right in sound recordings of the Copyright Act, are not subject to the DMCA compulsory license, and must be licensed directly from the copyright owner, usually the record company or artist.

All of this now leaves an artist with three types of copyright protection to consider for one piece of music, depending on how it is used:

- Traditional compulsory license via ASCAP, BMI, or SESAC for music broadcast on AM and FM radio stations.
- DMCA compulsory license via SoundExchange for music digitally distributed on Webcasts and cable or satellite subscription services.

- Voluntary (or direct) license via an individually-negotiated agreement for music to be downloaded on the Internet from a Website or an Internet jukebox; used in a movie, TV program, video, or commercial; or used in a compilation CD.

Secure Digital Music Initiative (SDMI)

The Secure Digital Music Initiative is a forum of more than 200 companies and organizations representing a broad spectrum of information technology and consumer electronics businesses, Internet service providers, security technology companies, and members of the worldwide recording industry working to develop voluntary, open standards for digital music. A DMAT mark is the trademark and logo specifying that products bearing this mark (including hardware, software, and content) meet SDMI guidelines. A list of SDMI members is available on the SDMI Website (http://www.sdmi.org).

These guidelines permit artists to distribute their music in both unprotected and protected formats—the choice is up to them. The SDMI guidelines also permit consumers to copy CDs onto their computers and digital recorders for personal use. In fact, the guidelines enable consumers to do so as many times as they wish—as long as they have the original disc.

The SDMI coalition has been slow to create a viable standard, citing disagreements over priorities between music and consumer electronics companies. It released a portable device specification and Phase I watermark in 1999, but is currently "on hiatus and intends to reassess technological advances at some later date" (SDMI, 2004). Digimarc, Microsoft, Real Networks, DivX Networks, and the Fraunhofer Institute (originator of the MP3 format) are all developing digital watermark technologies on their own.

Recent Developments

Digital Audiotape

Digital audiotape is a recording medium that spans the technology gulf between analog and digital. On one hand, it uses tape as the recording medium; on the other, it stores the signal as digital data in the form of numbers to represent the audio signals. A DAT cassette is about half the size of a standard analog cassette. Current DAT recorders support three digital sampling rates and do not use compression, allowing them to deliver excellent sound reproduction. However, as a tape format, DAT does not offer the random access capability of a disc-based medium such as a CD or MiniDisc (MD) (TargetTech, 2000). The technology of DAT, is closely based on that of video recorders, uses a rotating head and helical scan to record data on the tape. DAT has not been very popular outside of professional and semi-professional audio and music recording, although the prospect of perfect digital copies of copyrighted material was sufficient for the music industry in the United States to force the passage of the Audio Home Recording Act of 1992, which created the "DAT tax" (Alten, 2002).

Compact Disc

In the audio industry, nothing has revolutionized the way we listen to recorded music like the compact disc. Originally, engineers developed the CD solely for its improvement in sound quality over LPs and analog cassettes. After the introduction of the CD player, consumers became aware of the quick random-access characteristic of the optical disc system. In addition, the 12-cm (about 5-inch) disc was easy to handle compared with the LP. The longer lifetime for both the medium and the player strongly supported the acceptance of the CD format. The next target of development was to be the rewritable CD. Sony and Philips jointly developed this system and made it a technical reality in 1989. Two different recordable CD systems were established. One is the write-once CD named CD-R, and the other is the re-writable CD named CD-RW (Disctronics, 2002).

High-Density Compatible Digital (HDCD), developed by Pacific Microsonics and now owned by Microsoft Corporation, is a recording process that enhances the quality of audio from compact discs, giving an end result more acceptable to audiophiles than standard CD audio. HDCD discs use the least significant bit of the 16 bits per channel to encode additional information to enhance the audio signal in a way that does not affect the playback of HDCD discs on normal CD audio players. The result is a 20-bit per channel encoding system that provides more dynamic range and a very natural sound. Many HDCD titles are available and can be recognized by the presence of the HDCD logo (Disctronics, 2004). Microsoft has also included the HDCD playback technology in its Windows Media Series 9 Player.

Super Audio CD (SACD). When the CD was developed by Philips and Sony in the 1980s, the PCM format was the best technology available for recording. Nearly two decades later, professional recording capabilities have outgrown the limitations inherent in PCM's 16-bit quantization and 44.1 kHz sampling rate. Philips and Sony have produced an alternative specification called Super Audio CD that uses a different audio coding method, Direct Stream Digital (DSD), and a two-layer, hybrid disc format. Like PCM digital audio, DSD is inherently resistant to the distortion, noise, wow, and flutter of analog recording media and transmission channels. DSD samples music at 64 times the rate of a compact disc (64 × 44.1 kHz) for a 2.8224 MHz sampling rate. As a result, music companies can use DSD for both archiving and mastering (Super Audio CD, 2004).

The result is that a single hybrid SACD contains two layers of music—one layer of high-quality DSD-coded material for SACD playback and one layer of conventional CD encoded material for CD playback. The SACD makes better use of the full 16 bits of resolution than the CD format can deliver, and is backward-compatible with existing CD formats (Sony, 2002). In addition to the hybrid SACD, there are two other SADC variations—the Single Layer SADC (one high-density layer) and the Dual Layer SADC (two high-density layers for extra recording time) (Super Audio CD, 2004).

There are more than 600 SACDs available in a variety of music genres from a wide range of recording labels. And, as of January 2004, over two million SACD players had been sold worldwide (Super Audio CD, 2004).

MiniDiscs

MiniDiscs were announced in 1991 by Sony as a portable disc-based digital medium for recording and distributing "near CD" quality audio. There are two physically distinct types of discs:

premastered MDs, similar to CDs in operation and manufacture, and recordable MDs that can be recorded on repeatedly using a magneto-optical recording technology (Yoshida, 2000).

Magneto-optical disc recording technology has been used for computer data storage for several years. Sony updated the technology, developed a shock-resistant memory control for portable use, and applied a high-quality digital audio compression system called ATRAC (Adaptive Transform Acoustic Coding) to create the MiniDisc. With a diameter of 64mm, a MiniDisc can hold only one-fifth of the data stored by a CD. Therefore, ATRAC data compression of 5:1 is necessary in order to offer a CD-comparable 74 minutes of playback time (Sony, 2000). However, the use of compression also results in a slight reduction in sound quality.

MiniDisc's advantages over a tape format such as DAT include its editing capabilities and quick random access. MiniDisc's main advantage over the MP3 format is that MP3's sound quality is inferior to MD. MP3 compresses sound data in a 10:1 ratio, which results in a greater loss of sound quality (MD FAQ, 2002).

MiniDisc Long-Play (MDLP) is an encoding method for audio on MiniDisc that offers two long-play modes: 160 minutes of stereo audio (LP2) and 320 minutes of stereo audio (LP4). MDLP uses the encoding technique ATRAC3, which is also used in Sony's MemoryStick Walkman, Vaio Music Clip, and Network Walkman. Sound quality in LP2 is equal to standard MD, although LP4 quality may sometimes be inhibited by digital artifacts (MiniDisc.org, 2004b). To improve quality even further, ATRAC3plus technology has been developed by Sony. It achieves double the compression ratio of ATRAC3 with virtually no loss in sound quality (Sony, 2004).

NetMD. The one feature that had been limiting the popularity of the MiniDisc format was the necessity of recording music in real time. Rival MP3 products allow the user to download an hour's worth of songs in just a few minutes. In December 2001, NetMD players were introduced with prices for units from Sharp and Sony around $350. Since then, sales of these NetMD players have steadily grown as unit prices have dropped to below $250.

By employing a standard USB (universal serial bus) interface, music data can be transferred "drag and drop" from a PC to a NetMD product at high speed—five hours of audio in about 10 minutes, for example. In addition, the NetMD protocol enables a PC to control a NetMD product while editing music that has been recorded on a MiniDisc.

To ensure backward compatibility with existing MD products, NetMD supports both ATRAC and ATRAC3 audio compression technologies. The NetMD interface is also designed to support SCMS to prevent second-generation music/data digital copying (NetMD info, 2002).

Hi-MD. Launched in April 2004, Hi-MD is a nearly complete revamping of the original MiniDisc format. The most significant change is the introduction of Hi-MD media, which includes a new 1 GB blank disc in the existing MD form-factor, as well as a reformatting of existing MD media that doubles its capacity to 305 MB. The 1 GB disc can record up to 45 hours of audio using ATRAC3plus encoding and a new process called domain wall displacement detection (DWDD) to burn the audio on the disc (Canon, 2004). Discs are priced between $5 and $7 each.

By using the bundled SonicStage jukebox software, data can be transferred up to 100x real-time speed to the Hi-MD unit. Yet, like previous ATRAC3 MD players, the quality of the audio still falls just below CD quality (MiniDisc.org, 2004a). Hi-MD units are priced between $200 and $400.

MP3

Before MP3 came onto the digital audio scene, computer users were recording, downloading, and playing high-quality sound files using a coding scheme (codec) called .WAV. The trouble with .WAV files, however, is their enormous size. A two-minute song recorded in CD-quality sound would eat up about 20 MB of a hard drive in the .WAV format. That means that a 10-song CD would take up more than 200 MB of disk space.

The file-size problem for music downloads has changed, thanks to the efforts of the Moving Picture Experts Group (MPEG), a consortium that develops open standards for digital audio and video compression. Its most popular standard, MPEG, produces high-quality audio (and full-motion video) files in far smaller packages than those produced by .WAV. MPEG filters out superfluous information from the original audio source, resulting in smaller audio files with no perceptible loss in quality (MPEG, 1999).

Since the development of MPEG, engineers have been refining the standard to squeeze high-quality audio into ever-smaller packages. MP3—short for MPEG 1 Audio Layer 3—is the most popular of three progressively more advanced codecs, and it adds a number of advanced features to the original MPEG process. Among other features, Layer 3 uses entropy encoding to reduce to a minimum the number of redundant sounds in an audio signal. The MP3 codec can take music from a CD and shrink it by a factor of 12, with no perceptible loss of quality (MPEG, 1999).

To play MP3s, a computer-based or portable MP3 player is needed. Hundreds of portable MP3 players are available, from those with 1.5 GB to 4 GB microdrives to larger, 10+ GB hard drive-based models, sold by manufacturers such as Apple, Creative, iRiver, RCA, and Rio. The amount of available disc space is mostly relative to the way a person uses the portable player, either as a song selector or song shuffler. Song selectors tend to store all of their music on players with larger drives and select individual songs or playlists as desired, whereas song shufflers are more likely to load a group of songs on a smaller drive and let the player shuffle through selections at random.

Dozens of computer-based MP3 players are available for download. Winamp, still the most popular, sports a simple, compact user interface that contains such items as a digital readout for track information and a sound level display. This user interface can be customized by using "skins," which are small computer files that let the user change the appearance of the MP3 player's user interface.

Another intriguing part of the MP3 world is CD rippers. These are programs that extract—or rip—music tracks from a CD and save them onto a computer's hard drive as .WAV files. This is legal as long as the MP3s are created solely for personal use, and the CDs are owned by the user. Once the CD tracks have been ripped to the hard drive, the next step is to convert them to the MP3 format. An MP3 encoder is used to turn these .WAVs into MP3s.

All the copyright laws that apply to vinyl records, tapes, and CDs also apply to MP3. Just because a person is downloading an MP3 of a song on a computer rather than copying it from someone else's

CD does not mean he or she is not breaking the law. Prosecution of violators, through the efforts of the RIAA, is the recording industry's main effort to prevent unauthorized duplication of digital audio using MP3 technology.

The first era in Internet audio has undeniably belonged to the MP3 codec, the audio standard codified by MPEG 15 years ago. Thomson and Fraunhofer, the German companies that hold patents in MP3 technology, have long been collecting royalties from software and hardware companies that use the format.

Now, the MP3 codec, the technology most widely associated with unrestricted file swapping, is getting a makeover aimed at *blocking* unauthorized copying. As of this writing, Thomson and Fraunhofer are in the midst of creating a new digital rights management (DRM) add-on for the MP3 codec (Knight, 2004).

Their plan is aimed at pushing more deeply into the world of authorized music distribution through more pay-for-play and subscription services. Current services sell music, from 88¢ a song and up, wrapped in digital encryption—most in incompatible proprietary technologies. Thomson and Fraunhofer's DRM technology will be based on open standards. They will provide free use of the copy protection technology to any player manufacturer or song service that licenses MP3.

Competing with MP3, Microsoft, with its own digital audio codec, Windows Media Audio (WMA), has been a beneficiary of a similar type of interoperability. WMA is being used in over 500 different devices and by Musicmatch, Napster, Real Networks, Sony, Wal-Mart, and other song distributors. Additionally, Microsoft has incorporated its own DRM system into its WMA codec so that "content owners can now deliver music, videos, and other digital media content over the Internet in a protected format. It facilitates consumers to obtain digital media files legitimately, while maintaining the rights of the content owners" (Microsoft, 2004). Microsoft claims to have over 4.5 million WMA-supported devices in use in 2004.

In April 2004, Apple Computer incorporated MPEG4 Advanced Audio Coding (AAC) into QuickTime 6, iTunes 4, and iPod portable music players. AAC was developed by the MPEG group that includes Dolby, Fraunhofer, AT&T, Sony, and Nokia. The AAC codec builds upon new, state-of-the art signal processing technology from Dolby Laboratories and brings true variable bit rate (VBR) audio encoding to QuickTime and the iPod, which will now support AAC, MP3, MP3 VBR, Audible, AIFF, and WAV codecs.

According to Apple (2004), this means that AAC, when compared with the 10-year-old MP3 codec, provides:

- Higher-quality audio with smaller file sizes.
- Support for multichannel audio by providing up to 48 full frequency channels.
- Higher resolution audio by yielding sampling rates up to 96 kHz.
- Improved decoding efficiency by requiring less processing power for decodes.

Apple's Fairplay DRM standard, which has also won the big record labels' approval, and forms the heart of the company's number one ranked iPod player and iTunes Music Store, is not compatible with WMA or the proposed MP3 DRM (Music industry, 2004). This also applies to the popular Real Network RealAudio DRM format and Sony ATRAC DRM codec. This creates a five-way struggle—or opportunity for collaboration—among these primary players in the digital download arena. However, rather than develop a universal standard under the SDMI banner, companies are taking their technology to the marketplace to let consumers decide (Rost, 2004).

In addition to these three primary developers, there are also developers using new digital audio codecs. The most intriguing of the new codecs is Ogg Vorbis (sometimes just called Vorbis), which is an open source, patent-free audio codec developed as a replacement for MP3. Vorbis files (which have an .ogg extension) compress to a smaller size than MP3 files, reducing bandwidth and storage requirements. Also, according to some reports, Vorbis provides better sound quality than MP3 (Mitchell, 2003).

As with the original MP3 format, the new codec technology will have to be supported by software players and chipmakers before devices are able to play protected songs. Using DRM technology is a recognition of a dawning era in digital music, in which pay-per-song and subscription services are beginning to gain ground on file-swapping networks, and in which CDs themselves may ultimately be overtaken by digital downloads.

The growth in MP3 popularity has also had a positive impact on the price of music distributed on CD. Facing serious competition from pay-for-play and subscription Internet services, record labels have significantly lowered the prices for CDs for the first time in their history (Van Buskirk, 2003).

Wireless MP3. Another interesting MP3 development is the launch of the wireless MP3 player. The first use for a wireless MP3 player is fairly obvious: transferring music without physically connecting the player to a computer. This change is certainly convenient, but if wireless connections become common for MP3 players, the sky is (literally) the limit in terms of where your music can come from.

For example, perhaps a wireless MP3 user wants to sample the listening station, browse the racks of any record store, and scan the bar code of desired albums into the wireless MP3 player. Bypassing the cash register, she can leave empty-handed but connect her player to her computer once home. After the software identifies the song from the bar code, she could download the desired songs or albums or search for the lowest prices online.

SoniqCast's Aerio was the first wireless MP3 player to hit the market in 2004, complete with a 1.5 G hard drive, 802.11b Wi-Fi interface, USB interface, and FM tuner for just under $300. The player is also compatible with WMA and Musicmatch jukebox software (SoniqCast, 2004).

DVD-Audio

The initial version of DVD-Audio was released in April 1999, with discs and players appearing in the second half of 1999. DVD-Audio can provide higher-quality stereo than CD with a sampling rate of up to 192 kHz (compared with 44.1 kHz for CD). The standard DVD-Audio specification makes use of one or two layers (channel groups) of PCM multichannel stereo encoding. However, by using

Meridian Lossless Packing, a digital compression codec, DVD-Audio can deliver even more data: up to six channels of 96 kHz/24-bit surround-sound (compared with a standard CD's two channels of 44.1 kHz/16-bit data) (Disctronics, 2002). A typical DVD-Audio disc contains up to seven times the data capacity of a CD.

Home Theater

Since 1998, key innovations in digital audio processing and digital audio distribution have helped create a vibrant consumer marketplace for audio products. The primary home entertainment experience today is delivered by a quality home theater system. Home theater has grown increasingly affordable over the years, and a respectable system with all the necessary audio and video gear costs as little as $600. According to the Consumer Electronics Association, a home theater system is a TV set with a diagonal screen size of at least 25 inches, a video source such as a hi-fi/stereo VCR or DVD player, a surround-sound-equipped stereo receiver or compact shelf system, and four or more speakers.

According to a 2003 CEA survey, 31% of American households own a home theater system meeting this definition. At the very least, surround-sound puts the viewer in the center of the action, enveloping the viewer with the ambient background sounds (i.e., a driving rainstorm, a thunderous explosion, a fast-paced chase scene). It also enhances dialogue intelligibility and realism by channeling dialogue to a TV-top center-channel speaker, making the voices of on-screen actors come from the same direction as their images (CEA, 2004b).

Three companies have taken the lead in developing digital audio processing technologies, such as surround-sound and 3D audio, which are being used to enhance the media experience of the consumer: Q-Sound, Dolby, and DTS.

Q-Sound Digital Audio. Q-Sound has developed proprietary audio solutions that include virtual surround-sound, positional audio, and stereo enhancement for the consumer electronics, PC/ multimedia, Internet and healthcare markets. The company's audio technologies create 3D audio environments allowing users to receive stereo surround-sound from two, four, or 5.1 speaker systems. Q-Sound's key innovations are QSurround 5.1, Q3D, and QXpander. By creating virtual speakers, QSurround 5.1 reproduces spatially correct, multichannel output on regular two-channel equipment and enhances sound reproduction on five- and six-channel systems. Designed for professional audio recording and video game applications, Q3D places multiple individual sounds in specific locations outside the bounds of conventional stereo reproduction to provide a 3D listening environment. QXpander is a stereo-to-3D enhancement process available for both headphones and speakers. A robust algorithm allows it to process any stereo signal, making QXpander suitable for a broad range of consumer electronic applications (QSound Labs, 2004).

Dolby Digital 5.1 Audio. DolbyLabs revolutionized tape recording in the late 1960s and early 1970s with Dolby A-type (for professional applications) and Dolby B-type (for consumer applications) noise reduction. In the late 1980s and early 1990s, Dolby Surround and Dolby Pro Logic home theater systems entered the marketplace. These systems allowed home viewers to create the same four-channel setup found in theaters in the home (Dolby Labs, 2004).

Today's Dolby Digital 5.1 audio system takes the next step, providing six channels of digital surround-sound. Dolby Digital 5.1 delivers surround-sound with five discrete full-range channels—left, center, right, left surround, and right surround—plus a sixth channel for those powerful low-frequency effects (LFEs) that are felt more than heard in movie theaters. As it needs only about one-tenth the bandwidth of the others, the LFE channel is referred to as a ".1" channel (more commonly as the "subwoofer" channel) (Dolby Labs, 2004).

Dolby Digital 5.1 is the audio standard for the movie industry and for new digital media applications. Also, for DVD-video, Dolby Digital is called a "mandatory" audio coding format, meaning that a Dolby Digital soundtrack can be the only one on a disc. Discussed below, DTS, by comparison, is an "optional" coding format, meaning that the disc must have a mandatory-format soundtrack as well.

DTS Digital Audio. The other major developer of surround-sound technology, Digital Theater Systems, Inc. (DTS), uses an encode/decode system that delivers six channels (5.1) of 20-bit digital audio. In addition, DTS has developed a 6.1 surround-sound system, called DTS-ES, which is the only digital audio format capable of delivering 6.1 channels of discrete audio in the consumer electronics market. DTS-ES is also fully backward-compatible with previous 5.1 DTS decoders. DTS has also launched DTS 96/24, a technology for multichannel sound on DVD-Video, and is also fully backward-compatible with all DTS decoders. The "96" refers to a 96 kHz sampling rate (compared with the standard 48 kHz sampling rate), while "24" refers to 24-bit word length (DTS, 2004).

The audio industry continues to advance the state of the art in the Dolby and DTS digital surround formats by marketing 6.1 audio formats on DVD discs. Dolby Digital Surround EX (extended) and DTS ES (extended surround) are 6.1 formats that add a back center channel, which allows DVD makers to accurately place sound effects behind the listener, not just to the sides and front. Side-surround speakers still reproduce left and right surround information, but the new combination allows for effects that are supposed to be seamless.

The biggest industry push has been toward home-theater-in-a-box systems. During 2003, home-theater-in-a-box sales exceeded combined compact and component system sales for the first time ever (CEA, 2004b). Prices for these all-in-one systems range from $300 to over $3,000.

Hard-Drive Jukebox/Recorder

With computer hard drive prices steadily falling since 2002, and with computer technology becoming more crash-resistant and portable, manufacturers are rapidly producing portable digital recorders that use expansive hard drives (1.5 GB to 30+ GB) as their recording medium. The number of companies offering these jukeboxes rose from 10 in 2000 to over 30 in 2004, with the leaders being the Apple iPod, the Creative MuVo^2, and the Rio Karma. Storing music in MP3, AAC, or WMA, these devices archive hundreds of hours of songs, accessible by album title, song title, artist name, or music genre.

Some of the recorders feature built-in modems to download music from Web sites without the assistance of a PC. Some deliver streaming audio content from the Web to connected AV systems. For superior sound quality, the devices can be connected to cable modems or digital subscriber line (DSL) modems. Some hard-drive recorders also come with a built-in CD player, making it possible to rip songs from discs for transfer to the hard drive. Next-generation jukeboxes from Archos and RCA

Lyra are already being marketed as portable video players (PVPs) capable of displaying MPEG and MPEG4 video, JPEG pictures, MP3, MP3Pro, and WMA audio files (Portable video, 2004).

Current Status

The audio category is benefiting from rapid consumer adoption of digital products, in spite of declines across several products. MP3 players shot past all estimates for a record-setting year in 2003, despite continuing debates over home recording rights. Factory-to-dealer shipments of MP3 players totaled 3.8 million units during 2003, an increase of 121% compared with 2002, and they are expected to rise again in 2004 to more than 5.1 million units. Dollar revenues of MP3 players in 2003 leapt to $556 million, a 171% increase over 2002. CEA projects dollar revenues of MP3 players to increase another 27% to $706 million in 2004. Sales of portable CD/MP3 player combinations reached more than $223 million, an increase of 21% over 2002. In 2003, 62% of the dollars spent on MP3 players went toward higher-end units selling at $150 or more. Of that amount, 80% were used to buy hard-drive-based MP3 players, a market that Apple dominates with the iPod (CEA, 2004a).

Audio suppliers sold 43.3 million home, car, and portable CD players in 2003, according to CEA forecasts. On top of that, CEA expected suppliers in 2003 to ship more than 20 million home DVD players, all of which play CDs. For use in CD and DVD players, consumers bought 803.3 million CDs in 2002, according to RIAA. Although CD sales were down for the second year in a row (8.9% in 2002 and 6.4% in 2001), CDs remain a potent force in the music industry (CEA, 2004c).

DVD-Audio

The DVD-Audio format is seeing exponential growth. Currently, more than 700 DVD-Audio titles are available. Just over 150 DVD-Audio player models are available from more than 35 manufacturers. There are about 14 million DVD-Audio players already in the market, with player sales up 500% from 2002. An estimated two million computer-based DVD-Audio players have shipped, from companies such as Creative Labs and InterVideo. Additionally, DVD-Audio discs play on all DVD-Video players, as well as PS2 and Xbox consoles, although not with the enhanced audio heard when played on a DVD-A capable player (Chazen, 2004).

Factors to Watch

By the end of 2004, the two most important factors in digital audio will be:

1) Which digital audio format(s) consumers will adopt.

2) The protection of copyrighted material.

With a myriad of digital audio choices making it to the marketplace, consumers will be evaluating products on ease-of-use, audio quality, storage capacity, portability, and price. The successful worldwide adoption of the CD seems to be less of a challenge for new digital audio formats to supplant. Will consumers replace a perfectly good CD audio system with a new technology? Will they supplement their CD system with additional new technologies that fill specific needs such as Internet audio or home theater? Are there just too many new technologies for consumers to evaluate?

It appears that most music consumers are accepting and adapting fairly easily to the rapidly-growing number of pay-for-play download options that have appeared and evolved in the last year. Apple's iPod is leading the way with a 30% share of the portable MP3 player market, and its iTunes pay-for-play download site is selling over 2.5 million songs per week (Fried, 2004). A new pay-for-play Napster, now owned by Roxio, is leading the PC-only download service. Even Wal-Mart, the mass merchant giant and leading retailer of CDs, has joined in the pay-for-play market by teaming with Liquid Digital Media to offer WMA encoded songs for 88¢ each.

Players, software, and music appear to be properly positioned for interested users in terms of features and price, at the moment. The key factor for success may be download quality, speed, and wireless capability—similar to the situation that has developed in the cell phone and satellite TV industries where service providers give away previously pricey equipment while signing users to yearly service contracts.

It may just be the interoperability of DRM formats. A new consortium of media companies, anchored by Microsoft and Universal Media Group, established the Content Reference Forum (CRF) in 2004. CRF aims to publish standards that would use Internet-based references to identify content and the business agreements attached to them (CRF, 2004).

The group is hoping to make online content distribution more flexible and help break down barriers between incompatible formats and copy-protection technologies. As discussed previously, people who send files through file sharing networks, or via e-mail or instant messaging, are largely locked in to sending a specific file that may not be readable by people who lack the appropriate software or hardware.

Under the new technology, people would share the "content reference" file instead, which would point them to authorized versions of the content that would automatically fit whatever device or computer software the recipient is using. The groups behind the effort include ARM, ContentGuard, Macrovision, Microsoft, Nippon Telegraph and Telephone, Universal Music Group, and VeriSign.

In terms of piracy, a firestorm of controversy involving the Internet and music continues to be centered on the issue of unauthorized downloading of copyrighted music from two types of providers: MP3 hosting sites such as the "old" Napster and Aimster services and peer-to-peer (P2P) networks such as Grokster and Morpheus. The courts have ruled that MP3 hosting sites are illegal, while P2P networks are okay, for now.

Why? Napster and Aimster maintained a central directory server on their Web site to connect MP3 file swappers. This server enabled users to find uploaded music from other users—legal and pirated—and download it without paying the artists or the record companies. It also allowed the hosting site to monitor which users were online and what files they were sharing via the server. The district court found this to be illegal (*A&M Records v. Napster*, 2001).

Unlike users of MP3 hosting sites, persons using P2P networks download the P2P file sharing software (such as Morpheus or Gnutella), and then trade files on the network without going through a central server. This is accomplished because any computer running P2P file sharing software can search all of the other computers connected to the Internet that are also running the software and retrieve information that each user makes publicly available on his personal computer—not on a computer owned by the P2P network. This makes it virtually impossible to shut down the network without unplugging every connected computer. It also makes it difficult to control by legislation because there is no central server to restrict (Cha, 2000).

The only realistic way to shut down a P2P network is to pass legislation making the P2P file sharing software illegal to possess on a user's computer. As of mid-2004, there are a number of legislative efforts being considered by Congress (S. 1621, H.R. 2517, H.R. 2752, and H.R. 2885) and individual states such as California.

Another avenue may be through the courts. Several different cases are still pending that involve P2P networks (i.e., BearShare, Blubster, eDonkey2000, Gnutella, KaZaa, and Morpheus). In general, these cases—all of which are in civil court—revolve around claims for damages by major entertainment companies (such as record labels and movie studios). These companies, which own the copyrights to material that individual P2P software users have opted to make available to each other, claim that the P2P company that provided users with the software for sharing files should be liable for, in effect, "aiding and abetting" copyright infringement merely by virtue of providing the P2P file sharing software.

These charges are similar to those made in the early 1980s by major movie studios against Sony, the maker of the original Betamax VCR, seeking to outlaw the device because it facilitated home taping of TV movies without the copyright owners' permission. The U.S. Supreme Court ruled decisively against the studios. Thus far, different courts have reached different conclusions on these questions. These issues eventually may need to be settled by the Supreme Court as well.

The legal issues surrounding digital audio are not limited to file sharing networks. The RIAA assists authorities in identifying music pirates and shutting down their operations. In piracy cases involving the Internet, the RIAA's team of Internet specialists, with the assistance of a 24-hour automated Webcrawler, helps stop Internet sites that make illegal recordings available.

Based on the Digital Millennium Copyright Act's expedited subpoena provision, the RIAA sends out information subpoenas as part of an effort to track and shut down repeat offenders and to deter those hiding behind the perceived anonymity of the Internet. Information subpoenas require the Internet service provider (ISP) providing access to or hosting a particular site to provide contact information for the site operator. Once the site operator is identified, the RIAA will take steps to prevent repeat infringement. Such steps range from a warning e-mail to litigation against the site operator. The RIAA then uses that information to send notice to the site operator that the site must be removed. Finally, the RIAA requires the individual to pay an amount designated to help defray the costs of the subpoena process (U.S. Copyright Office, 2002b).

The wave of copyright lawsuits brought by the RIAA in early 2004 marked the first time the trade group has targeted individual computer users swapping music files over university networks. For example, in March 2004, the RIAA filed "John Doe" complaints against 89 individuals using networks at universities in Arizona, California, New York, Indiana, Maryland, Colorado, Pennsylvania, Tennessee, Wisconsin, and Washington. Lawsuits were also filed against 443 people using commercial Internet access providers in California, Colorado, Missouri, Texas, and Virginia. With the "John Doe" lawsuits, the RIAA must work through the courts to find out the identities of the defendants, which, at the outset, are identified only by the numeric Internet protocol addresses assigned to computers online (RIAA, 2004b).

Then, in April 2004, the RIAA brought several more copyright infringement lawsuits against named students at Princeton, Rensselaer Polytechnic University, and Michigan Technological Institute. These suits claim that the students "hijacked" their university's academic computer network to provide a pre-2004 Napster-like service that creates an "emporium of music piracy where copyright infringement is simplified down to the click of a computer mouse" (RIAA, 2004c).

Amidst this backdrop of legal wrangling, expect to see continued innovation, marketing, and debate in the next few years. Which technologies and companies survive will largely depend on the evolving choices made by consumers, the courts, and the continued growth and experimentation with the Internet and digital technology. Consumers seem to be very willing to move forward with the convenience and enjoyment being afforded by the new technologies of digital audio.

Bibliography

A&M Records v. Napster. (2001, April 3). Retrieved March 15, 2004 from http://caselaw.lp.findlaw.com/scripts/getcase.pl?court=9th&navby=case&no=0016401&exact=1.

Alten, S. (2002). *Audio in media,* 6th edition. Belmont, CA: Wadsworth.

Apple Computer, Inc. (2004, April). *MPEG4 Audio: AAC*. Retrieved April 22, 2004 from http://www.apple.com/mpeg4/aac/.

Canon. (2004, March). *Achieving 22 GB memory capacity on a 120mm disk*. Retrieved March 10, 2004 from http://www.canon.com/technology/detail/device/dwdd/.

Cha, A. (2000, May 18). E-power to the people. *The Washington Post*. Retrieved April 5, 2002 from http://www.washingtonpost.com/ac2/wpdyn?pagename=article&node=&contentId=A21559-2000May17.

Chazen, J. (2004, August 3). DVD-Audio penetration explodes—Over 700 albums available. *High Fidelity Review*. Retrieved March 15, 2004 from http://www.highfidelityreview.com/news/news.asp?newsnumber=10311668.

Consumer Electronics Association. (2003, September). *Sales of audio products show growth in September*. Retrieved March 14, 2004 from http://www.ce.org/press_room/press_release_detail.asp?id=10359.

Consumer Electronics Association. (2004a, April). *2004 sales of consumer electronics set new record*. Retrieved April 22, 2004 from http://www.ce.org/press_room/press_release_detail.asp?id=10384.

Consumer Electronics Association. (2004b, March). *Audio overview*. Retrieved March 13, 2004 from http://www.ce.org/publications/books_references/digital_america/audio/default.asp.

Consumer Electronics Association. (2004c, April). *Sales of audio products show growth during September*. Retrieved April 22, 2004 from http://www.ce.org/press_room/press_release_detail.asp?id=10359.

Consumer Reference Forum. (2004, March). *About the Content Reference Forum*. Retrieved March 13, 2004 from http://www.crforum.org/articles/about/overview.html.

Disctronics. (2002, March 30). *Audio coding*. Retrieved April 5, 2002 from http://www.disctronics.co.uk/technology/dvdaudio/dvdaud_audio.htm.

Disctronics. (2004, March). *HDCD*. Retrieved March 13, 2004 from http://www.disctronics.co.uk/technology/cdaudio/cdaud_hdcd.htm.

Dolby Labs. (2004, March). *Dolby Digital—General*. Retrieved March 13, 2004 from http://www.dolby.com/digital/diggenl.html.

DTS. (2004, March). *At a glance.* Retrieved March 13, 2004 from http://www.dtsonline.com/technology/at-a-glance.php.

Fried, I. (2004, March 15). Apple's iTunes sales hit 50 million. *C/Net News*. Retrieved March 15, 2004 from http://news.com.com/2100-1027-5173115.html?tag=nefd_hed.

Home Recording Rights Coalition (HRRC). (2000, April). *HRRC's summary of the Audio Home Recording Act.* Retrieved April 3, 2002 from http://www.hrrc.org/ahrasum.html.

Knight, W. (2004, March 2). MP3 creators to add copy protection. *New Scientist.* Retrieved March 13, 2004 from http://www.newscientist.com/news/news.jsp?id=ns99994731.

MD FAQ. (2002, April). *MiniDisc.org*. Retrieved April 5, 2002 from http://www.minidisc.org/faq_sec_4.html.

Microsoft. (2004, March 15). *Windows Media DRM FAQ.* Retrieved March 15, 2004 from http://www.microsoft.com/windows/windowsmedia/wm7/drm/faq.aspx#General1.

MiniDisc.org. (2004a). *Minidisc FAQ: Hi-MD topics.* Retrieved March 13, 2004 from http://minidisc.org/hi-md_faq.html.

MiniDisc.org. (2004b). *Mini-disc FAQ: MDLP mode topics*. Retrieved March 13, 2004 from http://www.minidisc.org/mdlpfaq.html.

Mitchell, G. (2003, June 25). *An introduction to compressed audio with Ogg Vorbis.* Retrieved March 15, 2004 from http://grahammitchell.net/writings/vorbis_intro.html#why_ogg_vorbis.

MPEG. (1999, December). *MPEG audio FAQ.* Retrieved April 3, 2000 from http://tnt.uni-hanover.de/project/mpeg/audio/faq/#a.

Music industry pushing Apple, Microsoft to make DRM "interoperable." (2004, February 2). *MacDailyNews*. Retrieved March 13, 2004 from http://www.macdailynews.com/comments.php?id=2065_0_1_0_C.

NetMD info! (2002, April). *Minidisco.com*. Retrieved April 5, 2002 from http://www.minidisco.com/minipages/netmdinfo.html.

Portable video players. (2004, March). *C/Net News*. Retrieved March 15, 2004 from http://reviews.cnet.com/4502-6499_7-0.html?tag=pvp.

QSound Labs. (2004, March). *3D audio timeline*. Retrieved March 13, 2004 from http://www.qsound.com/2002/technology/main1.asp.

Recording Industry Association of America. (2002, April). *Digital music laws.* Retrieved April 5, 2002 from http://www.riaa.org/Copyright-Laws-4.cfm#1.

Recording Industry Association of America. (2004a, March). *Royalty distribution.* Retrieved March 13, 2004 from http://www.riaa.com/about/collective/royalty.asp.

Recording Industry Association of America. (2004b). *RIAA brings new round of cases against illegal file sharers.* Retrieved April 22, 2004 from http://www.riaa.com/news/newsletter/032304.asp.

Recording Industry Association of America. (2004c). *Legal cases*. Retrieved April 23, 2004 from http://www.riaa.com/news/filings/pdf/pastcissues/PengFiledComplaint.pdf.

Rost, J. (2004, March 9). Interoperability: The top priority for digital music. *MEDIALINE*. Retrieved March 15, 2004 from http://medialinenews.com/articles/publish/article_491.shtml.

Secure Digital Music Initiative. (2004, March). *Frequently asked questions*. Retrieved March 13, 2004 from http://www.sdmi.org/FAQ.htm.

SoniqCast. (2004, March). *Specifications.* Retrieved March 13, 2004 from http://www.soniqcast.com/details.html.

Sony Electronics. (2000, April). *MiniDisc*. Retrieved April 3, 2002 from http://www.sel.sony.com/SEL/consumer/md/.

Sony Electronics. (2002, April). *SACD FAQS.* Retrieved April 5, 2002 from http://www.sel.sony.com/SEL/consumer/sacd/static/faqs.html.

Sony Electronics. (2004, March). *ATRAC3plus technology.* Retrieved March 13, 2004 from http://www.sony.net/Products/ATRAC3/tech/atrac3plus/index.html.

Super Audio CD. (2004, March). *Technical.* Retrieved March 13, 2004 from http://www.superaudio-cd.com/.

TargetTech. (2000, April). *DAT*. Retrieved April 3, 2000 from http://www.whatis.com/ dat.htm.

U.S. Copyright Office. (2002a). *Copyright law of the United States of America and related laws contained in Title 17 of the United States Code, Circular 92*. Retrieved April 5, 2002 from http://www.loc.gov/copyright/title17/circ92.html.

U.S. Copyright Office. (2002b). *The Digital Millennium Copyright Act of 1998: U.S. Copyright Office summary*. Retrieved April 4, 2002 from http://lcweb.loc.gov/copyright/legislation/dmca.pdf.

Van Buskirk, E. (2003, September 10). CD price drops coincide with MP3 hardware breakthroughs. *C/Net News*. Retrieved March 15, 2004 from http://reviews.cnet.com/4520-6450_7-5073810-1.html?tag=txt.

Watkinson, J. (1988). *The art of digital audio.* London: Focal Press.

Yoshida, T. (2000, April). *What are MiniDiscs?* Retrieved March 25, 2000 from http://www.minidisc.org/ieee_paper.html.

18

Digital Photography

Michael Scott Sheerin, M.S.*

I got a Nikon camera
I love to take a photograph
so mama don't take my Kodachrome away.

—Paul Simon

Sales of digital still cameras were estimated to be equal to sales of traditional still cameras in 2003, but projections for 2004 call for digital still cameras to outsell traditional still cameras for the first time (see Figure 18.1) (Photo Marketing Association, 2004). As of March 2004, industry giant Eastman Kodak stopped selling their 35mm still cameras and their Advanced Photo Systems (APS) cameras in the U.S. market. Looking at these factors, it is clear that the photo industry is in a transition stage as it converges with the computer industry and enters the age of digital imaging. Perhaps Paul Simon would have asked his mama not to take away his megapixels if he wrote *Kodachrome* today.

In addition to the increased sales of digital still cameras, industry experts are also noticing a shift in camera buyers. The early adopters of the digital still camera format were mostly professional photographers or avid enthusiasts and were predominantly male. This demographic has changed, as the digital still camera has entered the mass adoption phase with more than 42% of U.S. households owning one (Photo Marketing Association, 2004). Today's profile of the digital still camera user is typically a woman or a family with children.

---

* Assistant Professor, School of Journalism and Mass Communications, Florida International University (Miami, Florida)

Figure 18.1
U.S. Camera Sales, 1983-2003

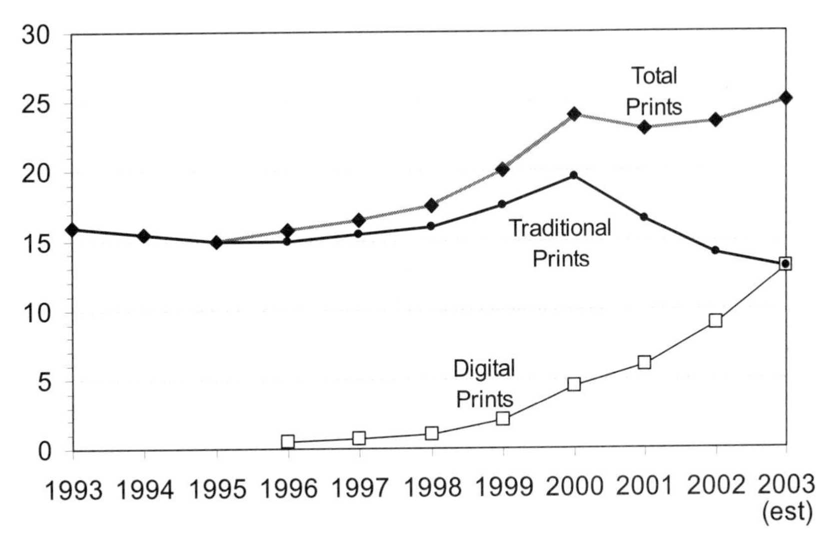

Source: PMA Marketing Research (2004)

Usage, defined as the number of prints made, is also on the rise. The high water mark of the number of photographic prints made in a year (30.4 billion) occurred in 2000, with the vast majority of those prints developed from film (Photo Marketing Association, 2004). Soon thereafter, the photo industry slumped, due to a sluggish economy and the decline of personal travel (or impact of 9/11). However, as travel has increased and the economy has picked up, the number of prints produced has also increased—just not from traditional cameras. As digital camera sales increase, so do the number of prints from digital cameras. Industry experts predict that the total number of prints (combined from both formats) will reach the year-2000 level by 2006 (Photo Marketing Association, 2004).

Figure 18.2
U.S. Amateur Film Sales, 1983-2003 (Millions)

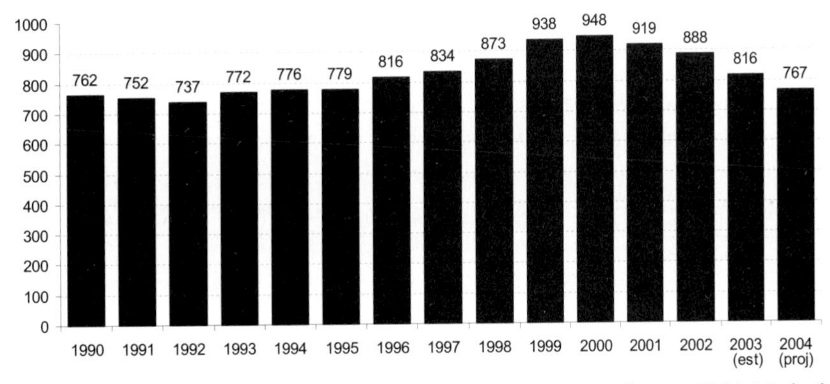

Source: PMA Marketing Research (2004)

Background

With digital imaging, images of any sort, be it family photographs or medical X-rays, are treated as data. This ability to take, scan, manipulate, disseminate, or store images in a digital format has spawned major changes in the communication technology industry. Digital still photography, the focus of this chapter, can be seen as the descendant of traditional photography. From the photojournalist in the newsroom to the magazine layout artist, digital imaging has changed print media. In fact, repercussions stemming from the ease with which digital photographs can be manipulated caused the National Press Photographers Association (NPPA) (2004) to update their code of ethics in 1991 to encompass digital imaging factors. Here's a brief look at how this photographic evolution occurred.

The first photograph ever taken is credited to Joseph Niepce, and it turned out to be quite pedestrian in scope. Using a process he derived from experimenting with the newly invented lithograph process, Niepce was able to capture the view from outside his Saint-Loup-de-Varennes country house in 1826 in a camera obscura (Harry Ransom Humanities, 2004). The capture of this image involved an eight-hour exposure of sunlight onto bitumen of Judea, a type of asphalt (Lester, 2000). Niepce named this process heliography, which is Greek for sun writing. Ironically, this was the only photograph of record that Niepce ever shot, and it still exists today as part of the Gernsheim collection at the University of Texas at Austin (Lester, 2000).

The next 150 years saw significant innovation in photography. Outdated image capture processes kept giving way to better image capture processes, from the daguerreotype (sometimes considered the first photographic process) developed by Niepce's business associate Louis Daguerre, to the calotype (William Talbot), wet-collodion (Frederick Archer), gelatin-bromide dry plate (Dr. Richard Maddox), and the continuous-tone panchromatic black-and-white and autochromatic color negative films of today. Additionally, exposure time has gone from Niepce's eight-hour exposure to 1/500th of a second or less.

Cameras themselves did not change that much after the early 1900s, until digital photography came along. Today's 35mm single lens reflex (SLR) camera works in principle like an original Leica, the first 35mm SLR camera introduced in 1924. Based on a relationship between film's sensitivity to light, the lens aperture (f-stop), and the shutter speed (time of exposure), the SLR camera allows photographers, both professional and amateur, to capture images using available light.

All images captured on these traditional SLR cameras has to be processed after the film's original exposure to light in order to see the image. Instant photography changed all that. Edwin Land invented the first instant photographic process in 1947. He used a Polaroid Land camera that produced a sepia colored print in about 60 seconds (Brasesco, 1996). This first camera, called Model 95, was sold in November 1948 at Jordan Marsh in Boston for $89.75 (About Polaroid, 2004). In 1972, Polaroid's SX-70 camera was introduced. Using Time-Zero film, this SLR camera was "fully automated and motorized, ejecting self-developing, self-timed color prints" (About Polaroid, 2004).

One of the main drawbacks of instant photography was that the images were not as sharp as the images derived from "negative" film cameras. Still, capturing instant photographs was something the public wanted, and Polaroid sales did quite well until the advent of the digital camera. In 2001, Polaroid declared voluntary bankruptcy, mainly due to increased competition from digital still cameras that also produced instant images (About Polaroid, 2004).

The first non-film camera produced analog images, not digital ones. In 1981, Sony announced a still video camera called the MAVICA, which stands for magnetic still video camera (Hammerstingl, 2000). It was not until 10 years later, in 1991, that the first digital still camera was introduced. Called the Dycam (and manufactured by a company called Dycam), it captured images in monochromatic grayscale only and had a resolution that was lower than most video cameras. It sold for a little less than $1,000 and had the ability to hold 32 images in its internal memory chip (Aaland, 1992).

In 1994, Apple released the Quick Take 100, the first mass-market color digital still camera. The Quick Take had a resolution of 640 × 480, equivalent to a TV image, and sold for $749 (McClelland, 1999). Complete with an internal flash and a fixed-focus 50mm lens, the camera could store eight 640 × 480 color images on an internal memory chip and could transfer images to a computer via a serial cable. Other mass-market digital cameras released around this time were the Kodak DC-40 in 1995 for $995 (Morgenstern, 1995) and the Sony Cyber-Shot DSC-F1 in 1996 for $500 (Karney, 1998).

The digital still camera works in much the same way as a traditional still camera. The lens and the shutter allow light into the camera, based on the aperture and exposure time, respectively. The difference is that the light reacts with an image sensor, usually a charge-couple device (CCD) or a complementary metal oxide semiconductor (CMOS) sensor. When light hits the sensor, it causes an electrical charge. The size of this sensor, and the number of picture elements (pixels) found on it determines the resolution, or quality, of the captured image. The thousands of pixels on any given sensor are referred to as megapixels. The sensors themselves can be different sizes. A common size for a sensor is 18 × 13.5mm (a 4:3 ratio), now referred to as the FourThirds System (Olympus Europa, 2002). In this system, the sensor area is approximately 25% of the area of exposure found on a traditional 35mm camera.

The pixel, also known in digital photography as a photosite, can only record light in shades of gray, not color. In order to produce color images, each photosite is covered with a series of red, green, and blue filters, a technology derived from the broadcast industry. Each filter lets specific wavelengths of light pass through, according to the color of the filter, blocking the rest. Based on a process of mathematical interpolations, each pixel is then assigned a color. Because this is done for millions of pixels at one time, it requires a great deal of computer processing. A digital still camera must "preview, capture, compress, filter, store, transfer, and display the image" in a very short period of time (Curtin, 2003).

This processing issue is one of the major drawbacks in digital photography, although it has been nearly eliminated in some of the newer, high-end cameras. In traditional photography, when the photographer pushes the button to take a picture, the film is immediately exposed, based on the shutter speed setting. In digital photography, there is computer processing time that delays the actual exposure from happening when the user pushes the capture button, especially if the camera uses auto focus and/or a built-in flash. Many professional photographers are used to dealing with a shutter speed of 1/250th of a second, so this "exposure shift" can cause the camera to record an image slightly different than the one expected.

Recent Developments

It has been written, "Unlike the evolution of other types of media, the history of photography is one of declining quality" (Pavlik & McIntosh, 2004, p. 113). Images shot with large format cameras in the 1880s are, in many ways, superior to the digital images of today. However, new image sensors are changing that. With the release of the Kodak DCS Pro 14N, which has an image capturing area of 35.8 × 23.8mm (the equivalent of the traditional 35mm frame) encompassing 13.5 megapixels, digital still photography has caught up to the resolution of 35mm traditional still photography.

Roger N. Clark (2002) concludes in his study of digital still versus film resolution that the resolution obtained from an image sensor of 10 megapixels or more is equal to or greater than the resolution of 35mm film. Not many people enjoy that quality of digital photography yet; a 2002 survey states that only 4% of all consumers who bought a digital still camera purchased one with 5 megapixels or more (Photo Marketing Association, 2003). It is important to note that two different camera models that have the same number of megapixels are not necessarily the same in terms of quality and price. According to Brad Tuckman (2004), president of KSC Studio, a large photography studio based in New York and South Florida, CCD image sensors tend to produce a superior image and are more expensive than CMOS image sensors.

This high-end image sensor, though superior in quality, is not the standard in digital still cameras. Introduced in the fourth quarter of 2002 by Kodak and Olympus Optical Co., Ltd., the FourThirds System (originally conceived by Fuji Photo Film Co., Ltd.) is considered the definitive standard SLR digital still camera format. Included in this definition is an open standard for lens mounts that assures the compatibility of all lenses manufactured for the FourThirds System, regardless of camera/lens type. Matsushita Electric Industrial Co., Ltd. (the parent company of Panasonic), Sanyo Electric Co., Ltd., and Sigma Corporation joined ranks with Kodak, Olympus, and Fuji in February 2004, forming the Universal Digital Interchangeable Lens System Forum (Matsushita Electric, 2004).

The lens used on any camera is critical to the clarity of the final image, whether traditional or digital. However, the optic systems for film and digital are not the same. Film can respond to light at high angles of incidence, while a "high angle of incidence can prevent sufficient light from reaching sensor elements at the periphery of a CCD and result in reduced color definition" in digital cameras (Olympus Europa, 2002). Knowing this, digital still camera manufacturers, concentrating on the FourThirds System, are now designing lenses to maximize the performance of the digital image sensor, resulting in a better, smaller lens that facilitates handling and functionality.

Philips has pushed the envelope on lens innovation with their demonstration of a new "fluid" lens that acts like the human eye. This new and very compact lens can change shape, thus changing focal length, overcoming the problem of the fixed focal length lens found in many standard digital still cameras. According to Phillips, the lens is suited for camera phones, endoscopes, home security systems, and optical storage drives, as well as digital cameras (Royal Philips Electronics, 2004). As of mid-2004, the lens was available in prototype, but was not yet used in commercial products.

A recent development for photojournalists is Canon's Digital Verification Kit (DVK-E2) that is designed to work with the EOS-1Ds and the new EOS-1D Mark II D-SLR cameras. The goal of the DVK-E2 kit is to validate digital photographs as original, non-manipulated images. Targeted at law enforcement and insurance companies, as well as the press, the DVK-E2 kit uses a secure mobile card that connects to a computer via a USB port and can detect a "single bit discrepancy in modification of an image since it was taken" (Canon validates, 2004).

For do-it-yourself consumers, who will make an estimated 3.5 billion prints from their digital still cameras in 2004, new consumer printers that produce quality digital photo prints have hit the market. Most have a front panel USB (universal serial bus) connection for quick connect to a digital still camera. Canon's new printers with ChromaPlus technology use eight inks (all in separate ink tanks so you only have to replace the empty one) to achieve native color (Canon's Bubble Jet, 2004). For the black-and-white enthusiast, HP Photosmart printers deliver black-and-white images that have been tested (under glass) to be fade resistant for 115 years (Third party, 2004). Because of the changing demographics of digital imaging users, Epson has introduced a printer geared toward women. Based in part on the fact that only 1% of women feel that technology devices are made with them in mind, Epson has released the PictureMate, a printer that works independent of a computer, can be easily carried (based on its lunchbox design), and can print 100 borderless 4 × 6-inch photo-quality images. All you need to do is insert a cartridge that contains the ink and glossy paper, plug in your camera to the printer, and print (Lewis, 2004).

On the lighter side of technology, Polaroid has enjoyed a sudden increase in awareness, thanks to a line in a 2004 mega-hit single *Hey Ya* by the band Outkast that included the line, "Shake it like a Polaroid picture." Trying to cash in on this new, "hip is retro" trend, the newly restructured company (Polaroid was acquired by an affiliate of One Equity Partners LLC in 2002) has signed agreements to have their cameras used at the *2004 Grammy Awards*, on *Saturday Night Live*, and at the *Vive Awards* show on Viacom's UPN. Being a privately-owned company, Polaroid does not release sales figures, so it remains to be seen if they profit from this newfound publicity. Polaroid did release a disclaimer stating that "its instant photographs no longer need shaking to dry" (Walker, 2004).

Current Status

It is projected that, by the end of 2004, over 15 million digital cameras will be in use, as the percentage of U.S. households owning digital still cameras exceeds 42% (Photo Marketing Association, 2004). Canon has obviously benefited from this growth, reporting a net sales growth of 8.8% for 2003, while also increasing their net income over the same period by 44.6%.

Currently, there are a wide variety of digital still cameras available, including many well-known brands from traditional camera manufacturers such as Nikon, Canon, Polaroid, and Kodak. Other manufacturers, such as Sony, are newcomers to the field, manufacturing still cameras of the digital variety only. The price of a digital still camera can vary greatly, from models with a low resolution of .1 megapixels that start at around $20, to professional models, such as the Canon EOS-1Ds, with a resolution of 11.1 megapixels, costing $7,000 or more (EOS-1Ds, 2004).

Traditional prints made up an estimated 83% of the total of prints processed in 2003, although that figure is projected to fall to 66% by 2006 (Photo Marketing Association, 2004). It has also been projected that prints made from digital pictures will exceed 10.6 billion by 2006, with 6.4 billion of these prints being produced at a retail store and another billion printed via online photo processing (Figure 18.3). It is interesting to note that traditional one-time-use camera sales continue to rise. Perhaps spurred by their ease of use and convenience, sales volume increased 7% in 2003, and it is projected to grow by 5% in 2004. However, the growth rate is slowing, down from a high of 17% in 2000 (Photo Marketing Association, 2004).

Figure 18.3
Volume of Prints Made (in Billions), Conventional vs. Digital

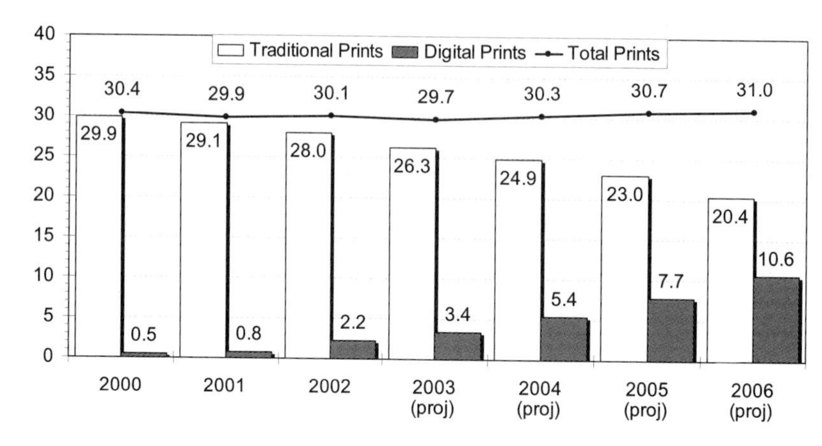

Source: PMA Monthly Processing Surveys (2004)

These figures show a shift in the business of print making from traditional film to digital prints, and many companies have overhauled their business model to reflect this. What looks like a loss in the traditional print business is a gain in the "print-from-digital" image business. In anticipation of this eventual crossover, Kodak installed do-it-yourself kiosks called Kodak Picture Maker in many retail photo departments. The Digital Imaging Marketing Association (DIMA) gave Kodak their 2004 DIMA Innovative Digital Product Award for the newest Kodak Picture Maker, the G3, which processes both digital and traditional formats. Kodak has grown their online photo fulfillment business, Ofoto, as well. Sales figures for Ofoto have more than doubled since 2001 as Kodak has built the "world's largest infrastructure for online photo processing" with 5,400 retail outlets in Western Europe and 20,000 in the United States (Eastman Kodak, 2004a). Although sales of their traditional 35mm cameras have ceased here, Kodak is experiencing sales growth of these cameras in other emerging markets, as 2004 sales are expected to increase 55% from 2003 numbers in overseas markets (Eastman Kodak, 2004b).

A direct offshoot of this digital growth is infoimaging, which has become a huge industry. Infoimaging is defined as the convergence of devices (hardware), infrastructure, and services/media, and is part image science and part information technology. The infoimaging industry, which uses imaging to improve both communication and commerce, has seen a growth of 8% to 10% annually. According to Kodak, a big player in infoimaging, "devices, infrastructure, and services/media are individually dynamic markets offering vast possibilities. But when linked together, the possibilities multiply exponentially, creating a burgeoning $385 billion opportunity" (Eastman Kodak, 2004c).

Factors to Watch

- *The image sensor battle, to be waged in both the consumer and professional markets.* New innovations, such as Foveon's X3 technology, and better CCD and CMOS sensors, based on either the FourThirds System or high-end 35mm systems, will produce less expensive cameras that have superior image quality.

- *Better raw data usage.* Scientists will continue to perfect the interpretation of the raw data the image sensor collects. As these software algorithms and mathematical equations are improved, the exposed image will better represent how the human eye perceives it. Factors that play a part in exposure interpolations are the light's color, luminance, degree of saturation, and latitude—the dynamic range of light on the sensor.

- *The newly introduced Philips fluid lenses will have a great impact on the digital still industry.* Watch for the adoption of these lenses on small image collecting devices, such as digital camera phones, for better image quality and focal length flexibility.

- *The wearable digital camera called the Sensecam Prototype V1.* Though not yet available outside of Microsoft's research lab in the United Kingdom, the Sensecam can record up to 2,000 images per day in addition to data streams (temperature, movement, light levels). Microsoft (2003) equates it to the "black box" of an airplane, recording the events of your life automatically. The camera senses movement, such as another person walking by, and records these images automatically (a manual hand signal can also trigger the camera). It will be used as a memory augmentation device, taking the computer one step closer into our lives. New prototypes will record audio and possibly physiological data such as a heartbeat. Whether the end user wants this "recorded life" remains to be seen.

- *The introduction of new products*, such as the FlashTrax by SmartDisk, to ease, as Douglas Rea of the Rochester Institute of Technology so eloquently calls it, the "digital constipation" caused by all our images (Lagesse, 2003). Surveys predict that Americans will shoot three times as many images with a digital camera than with a film camera. Ease-of-use, instant viewing, and the growing capacity of smart cards and memory sticks all contribute to this increase. The debate will continue on how many of these images will be preserved, how they will be preserved, and for how long.

- *Traditional photography will not fade away entirely.* Many professional photographers remain committed to traditional photography and will use both formats, depending on client need. Kodak will continue to introduce new 35mm film to the market, proven by the release of eight new 35mm film stocks from 2002 to 2004.

- Some theorists also note the *sociological implications* of ubiquitous digital photography, suggesting that it may help erode the perceived Utopian image of the American family. Sociologist Pierre Bourdieu, studying French families in the 1950s, called the camera a "festive technology," referring to the posed snapshots of extended family members (Lagessse, 2003). In Susan Sontag's book *On Photography*, she further illuminates the camera's role in disguising "the actual disappearance of the extended family as a functioning social unit: the portraits that include grandparents and relations are in fact the only moments at which such gatherings occur" (Murray, 1993). By capturing more images, the posed, perfect family portraits will give way to more candid snapshots, thus altering our perception of family life by representing a more realistic depiction.

- *Increased use of digital camera phones and the images derived from these phones.* Michael Polacek, vice president of imaging at National Semiconductor Corporation, predicts sales of digital camera phones to reach 100 million units in 2004, clearly outselling digital cameras. As third generation wireless networks propagate, sending images taken with these phones will increase. Japan is already experiencing a problem called digital shoplifting. Snapping images of new fashions from magazines in stores and sending them to friends is causing the Japanese Magazine Publishers Association to launch a national campaign against this "information theft" (Sommerville, 2003).

- *Privacy issues will arise as the dissemination of digital photos over the Internet increases.* Also, as the use of digital cameras by security and law enforcement agencies increases, the balance between "big brother" and personal rights will be tested.

Bibliography

Aaland, M. (1992). *Digital photography.* New York: Random House.

About Polaroid. (2004). *Polaroid Corporation.* Retrieved January 17, 2004 from http://www.polaroid.com/global/movie_2.jsp?PRODUCT%3C%3Eprd_id=845524441761320&FOLDER%3C%3Efolder_id=2825744883384418&bmUID=1076078667295&PRDREG=null.

Brasesco, J. D. (1996). The history of photography. *The photography network.* Retrieved January 20, 2004 from http://www.photography.net/html/history.html.

Canon's Bubble Jet i9950. (2004). Official press releases. *Digital Photography NOW.* Retrieved March 5, 2004 from http://dp-now.com/archives/000550.html.

Canon validates digital images with DVK-E2. (2004). *Consumer press releases.* Retrieved January 30, 2004 from http://www.canon.co.uk/For_Home/Product_Reviews/Consumer_Releases/DataVerificationKit.asp?ComponentID=156329&SourcePageID=26446#1.

Cartier-Bresson, H. (2004). The artist. *HCB Foundation.* Retrieved March 5, 2004 from http://www.henricartierbresson.org/hcb/home_en.htm.

Clark, R. N. (2002). Film versus digital information. *Clark Vision.* Retrieved March 4, 2004 from http://clarkvision.com/imagedetail/film.vs.digital.1.html.

Curtin, D. (2003). *Short course in choosing and using a digital camera.* Retrieved February 17, 2004 from http://www.shortcourses.com /choosing/how/03.htm.

Eastman Kodak. (2004a). Kodak Picture Maker G3 earns DIMA Innovative Digital Product Award at PMA. *News release.* Retrieved March 8, 2004 from http://www.kodak.com/eknec/ PageQuerier.jhtml?pq-path=2038&pq-locale=en_US.

Eastman Kodak. (2004b). Kodak to accelerate 35mm consumer film effort in emerging markets. *News release.* Retrieved January 20, 2004 from http://phx.corporate-ir.net/phoenix.zhtml?c= 115911&p=irol-newsArticle&ID=484120&highlight=.

Eastman Kodak. (2004c). *What is infoimaging?* Retrieved January 21, 2004 from http://www.kodak.com/eknec/PageQuerier.jhtml?pq-path=12/1245&pq-locale=en_US.

EOS-1Ds. (2004). Cameras. *DCVIEWS.* Retrieved January 30, 2004 from http://www.dcviews.com/cameras.htm.

Hammerstingl, W. (2000). *A little history of the digital (r)evolution*. Retrieved February 23, 2004 from http://www.olinda.com/ArtAndIdeas/lectures/Digital/ history.htm.

Harry Ransom Humanities Research Center at the University of Texas at Austin. (2004). *HRC Exhibitions and Events*. Retrieved March 3, 2004 from http://www.hrc.utexas.edu/ exhibitions/permanent/.

Karney, J. (1998). Sony DSC-F1, Mavica-FD5, Mavica-FD7. *PC Magazine*. Retrieved March 3, 2004 from http://www.pcmag.ru/archive/9805/059816.asp#l12.

Lagesse, D. (2003). Are photos finished? Digital makes memories easier to capture and share—But harder to hold on to. *U.S. News & World Report*, 66-69.

Lester, P. (2000). *Visual communication: Images with messages*. Belmont CA: Wadsworth.

Lewis, P. (2004). Finally, a cure for shooting pains. *Fortune*. Retrieved March 5, 2004 from http://www.fortune.com/fortune/peterlewis/0,15704,593143,00.html.

Matsushita Electric Industrial Co., Ltd., Sanyo Electric Co., Ltd. and Sigma Corporation to support FourThirds system standard for dedicated digital SLR camera systems. (2004). Official press releases. *Digital Photography NOW*. Retrieved February 17, 2004 from http://dp-now.com/archives/000556.html.

McClelland, D. (1999). Digital cameras develop. *Macworld*. Retrieved March 3, 2004 from http://www.macworld.com/1999/09/features/cameras/.

Microsoft, Inc. (2004). SenseCam: Personal image & data recall. *Microsoft Research*. Retrieved March 9, 2004 from http://research.microsoft.com/research/hwsystems/.

Morgenstern, S. (1995). Digital "cheese." *Home Office Computing*. Retrieved March 3, 2004 from http://cma.zdnet.com/texis/techinfobase/techinfobase/+-LewS1pecy 3qmwwwAo_qoWsK_v8_6zzmwwww1Fqrp1xmwBnVzmwwwzgFqnhw5B/ display.html.

Murray, K. (1993). Smile for the camera. *Visual Arts and Photography*. Retrieved March 8, 2004 from http://www.kitezh.com/texts/smile.htm.

National Press Photographers Association. (2004). Digital Manipulation Code of Ethics. *Business practices*. Retrieved January 23, 2004 from http://www.nppa.org/professional_development/business_practices/ digitalethics.html.

Olympus Europa GmbH. (2002). Olympus and Kodak agree to implement 4/3 system digital SLR camera standard. *Press information*. Retrieved February 17, 2004 from http://cf.olympus-europa.com/ FourThirds/ press_release.htm.

Pavlik, J., & McIntosh, S. (2004). *Converging media: An introduction to mass communications*. Pearson Old Tappan, NJ: Pearson Allyn & Bacon.

Photo Marketing Association International. (2003). *Photo industry 2003: Review and forecast*. Retrieved January 17, 2004 from http://www.pmai.org/new_pma/marketing_research.htm.

Photo Marketing Association International. (2004). *Photo industry 2004: Review and forecast*. Retrieved January 17, 2004 from http://www.pmai.org/new_pma/marketing_research.htm.

Royal Philips Electronics. (2004). Philips fluid lenses bring things into focus. *Press information*. Retrieved March 3, 2004 from http://www.research.philips.com/InformationCenter/Global/ FNewPressRelease.asp?lArticleId=2904&lNodeId.

Sommerville, Q. (2003). Japan's "digital shoplifting" plague. *BBC News*. Retrieved March 9, 2004 from http://news.bbc.co.uk/1/hi/world/asia-pacific/3031716.stm.

Third-party research confirms industry leading fade resistance and image quality. (2004). Official press releases. *Digital Photography NOW*. Retrieved March 5, 2004 from http://dp-now.com/archives/ 000516.html.

Tuckman, B. (2004, March 6). *Personal communication*.

Walker, A. K. (2004, March 5). New life develops for Polaroid. *Miami Herald*, 1C, 8C.

IV
Telephony & Satellite Technologies

Telephone revenues in the United States from local, long distance, and wireless services exceed those of all advertising media combined. Clearly, point-to-point transmission of voice, data, and video represents the single largest sector of the communications industry. The sheer size of this market has two effects: companies in other areas of the media want a piece of the market, and telephone companies want to grow by entering other media.

The Telecommunications Act of 1996 was designed to stimulate competition in the provision of these services, but, to date, comparatively little competition has emerged. Ironically, the advanced technology that promised new markets and revenues for incumbent and competitive service providers has been a major factor preventing a more competitive environment. The reason is the cost of the technology—those companies that made massive investments in new technologies have yet to receive a significant return, leaving the industry laden with debt and reeling from the aftermath of the technology bust.

The digital technology that generated this debt is, however, in place with the same potential to revolutionize tomorrow's communication as it had when it was conceived and purchased. Digital protocols have erased distinctions in the transmission process for video, audio, text, and data. Because all these signals are transmitted using the same binary code, any transmission medium can be used for almost any kind of signal, provided the needed bandwidth is available. Furthermore, the advance of digital compression technologies reduces the bandwidth needed to transmit a variety of signals, further blurring the lines dividing the transmission characteristics of communications media. Many of the organizational barriers remain, however, allowing division of these technologies and services into individual chapters.

The first chapter in this section discusses the basics of the U.S. telephone network, with a focus on organizational factors that are playing a much more important role in transforming the telecom landscape than technological change. The following chapter focuses on broadband networks. In addition to explaining how they work and how much information they can transmit, this chapter discusses a variety of organizational, economic, and regulatory factors that will influence when and how each becomes part of the telephone network. The application of these network technologies for home use is explored in Chapter 21. In exploring home network technologies, the role played by the Internet in the diffusion of these technologies is addressed.

Satellites are a key component of almost every communication system. Chapter 22 explains the range of applications of satellite technology, including the history of the technology and the range of equipment needed (on the ground and in space) for satellite communication. Chapter 23 then explores one of the most important applications of early satellite technology—distance learning—which has since evolved to encompass virtually every communication medium.

The many, rapidly-evolving forms of wireless telephony are reviewed in Chapter 24. The final chapter in this section discusses the range of teleconferencing technologies, from simple audio-conferencing to videoconferencing and videophone systems that are primarily designed to facilitate face-to-face communication over distances. It also discusses the rapid evolution of group-based videoconferencing systems and the continued failure of one-to-one videophones.

In studying these chapters, you should pay attention to the compatibility of each technology with current telephone technologies. Technologies such as cellular telephone are fully compatible with the existing telephone network, so that a user can adopt the technology without worrying about how many other people are using the same technology. Other technologies, including the videophone and ISDN, are not as compatible. Consumers considering purchase of a videophone or ISDN service have to consider how many of the people with whom they communicate regularly have the same technology available. (Consider: If someone gave you a videophone today, whom would you call?)

Markus (1987) refers to this problem as an issue of "critical mass." She indicates that adoption of interactive media, such as the telephone, fax, and videophone, is dependent upon the extent of adoption by others. As a result, interactive communication technologies that are not fully compatible with existing technologies are much more difficult to diffuse than other technologies. Markus indicates that early adoption is very slow, but once the number of adopters reaches a "critical mass" point, usage takes off, leading quickly to use by nearly every potential adopter. If a critical mass is not achieved, adoption of the technology will start to decline, and the technology will eventually die out.

One of the most important concepts to consider in reading this chapter (and the other chapters that include satellite technology) is the concept of "reinvention." This is the process by which users of a product or service develop a new application that was not originally intended by the creator of the product or service. For example, cellular telephone technology is being reinvented almost daily as enterprising individuals devise new uses for these ubiquitous communication instruments.

The final consideration in reading these chapters is the organizational infrastructure. Because of the potential risks and rewards, even the largest companies entering the market for new telephone services are hedging their bets with strategic partnerships and experimentation with multiple, competing technologies. In this manner, the investment and risk is spread over a number of technologies and partners, as just one successful effort could pay back all the time and money invested.

Bibliography

Markus, M. L. (1987). Toward a "critical mass" theory of interactive media: Universal access, interdependence, and diffusion. *Communication Research*, *14* (5), 491-511.

19

Local and Long Distance Telephony

David Atkin, Ph.D. & Tuen-yu Lau, Ph.D.*

The free-fall in telecommunications continues to dampen expectations—and job prospects—for a sector whose valuation has fallen by two-thirds since the end of 1999. The U.S. Department of Commerce definition of telecommunications includes elements ranging from local exchange and long distance/international services to cellular telephony and paging (the latter elements are explored in other chapters of this book). Scholars suggest that the telephone medium—focused on the delivery of analog voice services over copper wire for its first century—is being transformed by digital transmission of voice, video, and data services via fiber optic and wireless networks (Bates, et al., 2002).

As an engine for economic growth, the telephone can facilitate a global information super-highway, cultivating user skills in interactivity and scalability that are crucial to the operation of emerging technologies (Neuendorf, et al., 1998; 2002). The technology convergence generating this inertia flows from the highly deregulatory Telecommunications Act of 1996 (P.L. 104-104, 1996), notable for its removal of entry barriers between local, long distance, and cable service providers. Yet, despite cutthroat price competition within the long distance sector, competition in local telephony is still "on hold" and remains a "long distance" away.

This industry turmoil underscores its rapid transmogrification since the era of plain old telephone service (POTS), already one of the most ubiquitous and lucrative communication technologies in the

* David Atkin is Distinguished Faculty Research Professor and Assistant Chair, Department of Communication, Cleveland State University (Cleveland, Ohio). Tuen-yu Lau is the Director of the Digital Media Master's Program, School of Communications, University of Washington (Seattle, Washington).

world. In the United States, the $301.8 billion in gross revenues from local and long distance companies—30% of the world market—exceeds the combined revenues of all advertising media, even surpassing the gross national product (GNP) of most nations (Pelton, 2003). The profit potential of these two markets is the primary force behind the global revolution in telephony, as other media companies seek to enter this lucrative market and telephone companies engage other media. More than 95% of homes in the United States (104 million) have telephone service, with monthly average expenditures of $83 that can be broken down as follows: local service ($36), long distance ($12), and wireless ($35) (U.S. Telephone Association, n.d.). This chapter outlines the changing structure of the telephone industry and the influence of changing regulations on the conduct of telephone companies, including implications for cross-media competition.

We begin by examining the architecture of the telephone network. To understand the U.S. telephone system, it is helpful to know that the entire country is divided into 194 service areas known as LATAs (local access transport areas). Most phone calls made from one phone to another phone in the same LATA are handled by the local phone company. Any call made between LATAs, however, must go through a "long-distance" connection. After the breakup of the Bell system in 1984, local phone companies were forbidden to provide interLATA service; they can offer interLATA and interstate long distance service, however, in states where regulators deem local competition to be sufficient (discussed later in this chapter). Few companies are lining up to bypass the local operating companies because it can cost $3,000 to $5,000 per home to provide a hook-up.

The telephone network is often referred to as a "star" network because each individual telephone is connected to a central office with a dedicated circuit. As illustrated in Figure 19.1, the heart of the network is the central office that contains the switching equipment to allow any telephone to be connected to any other telephone served by the central office. To allow connections to other, more distant telephones, central offices are, in turn, connected to each other by two networks. Local phone companies have one network that interconnects all central offices within a LATA, and long distance companies have their own separate networks that connect central offices located in different LATAs.

Telephone switches are specialized, high-speed devices that route each call through the network to its destination. In the earliest days of the telephone network, a telephone operator made all of the connections by hand. Operators were then replaced by mechanical switches, which were faster and less expensive. The mechanical switches have since been replaced by electronic switches that add a host of functionality to the switching process, ranging from relaying the originating phone number to the destination (caller ID) to enabling a single phone line to connect with two or more other phone lines (three-way calling). These services are sometimes known as Class 5 services.

Today's telephone networks make extensive use of high-speed fiber optic cables to interconnect central offices. Fiber optics is also increasingly used in place of copper wire to allow more efficient connections between end users and the central office. Figure 19.2 illustrates a modern telephone network that uses fiber optics to aggregate signals from dozens of subscribers into a single connection to the central office.

Figure 19.1
Traditional Telephone Local Loop Network Star Architecture

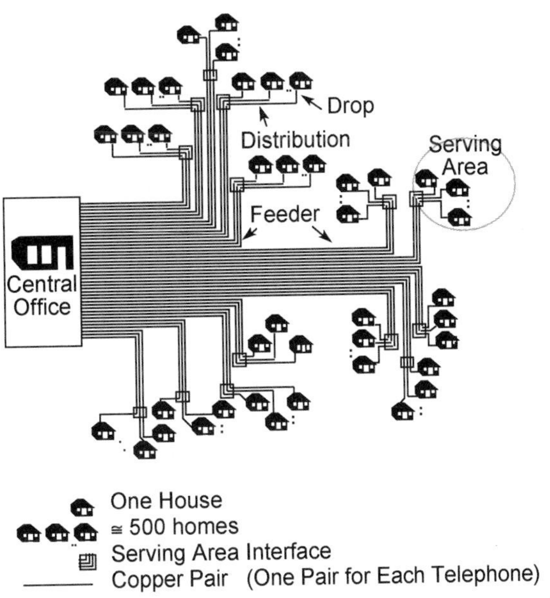

Source: Technology Futures, Inc.

Figure 19.2
Fiber-to-the-Feeder Local Telephone Network

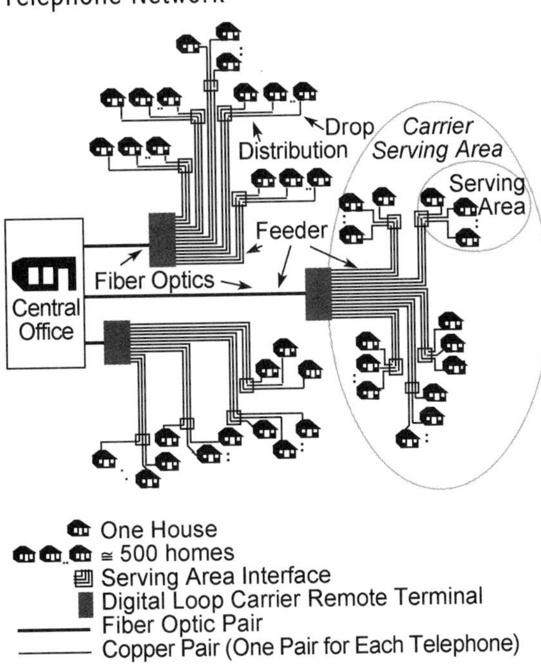

Source: Technology Futures, Inc.

Background

The Bell system dominated telephony for the century after Alexander Graham Bell won his patent for the telephone in 1876. After the original phone patent lapsed during the 1890s, the company came to be known as American Telephone & Telegraph (AT&T). This change ushered in a period of extensive competition—over 6,000 independent phone companies entered the fray, providing phone service and selling equipment (Atkin, 1996; Poole, 1983). However, concerns over gaps in service standardization and interconnection prompted government oversight of telephony in 1910 (Mann-Elkins Act, 1910).

Meanwhile, AT&T's acquisition of independent companies intensified after 1910, forming the building blocks for what later became known as the regional Bell operating companies (RBOCs). In response, the Justice Department threatened its first antitrust suit against AT&T in 1913 (*U.S. v. Western Electric, Defendant's Statement*). By the 1920s, however, Congress was actually in favor of a single monopoly phone system (Willis-Graham Act, 1921). In the meantime, AT&T worked to accommodate remaining independents, allowing interconnection with 4.5 million independent telephones in 1922 (Weinhaus & Oettinger, 1988). This industry rapprochement enabled Bell to focus energy on new ventures, such as "toll broadcasting" on radio stations (e.g., WEAF) (Brooks, 1976; Briggs, 1977).

Fearing telco domination of the nascent broadcast industry, Congress formalized a ban on telco-broadcast cross-ownership in the Radio Act of 1927 and the succeeding Communication Act of 1934. The 1934 Act also granted AT&T immunity from antitrust actions, in return for a promise to provide universal phone service; 31% of U.S. homes had a phone at that time (Dizard, 1999).

After investigating complaints concerning AT&T's market dominance in the 1930s, the Federal Communications Commission (FCC) endorsed the industry's structure, characterizing it as a "natural monopoly" (FCC, 1939). By the late 1940s, however, the Justice Department began to feel uneasy about the sheer magnitude of AT&T's empire. It initiated antitrust proceedings against the phone giant in 1948, which culminated in a 1956 Consent Decree. Under the decree, the government agreed to drop its lawsuit in return for an AT&T pledge to stay out of the nascent computing industry.

The following decades were marked by deregulation aimed at ending AT&T's notorious exclusive dealing practice, euphemistically known as the "foreign attachment" restriction. Under the guise of protecting its network from problems of incompatibility and unreliability, AT&T alienated several non-monopoly companies seeking to attach consumer-owned equipment to the network. In 1968, the FCC ruled against AT&T's ban of an acoustic coupler connecting radiotelephones. When allowing connection of this "Carterfone," the FCC and courts signaled that equipment not made by AT&T's manufacturing subsidiary, Western Electric, could be used in its network (*Carterfone*, 1968). During the following year, MCI was given permission to operate a long distance line, despite AT&T's objections (*Microwave Communications, Inc.*, 1969). In 1974, dissatisfied with industry conduct, the Justice Department initiated proceedings to dismember AT&T, an action reminiscent of its 1948 attempt. In particular, the complaint against AT&T (*U.S. v. AT&T*, 1982) alleged that the company:

1) Denied interconnection of non-Bell equipment to the AT&T network.

2) Denied interconnection of specialized common carriers with the Bell network.

3) Foreclosed the equipment market with a bias toward its Western Electric subsidiary.

4) Engaged in predatory pricing, particularly in the intercity service area (Gallagher, 1992).

This time, with the aid of several interested non-monopoly firms, the Justice Department was in a much stronger position to ultimately prevail. Thus, as demand for telecommunications services grew, the Bell system's regressive monopoly structure proved an impediment to growth and innovation (Jussawalla, 1993). After enjoying government protection during the first part of the century, the Bell monopoly lost favor with regulators for inefficient and anticompetitive market conduct.

In fairness, AT&T achieved the burdensome goal of providing universal service with the highest reliability in the world. The company employed one million workers in 1982, claiming that it lost about $7.00 a month on its average telephone customer (Dizard, 1999). While that plea may be debatable, such costs necessitated the practice of cross-subsidization that kept local phone service prices low by charging high rates for long distance service. Known as the behemoth that worked, AT&T was then the largest company in the world, subsuming 2% of U.S. GNP. The company carried over one billion calls per day at the time of divestiture (Dizard, 1999; *U.S. v. Western Electric*, 1980).

AT&T compensation issues notwithstanding, by 1982, even they viewed the regulated monopoly as an impediment to progress. The company's willingness to consent to divestiture was, arguably, as much a function of self-interest as exogenous government pressure. AT&T executives recognized that it was in their own interest to discontinue the "voice-only" monopoly in which the company had become encased. So they negotiated a divestiture settlement, or consent decree, known as the Modified Final Judgment (MFJ) in 1982 (*U.S. v. AT&T*, D.D.C., 1982). Concern over the impact of the MFJ prompted the presiding judge, Harold Greene, to maintain oversight over all aspects of the MFJ, including the divestiture, for the next decade.

The MFJ became effective in 1984, resulting in AT&T's local telephone service being spun off into seven new companies, known as the regional Bell operating companies, sometimes referred to as the "Baby Bells." The divestiture provided AT&T, which kept its long distance service and manufacturing arm (Western Electric), a convenient vehicle to exchange pedestrian local telephony for entry into the lucrative computer market. Perhaps more important, it enabled AT&T to cut most of its labor overhead without fear of union unrest or "bad press."

With recent deregulation allowing for the recombination of local and long distance services by a single provider, the experience with Ma Bell may provide a template for future industry conduct. For now, though, we examine the industry's post-divestiture conduct, and the implications of its entry into allied fields.

Recent Developments

After a series of FCC and court rulings relaxed restrictions on telco entry into cable (Telephone Company, 1992; *Chesapeake*, 1992), the RBOCs requested that the MFJ be vacated in 1994 (Atkin, 1996). In 1996, Congress and the President removed the ban as part of the Telecommunications Act. Since the law also removed the ban on telco purchases of cable companies in communities of over 35,000, new alliances between these industries became likely. Although U S WEST (now Qwest) was one of the most aggressive entrants into cable television in the 1990s, they split their phone and cable businesses into two companies just after the turn of the century.

The Telecom Act of 1996 also contained provisions designed to encourage competition in local telephony, leading to lower rates. Local rates would not decline, however, until AT&T, MCI, and other potential competitors obtained favorable terms for reselling Bell connections as their own services. New York City offered the first laboratory for local loop competition, with NYNEX competing against Tele-Communications, Inc. (TCI) and Cox's MFS Communications. By 1995, those cable-based competitors controlled over 50% of Manhattan's "special access" market, which involved the routing of long-distance calls to-and-from local lines (Atkin & Lau, 2003). The Telecom Act of 1996 also provided for universal service mandates, but allowed states and the FCC to decide how they should be funded.

With the new law resonating like the starting gun at a race, several companies issued threats about invading each other's markets. AT&T, for instance, issued a declaration of war on local phone companies everywhere, pledging to enter the market in all 50 states and win over at least one-third of the $100 billion sector within a few years. Their plan, which proved to be prohibitively expensive, was to offer service via alternative access providers while using the company's cable lines as a platform for speedy Internet access.

AT&T's bold move to deliver local service and become the nation's largest cable operator, bolstered by its acquisition of TCI and MediaOne, ended in 2002 when AT&T Broadband was sold to Comcast Cable. It seems, then, that the strategy of simultaneously pursuing an in-region and out-of-region service in the two industries did not succeed (Atkin, 1999). Telcos could also provide video programming in their own LATA, but the act prohibited joint telco-cable ventures in their home markets until 2004, when the FCC vacated the cross-ownership provision.

Despite this competitive rhetoric, telcos are more interested in pursuing markets in telephony than in cable. Before being allowed to enter long distance markets, RBOCs must prove that they have opened their local phone networks to new rivals, following a 14-point checklist contained in the 1996 Act. The FCC set rules for implementing RBOC entry into long distance in August 1996, promulgating 742 pages of guidelines. These rules were immediately challenged by several Baby Bells, and a U.S. District Court issued a ruling that the guidelines trampled on state's rights (Mehta, 1998). All such cases were consolidated in the Eighth Circuit Court of Appeals, which granted a stay of the FCC's order; the Supreme Court ultimately upheld the FCC's authority to review these applications on a state-by-state basis.

In December 1999, Bell Atlantic—which merged with GTE to become Verizon—became the first RBOC to win regulatory approval for its application to provide long distance service. Providing service to New York State, this arrangement represented the first time since AT&T's divestiture that consumers could receive both local and long-distance service from a former Bell operating company. Since then, most states have granted RBOC petitions for approval to offer long-distance telephone service to their local customers, with long distance service available from the RBOCs in over 36 states in 2003.

Given the limited telco entry into other industries, convergence will not prove as revolutionary as some might expect. As Bates, et al. (2002) note, new services can be grouped into several categories:

- Adding new types of content/uses that can utilize existing telephone networks.
- Value-added services complementing basic service.
- New mechanisms for delivering existing and emerging services.
- New uses which were not feasible under the technical limitations of the old networks, but which may be under new broadband networks.

Years after winning the right to provide information service, the RBOCs had little to show for their efforts (Atkin & Lau, 2003). In the meantime, AT&T, MCI, and the Baby Bells are actively promoting unlimited service plans that bundle local, long distance, and information applications through wireless and wireline channels for about $40 to $60 a month. Some companies are also planning to offer low-priced service via Internet broadband connections. The development of these alternatives, through wired and wireless channels, has led to a steady decline in the use of conventional landlines.

Beyond that, the RBOCs have engaged in projects involving information delivery (see Chapter 20). Despite blue-sky growth projections for video and data revenue— prompting telcos to assume $650 billion in debt worldwide—over 97% of fiber optic capacity went unused at the turn of the millennium (Zuckerman & Solomon, 2001).

Ancillary Services

Local and long distance service continues to dominate U.S. telco activities, generating $127.8 billion and $99.3 billion, respectively, in 2003 (USTA, n.d.). There has also been prolific growth in wireless ($74.7 billion revenue) and advanced services ($30 billion revenue), along with a host of other services, including call waiting, call forwarding, call blocking, prepaid phone cards, computer data links, and audiotext (or "dial-it") services. Long distance companies, in particular, bear little resemblance to the firms they were in the late 1990s, as their core landline business keeps shrinking.

One area of robust growth has been in an area involving mass-audience applications of the telephone—800 and 900 services (Atkin & LaRose, 1994; LaRose & Atkin, 1992). Virtually nonexistent prior to divestiture, these services generated 10 billion calls annually during the 1990s, with over 1.3 million 800 numbers earning billions in revenue for the industry (Atkin, 1995). Aside from dial-a-porn, audiotext presents hundreds of information options, including daily TV listings, national and international news, recordings of dead (and undead) celebrities, sports scores, horoscopes, and updates on soap operas (Atkin, 1993; 1995). The $1 billion audiotext industry may thus serve as a bridge technology to more advanced information services (Neuendorf, et al., 2002). Yet the Psychic Friends Hotline—which targeted low-income minority households with extensive promotions on cable—went out of business in 1998; they apparently did not see the end coming!

As telcos implement DSL (digital subscriber line—discussed in more detail in Chapter 20) technology and/or move to install fiber optics to the "last mile" of line extending to the home, providers will be able to offer a broad range of information and entertainment channels. There were 7.6 million U.S. DSL households in 2003.

In the meantime, the proliferation of new telephone services is having another effect: telephone companies are running out of available phone numbers. The traditional solution of splitting an area code into two and allocating a new area code is proving more difficult to implement as the geographical areas for these area codes becomes smaller and smaller. One solution being tested in many areas is an "overlay" of the new area code in the same area as the old one, requiring that 10 digits be dialed for every local call.

A related issue is local number portability (LNP). As competition emerges in the local telephone market, the FCC has been concerned that businesses and residences be able to keep their phone numbers when they choose to switch providers, adding slightly to the cost of local telephone service. The FCC implemented LNP for most U.S. cities in November 2003, allowing customers to keep their phone numbers when they switch providers.

The ability to transmit digital signals such as DSL on the same phone lines that carry ordinary POTS signals created another controversy, as competitive local exchange carriers (CLECs) argued that existing phone companies should share the copper wires from their switching offices to homes or businesses. The incumbent local exchange carriers (ILECs) argued against this, fearing a loss of up to half of their market share, which stood at 91% in 2002. This position delayed the introduction of digital services because of the lack of availability of enough existing telephone lines to allow additional service to everyone who wanted it. In early 2000, the FCC solved the problem by mandating line sharing among local phone companies. As a result, U.S. residents can now get multiple services (POTS and DSL) provided by different companies on the same phone line.

Presenting a textbook case of the regressivity of monopolists, DSL was developed by a Bell engineer in 1989 and languished for nearly a decade because the RBOCs did not want to sell against their other lucrative high-speed business Internet services (Dreazen, et al., 2002). The strong stock and bond markets of the 1990s helped fund CLECs in their efforts to increase DSL offerings, presenting a strong counterforce to the consolidated RBOCs. The number of local phone lines served by CLECs nearly doubled to 16.4 million (8.5% share) between 1999 and 2000 (Dreazen, 2001a). While there were 330 CLECs competing against the Bells at the end of 2000, only 150 remained a year later. After $50 billion in high-yield telecom bonds were issued in 1998 and 1999, investors balked at the CLEC upstarts in 2000, with ICG's CEO noting "We've gone from full spigot to a situation where every capital source has shut down at the same time" (Dreazen, et al., 2001, p. A10). This, along with regulatory uncertainty over broadband pricing, has prompted Bell slowness in upgrading to fiber. Only 23% of households have broadband in the United States—where much slower 56K modems reign supreme—compared with 73% of homes in Korea (Broadband fiasco, 2004).

Consolidation

Media merger and acquisition activity reached record levels in the wake of government deregulation during the 1990s, contributing to unprecedented levels of within-industry concentration. The impact of these mergers has been most dramatic in the area of local telephone service where, for instance, SBC is the local monopolist from Texas to Michigan.

SBC and Verizon account for 60% of phone lines in the United States, while Comcast Cable's merger with AT&T Broadband leaves the top three cable operators in control of 65% of the nation's cable market. This leaves the audience's essential communication lines under the control of powerful oligopolies. The forces of consolidation in telecommunications seem irresistible, given the large fixed costs, high barriers to entry, and low marginal costs of serving each additional subscriber. As Wolfe (2002) notes, "[c]ritics of media concentration will now wonder how much more wheeling and dealing can go on before there are but one or two juggernauts controlling every image, syllable, and sound of information and entertainment" (p. A18). Observers maintain that the Telecommunications Act of 1996 contributes to a "bigness" complex that threatens rate competition, service quality, public access to the media, freedom of speech, and democracy itself (Atkin, 1999).

Current Status

Declining stock valuations notwithstanding, the telephone companies remain in a formidable position to dominate video and information services, as they control more than half of U.S. telecommunications assets (Dizard, 1999). The industry handled 620 billion phone calls annually in the 1990s, and it has a long history of providing mass entertainment and information services (Atkin, 1995). The telecommunications sector accounts for one-twelfth of the economy, contributing $1 trillion to the "I.C.E." age economy (Pelton, 2003), a $4 trillion global colossus based on converging information, communication, and entertainment industries. Domestic phone revenue prospects may be dampened, however, by the Federal Trade Commission's 2003 implementation of a do-not-call registry designed to stop telemarketing.

While the telcos are anxious to compete in long distance, long distance companies are concerned that local companies still command 98% of local revenues over their regions. RBOCs retain control of over 90% of the 155 million phone lines in the United States (Dreazen, 2001a); they have "bottleneck control of the local loop and they'll be able to squash competition" (Berniker, 1994, p. 38). Given that MCI WorldCom paid $0.46 of every dollar it earned to the RBOCs in access charges, they have an interest in entering the $25 billion access charge business. Local phone companies receive about $8.5 billion annually for connecting long-distance calls to the local phone network. This local exchange bottleneck helped motivate GTE to enter into its $70 billion merger with Bell Atlantic to create Verizon (Mehta, 2000).

At the time of divestiture, the RBOCs represented roughly 75% of AT&T's assets. Between 1940 and 1980, the number of independent companies decreased by 77%, owing chiefly to consolidation among these companies, although some were acquired by AT&T (Weinhaus & Oettinger, 1988). By 1982, there were 1,459 independent telephone companies, producing 15.9% of the industry's revenues. Merger activity has accelerated since then, but hundreds of phone providers remain. Shifting power dynamics between local and long distance have been so sweeping that, in April 2004, the Dow Jones Stock Exchange removed AT&T from its list of 500 leading economic indicators, only to replace it with a Baby Bell progeny—Verizon—which had almost doubled the $34 billion annual revenue of its declining parent.

Although the Telecommunications Act of 1996 returned telephone regulation to the federal government, the Baby Bells were previously governed by the judiciary through Judge Harold Greene's review of the MFJ. The MFJ imposed several line-of-business restrictions on the RBOCs, including the manufacture of telecommunications products, provision of cable or other information services, and provision of long distance services (*U.S. v. A.T.& T.*, D.D.C. 1982).

The FCC reified telco/cable cross-ownership restrictions out of fear that telcos would engage in the predatory pricing characteristic of capital-intensive industries, and use their natural monopoly over utility poles and conduit space to hinder competition with independent cable operators. Under these rules, former RBOCs were banned from providing cable service outside of their local access transport area (Telephone Company, 1992). Although these terms of conduct have been subsequently deregulated, the "home court" advantage that RBOCs enjoy—in delivering a range of services—remains a concern. For example, in 2001, a federal appeals court overruled the FCC, which had approved SBC's application to sell long distance service in Oklahoma and Kansas, because the company had overcharged rivals who use their network (Dreazen, 2001b).

Factors to Watch

As the recent travails of AT&T suggest, the long distance sector has been open to competition for years, and may even be doomed as Internet telephony and other players push prices ever closer to zero. Aside from this convergence, telephone service will also be characterized by greater bandwidth capability, mobility, and globalization. As these trends unfold, there will be an intensification of the debate over whether Internet telephony should be subject to the same taxes and regulation as POTS (as advocated by the telcos). Telephone companies are less anxious to see regulation applied to their television transmissions over the public switched telephone network (PSTN), claiming that they should be exempt from taxes levied in the form of cable franchise fees even when these competing wires deliver the same services. An appeals court ruled in 2003 that cable Internet is a telecommunications service and should thus be subject to far more regulation (Broadband fiasco, 2004).

This regulatory standardization may facilitate the rollout of voice over Internet protocol (VoIP) (see Figure 19.3), a phone service that flows through the Internet and promises unlimited calling at discounted rates (Funk & Seper, 2004). Customers with high-speed Internet can make unlimited calls in North America for as little as $35 per month, using providers from AT&T, SBC, and Verizon to Cox Cable and Vonage Holdings Co. Although VoIP accounts for only 0.2% of the nation's phone lines as of early 2004, it has grown 640% worldwide, accounting for 12.8% of global phone traffic.

The industry's basic service market—local exchange service—is thus in transition from a "natural regulated monopoly" to an "open competitive marketplace" (Bates, et al., 2002, p. 92). The catalyst, the Telecommunications Act of 1996, was passed amidst heady optimism concerning the benefits of deregulation for competition, consumer prices, the construction of an information superhighway, and resulting job creation in the coming information age. This radical deregulation was designed to encourage unprecedented media cross-ownership and competition by enabling cable companies to enter local telephone markets and local and long distance companies to enter each other's markets.

Shortly after its passage, however, the act encountered heavy criticism in the popular press (Schiller, 1998), as even prominent congressional supporters sought hearings to investigate problems with industry conduct that the measure was designed to remedy (e.g., service, pricing) (Atkin, 1999). Preliminary data suggest that the act has not lived up to its promise. Although rates in the competitive long distance sector have declined 13.1% since 1996, cable rates have risen 31.9% and local phone rates—still regulated—have risen 12.1% (Solomon & Frank, 2001). Even in the beleaguered long distance area, rates for per-minute and bundled plans increased in 2003. Given recent declines in landline revenue, and WorldCom's bankruptcy filing following an $11 billion accounting scandal, long distance companies raised several monthly fees and per-minute rates in 2003.

Figure 19.3
Voice over Internet Protocol

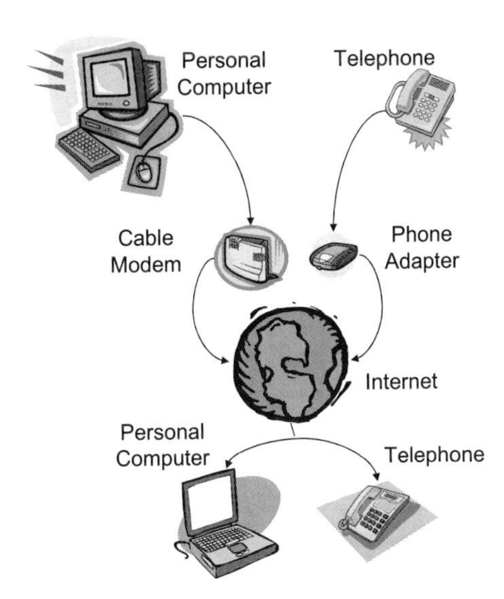

Source: Technology Futures, Inc.

Some eight years after the act's passage, telecommunications heavyweights are retreating to their neutral corners, as the proportion of customers who receive phone service over cable TV lines remains under 1%. Gene Kimmelman, co-director of Consumers Union, concludes, "It's an abysmal failure so far. The much-ballyhooed opening of markets to competition was a vast exaggeration" (Schiller, 1998, p. A1). Local telephone costs are likely to increase further in the wake of an appeals court ruling, in March 2004, that struck down FCC rules governing how RBOCs open their networks to competitors; expected increases in access payments could endanger several carriers.

Consumers are thus far paying more for telecommunications services in the new era of deregulation ushered in by the act. While former FCC Chair Reed Hundt expressed dismay that competition was delayed by legal wrangling and industry "détente," he suggested that higher rates might be a necessary first step toward competition (Mills & Farhi, 1997), with longer-term trends favoring convergence, competition, and lower prices. One competitive bright spot involves a group of smaller companies, such as Teligent, that are building their own advanced, high-speed communications facilities to give them direct access to their customers, bypassing remnants of the old Bell system. A related issue involves the ongoing redefinition of "universal service" mandates to include some form of lifeline access to video and information services (Auferheide, 1999).

While politicians might debate the need to maintain regulatory oversight until competition takes root, few would dispute that competition (and lower prices) will eventually be in the offing for telephony. The decade following divestiture saw competitive market forces dramatically reshape the telecommunications landscape. Consumer long distance rates dropped from an average of $0.40 per minute in 1985 to just $0.14 per minute by 1993 (Wynne, 1994). New long distance carriers helped spawn thousands of new jobs, although the sector has lost nearly 500,000 jobs since 2001, and market capitalization is down $2 trillion from its peak in the late 1990s (Broadband fiasco, 2004).

The consolidation of three former "Baby Bells" (Southwestern Bell, Pacific Telesis, and Ameritech) into SBC, in conjunction with the consolidation of NYNEX, Bell Atlantic, and GTE into Verizon, raises the issue of whether the remaining Baby Bells (Qwest, formerly U S WEST, and BellSouth) can go it alone. Although their territories range in profitability from the fast-growing south to the depopulated west, there is still plenty of opportunity for all RBOCs in the local phone industry.

Industry leader AT&T remains effectively blunted from access into local service. Meanwhile, more than 100 smaller long-distance players (e.g., Metromedia) compete in the long-distance market, but their numbers have been thinned by the recent industry downturn. Table 19.1 reviews the top 10 worldwide telecom mergers and acquisitions, the five biggest of which rank among the 10 largest business deals in the 20th century.

SBC's growth and aggressive approach to local loop competition, in particular, seems to be reawakening fears about the dysfunctional conduct of local telephone monopolists. Following its acquisition of Ameritech, the company was besieged with complaints of poor service—racking up $188 million in penalties since 1999—along with charges of anticompetitive behavior (Waters, 2001). This new consolidation, combined with obvious telco cross-subsidization concerns, presents a basis for continued vigilance by regulators.

AT&T's 2002 sale of its cable television assets to Comcast suggests that the promised "revolution" in these industries will probably not be created by telco entry into the video marketplace. This is also bad news for various independent programmers, who supply specialty entertainment and information services. Speculation now centers on the possibility that the conservative FCC—whose Chair William Powell opposes ownership caps—may allow a Bell company to acquire a long distance firm such as Sprint (Dreazen, et al., 2002).

In sum, the industry-wide devaluation, consolidation, and retreat from competition present regulators with the challenge of preventing undue concentration of essential wired communication capabilities into ever fewer hands. Even so, the definition of telephony continues to evolve beyond a fixed line to include such methods as prepaid phone cards, which can also be used as promotional and marketing tools. As Bates, et al. (2002) note, "[W]hat we think of today as the telephone is increasingly just one piece of telecommunications goods and services being offered by an ever-increasing variety of suppliers" (e.g., Internet-flavored structures) (p. 199). To maximize innovation, quality, and diversity in the information grid, telephony's vast capital and human resources should be channeled into allied fields in a way that augments (rather than depletes) the existing cast of players.

Despite sluggish growth prospects in the U.S. market, the share of the global telecommunications market being claimed by developing countries is growing, and global trade will increase under new World Trade Organization guidelines. The rate of information continues to expand 300,000 times faster than the world's population—some 500-fold in the past 30 years—as the speed and scope of telecommunications investment stimulates global economic development (Pelton, 2003). The initial liberalization in the U.S. market has prompted other countries (e.g., Japan) to follow suit, opening monopolies up to competition and "greasing the wheels" of global commerce. Thus, even as POTS continues to evolve, it should remain at the center of the emerging information economy.

Table 19.1
Top Telecom Mergers through 2003

Target	Acquirer	Value (in Billions)	Announced
Ameritech	SBC Communications	$72.4	5/11/98
GTE	Bell Atlantic	$71.3	7/28/98
AirTouch Communications	Vodafone Group	$65.9	1/18/99
U S WEST	Qwest Communications	$48.5	6/14/99
AT&T Broadband	Comcast	$44.0	12/21/01
MCI Communications	WorldCom	$43.4	10/1/97
Orange	Mannesmann	$35.3	10/20/99
Telecom Italia	Olivetti	$34.8	2/20/99
NYNEX	Bell Atlantic	$30.8	4/22/96
TCI Cable	AT&T	$30.0	6/24/98
Pacific Telesis	SBC Communications	$22.4	4/1/96

Source: Federal Communications Commission

Bibliography

Atkin, D. (1993). Indecency regulation in the wake of Sable: Implications for telecommunications media. *1992 Free Speech Yearbook*, *31*, 101-113.

Atkin, D. (1995). Audio information services and the electronic media environment. *The Information Society*, *11*, 75-83.

Atkin, D. (1996). Governmental ambivalence towards telephone regulation. *Communications Law Journal*, *1*, 1-11.

Atkin, D. (1999). Video dialtone reconsidered: Prospects for competition in the wake of the Telecommunications Act of 1996. *Communication Law and Policy Journal*, *4*, 35-58.

Atkin, D., & LaRose, R. (1994). Profiling call-in poll users. *Journal of Broadcasting & Electronic Media*, *38* (2), 211-233.

Atkin, D., & Lau, T. (2003, May). *Still on hold: Prospects for competition in the wake of the Telecommunications Act of 1996*. Paper presented at the International Communication Association, San Diego, CA.

Auferheide, P. (1999). *Communications policy and the public interest: The Telecommunications Act of 1996*. New York: Guildford.

Bates, B., Jones, K., & Washington, K. (2002). Not your plain old telephone: New services and new impacts. In C. Lin and D. Atkin (Eds.). *Communication technology and society: Audience adoption and uses*. Cresskill, NJ: Hampton, pp. 91-124.

Berniker, M. (1994, December 19). Telcos push for long-distance entry. *Broadcasting & Cable*, 38.

Briggs, A. (1977). The pleasure telephone: A chapter in the prehistory of the media. In I. de Sola Pool (Ed.). *The social impact of the telephone*. Cambridge, MA: MIT Press.

Broadband fiasco. (2004, February 11). *The Wall Street Journal*, A18.

Brooks, J. (1976). *Telephone: The first hundred years*. New York: Harper & Row

Carterfone. In the matter of use of the Carterfone device in message toll telephone service. (1968). FCC Docket Nos. 16942, 17073; Decision and order, 13 FCC 2d 240.

Chesapeake and Potomac Telephone Co. v. U.S. (1992), Civ. 92-17512-A (E.D.-Va).

Dizard, W. (1999). *Old media, new media*, 3rd edition. New York: Longman.

Dreazen, Y. (2001a, May 22). Bells' rivals double local market share. *The Wall Street Journal,* B6.

Dreazen, Y. (2001b, December 31). Court says SBC overcharges competitors, tells FCC to review long-distance decision. *The Wall Street Journal*, A2.

Dreazen, Y., Ip, G., & Kulish, N. (2002, February 25). Why the sudden rise in the urge to merge and form oligopolies? *The Wall Street Journal*, A10.

Federal Communications Commission. (1939). Investigation of telephone industry. *Report of the FCC on the investigation of telephone industry in the United States*, H.R. Doc. No. 340, 76th Cong., 1st Sess. 602.

Funk, J., & Seper, C. (2004, February 8). Internet phone service about to come calling. *Cleveland Plain Dealer*, A1, 19.

Gallagher, D. (1992). Was AT&T guilty? *Telecommunications Policy*, *16*, 317-326.

Jussawalla, M. (1993). *Global telecommunications policies: The challenge of change*. Westport, CT: Greenwood Press.

LaRose, R., & Atkin, D. (1992). Audiotext and the re-invention of the telephone as a mass medium. *Journalism Quarterly*, *69*, 413-421.

Mann-Elkin's Act. (1910). Mann-Elkins Act, Pub. L. No. 218, 36 Stat. 539.

Mehta, S. (1998, January 5). Baby Bells cautious on quick entry to long-distance market after ruling. *The Wall Street Journal*, B8.

Mehta, S. (2000, January 20). Bell Atlantic and GTE file with FCC to split off GTE's Internet backbone. *The Wall Street Journal*, A4.

Microwave Communications, Inc. (MCI). (1969). FCC Docket No. 16509. Decision, 18 FCC 2d 953.

Mills, M., & Farhi, P. (1997, January 27). A year later, still lots of silence: Dial up the Telecommunications Act and get bigger bill and not much else. *The Washington Post*, 18.

Neuendorf, K., Atkin, D., & Jeffres, L. (1998). Understanding adopters of audio information services. *Journal of Broadcasting & Electronic Media, 42*, 80-95.

Neuendorf, K., Atkin, D., & Jeffres, L. (2002). Adoption of audio information services in the United States. In C. Lin and D. Atkin (Eds.). *Communication technology and society: Audience adoption and uses*. Cresskill, NJ: Hampton, pp. 125-152.

Pelton, J. (2003). International telecommunications. In K. Anowkwa, C. Lin, & M. Salwen, (Eds.). *International communication: Theory and cases*. New York: Wadsworth, pp. 267-283.

Poole, I. (1983). *Forecasting the telephone: A retrospective technology assessment*. Norwood, NJ: Ablex.

Schiller, Z. (1998, February 9). Local phone competition is still just a promise. *Cleveland Plain Dealer*, A1, A6.

Solomon, D., & Frank, R. (2001, December 21). Comcast deal cements rise of an oligopoly in the cable business. *The Wall Street Journal*, 1.

Telecommunications Act of 1996, 104 Pub. L. 104, 110 Stat. 56, 111 (1996) (codified as amended in 47 C.F.R. S. 73.3555).

Telephone Company/Cable Television Cross Ownership Rules. (1992). 47 C.F.R. 63.54-63.58. *Second report and order, recommendations to Congress and second further notice of proposed rulemaking*. 7 FCC Rcd. 5781.

U.S. Telephone Association. (n.d.) *Telecom statistics*. Retrieved March 13, 2004 from http://www.USTA.org/index.php?urh=home.news.telecom_stats.

U.S. v. A.T.& T., 552 F. Supp. 131, 195 (D.D.C. 1982), *aff'd sub nom. Maryland v. U.S.*, 460 U.S. 1001 (1983).

U.S. v. Western Electric, Defendant's Statement, U.S. v. Western Electric Co. (1980). Civil Action No. 74-1698, pp. 169-170.

Waters, R. (2001, July 18). Back to the future. *Investor's Business Daily*, 15.

Weinhaus, C., & Oettinger, A. (1988). *Behind the telephone debates*. Norwood, NJ: Ablex.

Willis-Graham Act. (1921). Pub. L. No. 15, 42 Stat. 27.

Wolfe, M. (2002, Feb. 21). Here comes another wave of mergers. *The Wall Street Journal*, A18

Wynne, T. (1994, November 16). An earshocking proposition. *Cleveland Plain Dealer*, 11-B.

Zuckerman, G., & Solomon, D. (2001, May 11). Telecom debt debacle could lead to losses of historic proportions. *The Wall Street Journal*, A1.

20

Broadband Networks

Lon Berquist, M.A.*

Broadband technology has been adopted by consumers throughout the world at a remarkable rate. Worldwide, there are more than 100 million broadband connections (Point Topic, 2004). In the United States, broadband has become one of the most rapidly-adopted consumer technologies ever, with early market penetration growing more swiftly than either personal computers or cell phones (Beardsley, et al., 2003). By 2003, 24% of U.S. households had broadband connections (Pew Internet, 2004).

Background

Broadband delivers voice, video, and data over a variety of transmission media. The terms "high-speed," "advanced telecommunications," "advanced services," and "broadband" are often used interchangeably to describe telecommunications networks with considerable transmission capability.

The International Telecommunications Union (ITU), an international telecommunications standards-making body, has defined broadband as transmission capacity that is faster than 1.5 megabits per second (Mb/s) (ITU, 2003). However, in the United States, the Federal Communications Commission (FCC) defined broadband in the Telecommunications Act of 1996 as an advanced telecommunications service with a bidirectional transmission speed of more than 200 kilobits per second (Kb/s) (FCC, 2004)

* Telecommunications and Information Policy Institute, University of Texas at Austin (Austin, Texas).

Although a transmission speed of 200 Kb/s is almost four times faster than 56 Kb/s dial-up modems, it is far below the internationally recognized definition of broadband. It is likely that the U.S. definition of broadband will evolve as network capabilities improve, and the ever-increasing bandwidth needs of complex Internet content require even greater transmission speeds.

For those researching broadband development, the most crucial aspect of broadband is not the technology itself, but the content or services it can provide. In addition to broadband's ability to enhance transmission speeds for traditional Internet use, it offers the potential for greater multimedia-based entertainment and information programming, improved capability for telework, better potential for e-government, and increased possibilities for innovative services yet to be developed. Also, the "always-on" nature of most broadband access simplifies the process of connecting to the Internet.

The deployment of telecommunications technology has long been touted as an economic development tool to promote regional and national economies, and policymakers recognize the importance of broadband networks as a means to sustain current businesses and attract new businesses or industries. A broad range of industries desire broadband deployment to enhance business efficiencies and to develop new and innovative products and services for consumers.

At the same time, the use of broadband for education (both K-12 and higher education), healthcare (telemedicine), and online government services offers more incentives for developing policies promoting broadband deployment.

The economic downturn that began in 2001 has had a direct impact on the telecommunications industry and the deployment of broadband. Despite the initial rapid adoption of broadband, the pace of growth has slowed. Although there was a steady increase in Internet connectivity throughout the 1990s, the pace of Internet growth has declined as well (Angwin, 2001). In the second half of 2003, broadband connections grew 27%, while dial-up connections (narrowband) showed no growth (see Table 20.1). In contrast, the amount of information transmitted on the Internet continues to double each year. According to industry consultant IDC, Internet traffic volume will grow from 180 petabits (180 million gigabits) per day in 2002, to 5,175 petabits per day in 2007 (Legard, 2003). Broadband networks are essential to meet the increasing demands of Internet traffic.

Table 20.1
U.S. Broadband and Narrowband Growth

Speed	May 2003	November 2003	Growth
Broadband Users	38,957,000	49,465,000	27%
Narrowband Users	69,647,000	69,609,000	0%

Source: Nielsen/NetRatings (December 2003)

Broadband Technologies

Current broadband technologies utilized throughout the United States include cable modems, digital subscriber line (DSL), fixed wireless broadband (FWB), wireless local area networks

(WLANs), fiber-to-the-home (FTTH), satellite broadband, and the emerging broadband over power line (BPL) technology. Sixty-one percent of consumers, however, still access the Internet by using dial-up modems, while 39% of Internet users utilize broadband technologies. Fifty-four percent of broadband users access the Internet via cable modems, 42% use DSL, and 3% use wireless or satellite for broadband access (Pew Internet, 2004).

In their 2003 study of Internet use, the Pew Internet and American Life Project (2003) discovered that users of broadband technologies are much more active in their use of the Internet than dial-up users. This is particularly true of data-intensive activities such as viewing streaming multimedia and downloading music (see Table 20.2).

Table 20.2
Internet Activities Among Broadband and Dial-Up Users

Daily Internet Activities	Broadband Users	Experienced Dial-Up Users	Dial-Up Users
News	41%	35%	23%
Research for Work	30%	30%	15%
Participation in Group	12%	11%	4%
Content Creation	11%	9%	3%
Stream Multimedia	21%	13%	7%
Download Music	13%	3%	3%

Source: Pew Internet & American Life Project (2003)

Recent Developments

Cable Modems

A majority of U.S. consumers receive broadband connectivity through their local cable television system. According to the National Cable Television Association (NCTA), cable modem service is available to 90 million U.S. homes (NCTA, 2004b), with over 15 million households subscribing to service (Leichtman Research Group, 2004).

The largest multiple system operators (MSOs) dominate the cable modem market. Comcast's merger with AT&T Broadband in 2002 allowed it to surpass Time Warner Cable as the dominant MSO in the cable broadband market, capturing almost 5.3 million subscribers. Time Warner has 3.2 million subscribers, followed by Cox with over 1.9 million subscribers, Charter with close to 1.6 million, and Cablevision providing cable broadband service to over one million cable modem subscribers (see Table 20.3).

Installation of cable modem service runs from $50 to $200, with a monthly service fee of $30 to $50. A typical cable modem will provide high-speed Internet connections of up to 1.5 Mb/s downstream and 128 Kb/s upstream; however, the downstream bandwidth is shared with cable subscribers on the same distribution node, which can impact download speeds among simultaneous users. Cable's dominance of the broadband market can be explained by the fact that a cable system can provide ubiquitous broadband service throughout its network, while the competing telephone system's DSL service is distance sensitive and not always available to consumers desiring broadband service.

Table 20.3
U.S. Cable Modem Market, 2003

Cable MSO	Cable Modem Subscribers
Comcast	5,283,900
Time Warner Cable	3,228,000
Cox	1,988,527
Charter	1,565,600
Cablevision	1,057,020
Adelphia	960,000
Bright House	625,000
Mediacom	280,000
Insight	230,000
RCN	200,000
CableOne	133,800
Other	225,000
Total	15,776,847

Source: *Cable Datacom News* (2004)

Cable modems operate under the open standard Data Over Cable Services Interface Specifications (DOCSIS). The first generation DOCSIS 1.0 certified modems enabled cable systems that had been upgraded with two-way hybrid fiber/coax (HFC) networks to offer Internet service. Increasingly DOCSIS 1.1 modems are being deployed, improving quality of service (QoS) requirements that allow tiered data, voice, and multimedia services. The next generation modems, DOCSIS 2.0, are being field tested to add even greater capacity (up to 30 Mb/s) and symmetric services that will enhance data, video, and voice services (Eshad & Melamed, 2004).

In addition to offering broadband connectivity to the Internet, the cable industry provides telephone service for roughly 2.5 millions subscribers (NCTA, 2004a). Most cable telephony is provided using traditional circuit switching over the cable system's network; however, the cable industry has been an early provider of Internet protocol (IP) telephony that leverages the packet switching capability of their networks to offer voice over IP (VoIP). In 2003, Cablevision Systems had almost 30,000 subscribers paying $34.95 per month for VoIP in their New York market (Breznick, 2004).

In March 2002, the FCC ruled that cable modem service, rather than being a typical cable television service or a regulated telecommunications service, was an unregulated "information service" and thereby not subject to open access requirements. In 2003, the Ninth Circuit U.S. Court of Appeals ruled that cable modem service was, instead, a telecommunications service—disappointing both the FCC and local municipalities who preferred a cable service classification (Levine, 2003). It is unclear whether the FCC or local government alliances will appeal the decision to the U.S. Supreme Court, but, in the meantime, local governments, which charge cable operators franchise fees based on their cable subscription revenues, are prohibited from collecting franchise fees from cable modem revenue.

DSL

Digital subscriber line technology makes use of existing telephone networks to transmit high-speed data. Incumbent local exchange carriers (ILECs), such as SBC and Verizon, are the dominant providers of DSL service, with competitive local exchange carriers (CLECs) holding less than 10% of the broadband market (Ferguson, 2004). As of late 2003, there were over 9 million DSL subscribers nationwide (see Table 20.4).

Table 20.4
U.S. DSL Subscribers, 2003

DSL Provider	DSL Subscribers
SBC	3,516,000
Verizon	2,319,000
BellSouth	1,462,000
Qwest	637,000
Covad	517,000
Sprint	304,000
ALLTEL	153,028
Cincinnati Bell	99,000
Century Tel	83,400

Source: Leichtman Research Group (2004)

Digital subscriber line technology is utilized by telephone systems in order to transmit data over ordinary copper phone lines. DSL's advantage is that it offers a dedicated line from the central office to the residence, providing both voice service and data with typical rates of 1.5 Mb/s downstream (although offering slower rates upstream). DSL service is similar to cable modems in that transmission is asymmetrical, with a greater downstream capacity, adequate for most Web browsing and Internet use.

Installation costs for DSL can range from $50 to $200, with a monthly service fee of $30 to $70 depending on the desired transmission speed and service. Telephone companies have lowered prices to compete with the more successful cable modem, and they have begun to offer varying tiers of service to provide pricing and speed options to customers (Musgrove, 2003). As of mid-2004, SBC charges $34.94 per month for download speeds between 384 Kb/s and 1.5 Mb/s, and $44.99 per month for download speeds between 1.5 Mb/s and 3 Mb/s. Bundling services has also become a marketing strategy for phone companies—promoting discounted packages combining local and long distance phone service with cellular phone service and DSL (Hu, 2004b).

The increase in data rates for DSL is made possible by enhancements to traditional asymmetric digital subscriber line (ADSL) technology. DSL extension technology has been introduced to extend the reach of DSL service within the telephone network. Typically, a DSL customer must reside within approximately 15,000 feet from the digital subscriber line access multiplexer (DSLAM) located in the phone company's central office. Extenders can provide DSL service up to 25 miles from the DSLAM (FCC, 2002).

In addition, new standards have been introduced such as ADSL2 that increase the reach of DSL service to customers and boost downstream data rates to over 12 Mb/s. ADSL2+ provides data rates of up to 25 Mb/s (Leblanc, 2003). As phone companies such as Qwest Communications build networks that deploy high-speed fiber optics closer to homes, very-high-data-rate DSL (VDSL) technology is being deployed that offers transmission capability of 45 Mb/s (Krim, 2003a).

In 2003, the FCC voted to modify provisions in the Telecommunications Act of 1996 that required ILECs to share their networks with CLECs. The revised rules allow ILECs to avoid sharing their fiber-optic networks with competitors and ends requirements for leasing DSL lines to competing providers at discounted rates. The FCC order sustained the requirement that ILECs lease traditional phone lines to CLECs at a significant discount (Krim, 2003b), but this provision was subsequently overturned by the U.S. Court of Appeals.

Wireless

Wireless Internet service providers offer broadband wireless access via fixed wireless broadband technologies such as multipoint distribution services MMDS and LMDS, and new generations of wireless broadband over regulated and unregulated frequencies.

The FCC provides licenses in fixed wireless frequencies (2 GHz to 42 GHz) for multichannel multipoint distribution service and instructional television fixed service. Originally developed for analog video distribution, they are now being utilized for data transmission with data rates of 128 Kb/s to 10 Mb/s over 50 kilometers. Local multipoint distribution service offers transmission speeds of up to 100 Mb/s (shared among users) over a distance of one to three kilometers. Because of their line-of-sight requirements and potential signal interference due to bad weather, subscription to fixed wireless broadband has been limited to roughly 150,000 subscribers (FCC, 2002). Despite the limitations of fixed wireless, major telecommunications providers such as Clearwire, BellSouth, Sprint, and Verizon continue to test wireless service in localized markets (Martin, 2003).

Free space optics (FSO) utilizes lasers to transmit data point-to-point through the air at data rates up to 1.55 Gb/s, although its line-of-sight signal is limited to less than a kilometer (FCC, 2002).

There is momentum building for wireless metropolitan area networks (WMANs) with the development of the IEEE 802.16 standard commonly referred to as WiMAX (worldwide interoperability for microwave access). WiMAX transmits point-to multipoint data in the 10 GHz to 66 GHz range at data rates up to 120 Mb/s over 50 kilometers.

Combined with the growing use of Wi-Fi local networks, wireless broadband promises to free Internet users from the tether of wireline connectivity. (See Chapter 21 for more information on wireless local area networks.)

Broadband Satellite

Broadband Internet service via satellite is offered primarily by Hughes Electronics' DIRECWAY with close to 180,000 subscribers and StarBand Communications, recently surviving a bankruptcy scare, claiming 40,000 subscribers. Transmission speeds for residential broadband satellite typically range from 400 Kb/s to 500 Kb/s downstream and 128 Kb/s to 256 Kb/s upstream. The broadband satellite dish and modem retail for around $600, with monthly fees ranging from $40 to $70.

Financial woes continue to plague the direct broadcast satellite (DBS) industry, delaying the launch of new competing systems. Wildblue Communications, financed by INTELSAT, Liberty Media, Telesat, Kleiner Perkins Caulfield and Byers, Arianespace, and the National Rural Telecommunications Cooperative (NRTC), is set to launch in 2004 along with SES Americom's American2Home DBS and broadband service (Labrador, 2004). Wildblue offers downstream speeds of up to 1.5 Mb/s and upstream speeds of up to 256 Kb/s.

Hughes has introduced the Internet protocol over satellite (IPoS) standard that was ratified as a U.S. Telecommunications Industry Association standard in November 2003 (Hu, 2004a). The standard specifies the protocols for the transmission of IP packets via satellite. In 2004, Hughes is introducing its next-generation broadband satellite service, SPACEWAY, that will provide services to small businesses and eventually residential customers with download speeds as high as 50 Mb/s and upload rates as fast as 16 Mb/s.

Late in 2003, News Corporation acquired 34% of Hughes to gain a controlling interest in the DBS service, after an attempt by EchoStar to purchase Hughes was thwarted by the FCC and the U.S. Justice Department due to antitrust concerns (Silverstein, 2003).

Fiber Optics

Both DSL and cable modem service rely on fiber optic cables for the bulk of the backbone of their networks (the middle mile), but the last mile of transmission media consists of coaxial cable for cable modem service and twisted pair copper wire for the telephone system's DSL service. Gradually, fiber optic cable is reaching closer to the home, and eventually fiber-to-the-home installations will become an accepted residential broadband connectivity option.

Currently, however, there are few FTTH networks. The FCC estimates there are 460,000 direct-to-subscriber fiber links throughout the United States, but only 0.6% are connected to residences (FCC, 2002).

Fiber-to-the-home, also described as fiber-to-the-premises (FTTP), currently uses passive optical network (PON) technology to provide speeds of 622 Mb/s to 2.5 Gb/s (shared among network users) at an almost unlimited distance from the central office. Surprisingly, many examples of FTTH networks can be found in small rural towns, forgotten by cable operators and telephone companies (Johnston, 2002). In some instances, cities have formed partnerships, such as the Utah Telecommunications Open Infrastructure Agency (UTOPIA), to develop a publicly-owned fiber network.

ILECs have recently enacted a number of pilot projects to assess the market for fiber optic networks. SBC's fiber project in Mission Bay (California) offers 1.5 Mb/s speeds for $26.95 per month, and 6 Mb/s for $139.95 per month. SBC has announced it will develop fiber projects in five other U.S. sites. Verizon Communications has taken an even stronger step with plans to put fiber in one million homes in California and eight other states to determine the best operational, marketing, and technology practices for fiber optic networks (Granelli, 2004).

The FCC decision to discontinue network sharing requirements was meant to encourage ILEC network upgrades and investment in fiber optic networks, and thus far, the ILECs have responded. The development of publicly-managed fiber networks, however, suffered a set-back in early 2004 as the U.S. Supreme Court ruled that state governments are allowed to prohibit municipalities from building public telecommunications networks (Gardner, 2004).

Broadband over Power Line

For residences with limited access to advanced telephone systems or upgraded cable television systems, there is an emerging option for broadband utilizing the existing ubiquitous infrastructure of electric power lines. BPL is being tested in a number of primarily rural regions. BPL allows homes to connect to the Internet by plugging adapters into existing electrical outlets. A pilot program initiated by an electric cooperative in central Virginia delivers Internet access at 256 Kb/s for $30 per month (Robinson, 2004).

The FCC is examining BPL as an option for broadband telecommunications; however, its primary focus is to discover if radio frequency emissions from BPL interfere with existing licensed radio spectrum (Gross, 2004). Although it is unlikely BPL will be a direct competitive threat to DSL or cable modems, it offers a unique solution for regions of the country desiring broadband Internet connectivity, but lacking the needed telecommunications infrastructure.

Factors to Watch

Despite the recent growth of broadband adoption, the U.S. government, the high-technology industry, and consumer groups have continued to suggest that more initiatives are needed to promote broadband deployment and use. Current U.S. policy, based on a limited role for government, is premised on confidence that market competition will necessarily lead to broadband accessibility.

The FCC has determined that the current level of broadband deployment is reasonable and timely; however, they are reevaluating their definition of broadband and will examine the availability of broadband in different market segments (urban versus rural), as well as the economic considerations that support broadband deployment (FCC, 2004). They will also explore how the country's deployment of broadband impacts the U.S. role in the global economy.

The United States ranks 10th worldwide in broadband deployment (see Table 20.5). Korea, with the highest broadband connectivity per capita, has subsidized broadband deployment with a number of government initiatives, including building a nationwide fiber optic network (Belson & Richtel, 2003). A recent study of "e-readiness"—measuring a nation's openness to Internet-based opportunities—found that the U.S. rated sixth due to its comparatively lower rate of broadband adoption (Chabrow, 2004). To enhance America's competitive stance with other nations, President Bush has called for universal broadband service by 2007 (Haley, 2004). Broadband is likely to attract the attention of more policymakers as court challenges and FCC decisions lead to more regulatory uncertainty.

Table 20.5
International Broadband Penetration

Country	Broadband Penetration*
Korea	23.17
Canada	13.27
Iceland	11.22
Denmark	11.11
Belgium	10.34
Netherlands	9.20
Sweden	9.16
Switzerland	9.13
Japan	8.60
United States	8.25

* Broadband access per 100 inhabitants

Source: OECD (2003)

With the current business and policy environment failing to require open access to cable television systems, and with the recent FCC decision ending the broadband line sharing requirements for ILECs, the hoped for intra-modal competition touted in the Telecommunications Act of 1996 seems unlikely to occur. Instead, inter-modal competition or competing technologies appear to be the primary source of market competition that may serve consumers in advancing broadband technology and keeping broadband prices in check. Worldwide, DSL subscribership surpasses cable modem use, and despite cable modem dominance in the United States, DSL use is growing rapidly and is likely to supplant cable modems as the most popular U.S. broadband technology (see Table 20.6). Technological competition will force vendors to push the capabilities of their technologies, providing even greater transmission speeds and services.

Table 20.6
U.S. Broadband Market

Broadband Type	March 2003	March 2004
Cable	67%	54%
DSL	28%	42%
Wireless/Fixed Sat.	4%	3%

Source: Pew Internet & American Life Project (2004)

Policymakers have traditionally concentrated on the supply-side aspects of broadband—focusing on the deployment and availability of broadband. More recently, the demand side has received attention, with promoters of broadband suggesting that broadband would quickly reach critical mass with the introduction of a "killer application" that would entice consumers to subscribe. Some insist that Hollywood needs to produce more online entertainment to attract consumers to broadband (Shiver, 2002). Streaming music and video have certainly encouraged broadband connectivity and led to new online companies and services. AtomFilms credits broadband for the increase in its viewers, with three million users watching its short films online every month (Chmielewski, 2004). However, Hollywood is reluctant to showcase much of its content online due to copyright concerns (Lessig, 2002).

As broadband adoption increases, there will be a greater supply of consumers for bandwidth-intensive applications such as streaming video over the Internet. It is uncertain which new services will lure the current dial-up users to broadband. Looking to the future, VoIP has been hyped as the next crucial application for broadband, with Juniper Research reporting that the VoIP market will become the main revenue source for broadband service providers by 2009. It would certainly be ironic to discover the killer application for broadband is similar to the voice communications application introduced by Alexander Graham Bell in 1876.

Bibliography

Angwin, J. (2001, July 16). Has growth of the net flattened? *The Wall Street Journal*, B1.

Beardsley, S., Doman, A., & Edin, P. (2003). Making sense of broadband. *McKinsley Quarterly*, *2*, 78-87.

Belson, K., & Richtel, M. (2003, May 5). U.S. broadband dream is alive in Korea. *The New York Times*, C1.

Breznick, A. (2004, April). Cablevision races to early lead in cable VoIP. *Cable Datacom News*. Retrieved April 10, 2004 from http://www.cabledatacomnews.com/apr04/apr04-4.html.

Chabrow, E. (2004, April 23). U.S. Internet leadership is slipping. *Information Week*. Retrieved April 26, 2004 from http://www.informationweek.com/story/showArticle.jhtml?articleID=19200096.

Chmielewski, D. (2004, March 15). Broadband's bounty. *San Jose Mercury News*, 1F.

Eshed, E., & Melamed, O. (2004, January). DOCSIS 2.0: Where upstream network and application needs meet. *Communications Technology*. Retrieved April 1, 2004 from http://www.broadband-pbimedia.com/ct/archives/0104/0104_docsis.html.

Federal Communications Commission. (2002, February 6). *Inquiry concerning the deployment of advanced telecommunications capability to all Americans in a reasonable and timely fashion and possible steps to accelerate such deployment pursuant to Section 706 of the Telecommunications Act of 1996*, CC Docket No. 98-146, (Third Report).

Federal Communications Commission. (2004, March 11). *Inquiry concerning the deployment of advanced telecommunications capability to all Americans in a reasonable and timely fashion and possible steps to accelerate such deployment pursuant to Section 706 of the Telecommunications Act of 1996*, GEN Docket No. 04-54, (Fourth Notice of Inquiry).

Ferguson, C. (2004). *The broadband problem: Anatomy of a market failure and a policy dilemma*. Washington, DC: Brookings Institution Press.

Gardner, W. (2004, March 25). Court blocks municipalities from offering phone service. *Information Week*. Retrieved April 26, 2004 from http://www.informationweek.com/showArticle.jhtml?articleID=18402660.

Granelli, J. (2004, April 19). SBC: Fiber optics' future is focus of test project. *The Los Angeles Times*, C1.

Gross, G. (2004, February 13). FCC moves ahead with power-line broadband rules. *Computerworld*. Retrieved April 14, 2004 from http://www.computerworld.com/managementtopics/outsourcing/isptelecom/story/0,10801,90212,00.html.

Haley, C. (2004, March 31). Bush calls for universal broadband by 2007. *Internet News*. Retrieved April 22, 2004 from http://www.internetnews.com/xSP/article.php/3333711.

Hu, J. (2004a, March 11). Satellite seeks broadband re-entry. *C/Net News*. Retrieved April 14, 2004 from http://news.com.com/2100-1034-5172088.html.

Hu, J. (2004b, February 3). SBC changes DSL pricing. *C/Net News*. Retrieved April 14, 2004 from http://news.com.com/2100-1034-5152274.html.

International Telecommunications Union. (2003, September). Birth of broadband. *ITU Internet Reports*.

Johnston, F. (2002). Taking fiber home. *Public Power, 60* (5).

Krim, J. (2003a, February 7). Copper lines regaining luster. *The Washington Post*, E01.

Krim, J. (2003b, February 21). FCC delivers mixed vote on competition. *The Washington Post*, A01.

Labrador, V. (2004, February). Whither broadband? What's in store for in 2004. *Satmagazine.com*, 12-14.

Leblanc, P. (2003, January 20). ADSL2 boosts data rate and reach. *Network World*, 4-5.

Legard, D. (2003, March 6). Internet traffic to double each year. *InfoWorld*. Retrieved on March 10, 2003 from http://www.infoworld.com/article/03/03/06/HNnettraffic_1.html.

Leichtman Research Group. (2004). *Broadband Internet grows to 25 million in the U.S.* Retrieved April 12, 2004 from http://www.leichtmanresearch.com/press/030804release.html.

Lessig, L. (2002, January 8). Who's holding back broadband? *The Washington Post*, A17.

Levine, S. (2003, November). Court muddies the broadband debate. *America's Network*, 14.

Martin, M. (2003, January 20). Broadband wireless service drawing renewed interest. Network World, 27.

Musgrove, M. (2003, February 2). Broadband broadens its pitch. *The Washington Post*, H01.

National Cable Telecommunications Association. (2004a). *Balancing responsibilities and rights: A regulatory model for facilities-based VoIP competition*. Retrieved April 10, 2004 from http://www.ncta.com/pdf_files/whitepapers/VoIPWhitePaper.pdf.

National Cable Telecommunications Association. (2004b). *Statistics and resources*. Retrieved April 10, 2004 from http://www.ncta.com/Docs/PageContent.cfm?pageID=86.

Nielsen/NetRatings. (2004, January 4). *Fifty million Internet users connect via broadband.* Retrieved April 1, 2004 from http://www.nielsen-netratings.com/pr/pr_040108_us.pdf.

Organisation for Economic Cooperation & Development. (2001). *The development of broadband access in OECD countries*. Report of the Directorate for Science, Technology and Industry, Committee for Information, Computer and Communications Policy.

Pew Internet & American Life Project. (2003, May 28). *Broadband adoption at home: A Pew Internet Project data memo*. Retrieved April 22, 2004 from http://www.pewinternet.org/reports/toc.asp?Report=90.

Pew Internet & American Life Project. (2004, April). *55% of adult Internet users have broadband at home or work: A Pew Internet Project data memo*. Retrieved April 22, 2004 from http://www.pewinternet.org/reports/toc.asp?Report=120.

Point Topic. (2004, March 23). *World broadband statistics: Q4 2003*. Retrieved April 1, 2004 from http://www.point-topic.com/contentDownload/freeforyou/world%20broadband%20statistics%20q4%202003.pdf.

Robinson, B. (2004, April 1). Va. Power co-op lights up broadband. *Federal Computer Week.* Retrieved April 16, 2004 from http://www.fcw.com/geb/articles/2004/0329/web-power-04-01-04.asp.

Shiver, Jr., J. (2002, September 24). More engaging online content urged. *The Los Angeles Times,* C5.

Silverstein, S. (2003, December 23). News Corp. in control of Hughes. *Space News*. Retrieved April 1, 2004 from http://dev.space.com/spacenews/marketmonitor/hughes_122303.html.

21

Home Networks

Jennifer H. Meadows, Ph.D. & August E. Grant, Ph.D.*

The idea of setting up a computer network for home or personal use seemed unusual just a few years ago. However, with the increased use of computers in the home and the increasing number of homes with more than one computer, the idea of a home network is not only possible, in many cases, it is desirable. At the same time, the use of portable laptop computers has exploded and, with that, has come the desire to access the Internet anywhere, anytime. Wireless network technology, such as Wi-Fi, is now being employed in a variety of public places, institutions, and business to give laptop users with a wireless card Internet access on the go.

Most homes used to have only one personal computer with one Internet connection. With the rapid growth of personal computer ownership came the rapid growth of multiple computer households. With multiple computers in a household comes the problem of sharing the Internet connection, as well as the various computer peripherals such as printers and scanners. A home network solves that problem by allowing multiple computers to share an Internet connection, as well as those peripherals. As discussed below, this sharing is only the beginning of what a home network can do.

One of the primary drivers behind home networking is the increase in broadband access to the home. Ironically, the key attribute for home networking of these broadband services is not the speed, but the fact that the connection is always on. Once a broadband user becomes accustomed to instant access, without spending any time to dial up or connect, the utility of having this access available to all computers in a home becomes even more salient. The value of this connection was underscored by a 2002 survey of computer users indicating that the always-on connection was more important than

---

* Jennifer Meadows is an Associate Professor in the Department of Communication Design, California State University, Chico (Chico, California). Augie Grant is Associate Professor in the College of Mass Communications and Information Studies, University of South Carolina (Columbia, South Carolina).

speed for people considering upgrading from dial-up to broadband Internet access in their homes (Demand for, 2002).

The key device in most home networks is a residential gateway, sometimes known as a cable/DSL router. These are devices that interconnect all of the computers and other devices that use IP (Internet protocol) data streams to create the home network, in turn connecting the home network to the outside broadband connection, thereby allowing different streams of information to be routed intelligently throughout the home. The capability of routing any type of data stream to set-top boxes, telephones, and other devices will eventually allow audio, video, and telephone signals to be distributed throughout the home in the same manner as computer data streams are routed.

Here are some possible applications of a home network and residential gateway:

- You order a boxing match that is sent to the downstairs television set, while Internet radio is sent to one receiver upstairs, and another household member plays online video games—all at the same time.
- Telephone calls to your teenager are routed automatically to her bedroom, while business calls go to the home office.
- The dishwasher breaks and uses the home network to contact the repair facility before you know something is wrong.
- On your way home from a trip, you hear of a hard freeze coming to your area. You contact your home network from your wireless Web phone and turn the heat on so the house is comfortable by the time you get home and the pipes will not freeze.

With such networking, computer use in the home becomes more efficient. With wireless networking technologies such as Wi-Fi, the utility of a portable computer greatly expands. The laptop user is no longer tied to a particular location for Internet access. Now, a variety of locations including coffee shops, airports, hotels, and schools have installed wireless Internet access for computer users who may or may not pay a usage charge. Business travelers no longer have to deal with slow, awkward dial-up access in their hotel rooms and airport lounges. Students can escape the busy computer lab and noisy dorm room, and conduct research on the Web from the quiet coffee shop. All the user needs is a compatible wireless networking card installed in his or her computer.

This chapter will briefly review the development of home networks, residential gateways, and wireless networks; discuss the types and uses of these technologies; and examine the current status and future developments of these exciting technologies.

Background

Networking was once thought to be exclusively the domain of the office or institution, not the home or public spaces. Those few homes that did have networks typically had a traditional Ethernet network that required an expensive type of telephone wiring called Category 5 (Cat 5). Such a network needed a server, hub, and router. The network needed to be administered, requiring one household member to have computer network expertise. The installation and administration of these early home networks required a major allocation of time, money, and effort on the part of the user. Clearly, they were not for the average home computer user, because the user needed extensive computer and networking expertise, as well as a means of financing the network.

Several factors changed the environment to allow home networks to take off:

- *The Internet.* As discussed in Chapter 13, Internet use and growth has been exponential. More and more people want access, and competition for Internet access within the home is quickly becoming a common occurrence.

- *Computer and peripheral sales.* The number of multiple computer homes is rapidly growing, facilitating a need for shared peripherals and Internet access. Who wants to buy several printers and have multiple Internet service provider (ISP) accounts when all the computers in the home can share these resources?

- *Broadband.* The availability of broadband connections such as digital subscriber line (DSL) and cable modems is growing. These connections carry much more information than a traditional phone line, allowing for faster and enhanced Web browsing as well as delivery of telephony, video, and audio services.

- *New consumer electronics devices.* There is a plethora of new consumer electronics devices that work in concert with the Internet such as Internet-ready video game consoles, Internet audio receivers, MP3 players, and even Internet-enhanced appliances. Technologies such as direct broadcast satellites and digital video recorders also can be used to great benefit on home networks. (See Chapters 4 and 16.)

The five major uses for home networks include resource sharing, communication, home control, scheduling, and entertainment. Sharing one broadband connection and computer peripherals within the home is an example of resource sharing. Communication is enhanced when one computer user can send files to another computer in the house for review or, perhaps, to send reminders. A family can keep a master household schedule that can be updated by each member from his or her computer. Home control includes being able to remotely monitor security systems, lighting, heating and cooling systems, etc. Being able to route digital video and audio to different players within the home is an example of entertainment. The residential gateway allows the home network to have these multiple functions.

Home Networking

There are four basic types of home networks:

- Traditional.

- Phone line.

- Power line.
- Wireless.

When discussing each type of home network, it is important to consider the transmission rate, or speed, of the network. High-speed Internet connections and digital audio and video require a faster network. Regular file sharing and low-bandwidth applications such as home control may require a speed of 1 Mb/s (Megabits per second) or less. The MPEG2 digital video and audio from DBS (direct broadcast satellite) services requires a speed of 3 Mb/s, DVD (digital videodisc) requires between 3 Mb/s and 8 Mb/s, and compressed high-definition television (HDTV) requires around 20 Mb/s.

Traditional networks use Ethernet, which has a data transmission rate of 10 Mb/s to 100 Mb/s. There is also Gigabit Ethernet, used mostly in business, that has transmission speeds up to 1 Gb/s. Ethernet is the kind of networking commonly found in offices and universities. As discussed earlier, traditional Ethernet has not been popular for home networking because it is expensive and difficult to use. To direct the data, the network must have a server, hub, and router. Each device on the network must be connected, and many computers and devices require add-on devices to enable them to work with Ethernet. Thus, despite the speed of this kind of network, its expense and complicated nature make it somewhat unpopular in the home networking market, except among those who build and maintain these networks at the office.

Some new housing developments come "Internet-ready" and include Cat 5 wiring with routers and hubs. For example, the properties at the Irvine Ranch (Irvine, California) include homes, condos, and apartments that come with home networks (The Irvine Company, n.d.). Homes under construction are more likely to have network wiring built in. The extra wiring is typically offered as a premium enhancement to the home at prices ranging from a few hundred dollars (for simple wiring and no hardware) to tens of thousands of dollars (for complex systems that route any type of data, audio, video, or telephone signals). This option is increasingly popular with builders: 74% of respondents in an In-Stat/MDR study said that they offered structured wiring in the homes they build, with 30% including the wiring as standard (In-Stat/MDR, 2003).

New homes represent a small fraction of the potential market for home networking services and equipment, so manufacturers have turned their attention to solutions for existing homes. These solutions almost always are based upon "no new wires" networking solutions that use existing phone lines or power lines, or they are wireless.

Phone lines are ideal for home networking. This technology uses the existing random tree wiring typically found in homes and runs over regular telephone wire—there is no need for Cat 5 wiring. The technology uses frequency division multiplexing (FDM) to allow data to travel through the phone line without interfering with regular telephone calls or DSL service. There is no interference because each service is assigned a different frequency. The Home Phone Line Networking Alliance (HomePNA) has presented three open standards for phone line networking. HomePNA 1.0 provided data transmission rates up to 1 Mb/s and was replaced by HomePNA (HPNA) 2.0, which allows data transmission rates up to 10 Mb/s and is backward-compatible with HPNA 1.0. HomePNA 3.0 provides data rates up to 128 Mb/s and is backward-compatible with both HPNA 1.0 and 2.0. In addition to the presence of a router or residential gateway that has HPNA, each device on an HPNA network must have an adaptor that connects the device to a phone jack. The most common adaptors are internal cards that plug into expansion slots of computers and external adaptors that connect between the phone jack and a computer's USB (universal serial bus) port.

The developers of power line networking realized that all of the devices a consumer might want to network are already plugged into the home's electrical outlets. In addition, there are almost always power outlets in the rooms where a networked device would be placed. Why not network over these power lines? Because the power line networking industry was a few years behind its phone line and wireless competitors, many of the companies in this industry banded together to form the Homeplug Powerline Alliance in 2001 (Handley, 2002). The primary purpose of the organization was to create an open specification for power line networking, allowing the industry to more effectively compete with other networking systems (Gardner, 2001).

Homeplug introduced their v1.0 specification in June 2001 (Gardner, et al., 2002). Homeplug's specification uses OFDM (orthogonal frequency division multiplexing) for transmission (Gardner, et al., 2002). In the past, power line networking technologies have been hampered by a lack of standards in addition to immature technologies and regulatory issues. Homeplug v1.0 is intended to overcome these problems. With power line networking, household electrical wires are used for data transmission, with data rates up to 14 Mb/s as of mid-2004. With these products, a gateway is plugged into an electrical outlet and a modem. Network interface adapters are then connected between electrical outlets and the USB, Ethernet, or parallel ports of a networked device such as a computer.

The data speed available over power line networking has been limited by the hostile nature of electrical lines. Data rates are affected by change and flux created by power surges, lightning, and brown outs. Potential interference is another problem with power line networking. Household appliances that draw a great deal of electricity, such as hair dryers, vacuum cleaners, microwave ovens, and drills, can introduce noise to the network.

One of the most talked about types of home network is wireless. There were four major types of wireless home networking technologies competing for the lead: Wi-Fi (otherwise known as IEEE 802.11b and 802.11a), HomeRF, and Bluetooth.

Wi-Fi, HomeRF, and Bluetooth are based on the same premise: Low-frequency radio signals from the instrumentation, science and medical (ISM) bands of spectrum are used to transmit and receive data. The ISM bands, around 2.4 GHz, are not licensed by the Federal Communications Commission (FCC) and are used mostly for microwave ovens and cordless telephones. (802.11a, on the other hand, operates at 5 GHz.)

Wireless networks are configured with a receiver that is connected to the wired network or gateway at a fixed location. Transmitters are either within or attached to electronic devices. Much like cellular telephones, wireless networks use microcells to extend the connectivity range by overlapping to allow the user to roam without losing the connection (Wi-Fi Alliance, n.d.-a.).

The most common standards for wireless networking fall under the umbrella moniker, 802.11, also known as Wi-Fi. These standards each use a different one-letter suffix, e.g., 802.11b, 802.11a, and 802.11g. IEEE 802.11b was the specification for wireless Ethernet. It can transfer data up to 11 Mb/s and is supported by the Wi-Fi Alliance. 802.11b operates at 2.4 GHz and utilizes direct sequence spread spectrum technology. 802.11b has been widely adopted worldwide for both enterprise and home markets. The inhibiting factors to its adoption include limited bandwidth, radio interference, and security. Radio interference affects the network because the 2.4 GHz band used by 802.11b is home to a myriad of technologies including cordless phones, microwave ovens, and the competing but incompatible Bluetooth standard. At the same time as the 802.11b specification was released, the 802.11a specification was released. 802.11a operates within a higher frequency band, 5 GHz, using OFDM. With data throughput up to 54 Mb/s, 802.11a clearly has a speed advantage over 802.11b's 11 Mb/s and does not have the RF (radio frequency) interference problems that come with operating at 2.4 GHz. Because it operates at a higher frequency, 802.11a only works at shorter distances than 802.11b. A third standard, 802.11g, uses the same frequencies as 802.11b (2.4 GHz) with the same 54 Mb/s capacity of 802.11a.

Wi-Fi is not only used in the home. Increasingly, it was been adopted by business and institutions to create hotspots—places where users with the appropriate wireless card and account can access the network. Hotspots began to appear in places primarily for business travelers such as in airports and hotels. As the popularity of these hotspots grew, so did the availability. They can now be found in coffee shops, restaurants, student unions, and McDonald's and the San Francisco Giants' SBC Park.

HomeRF, like Wi-Fi, was developed primarily for the home networking market. Although its transmission speeds were comparable to Wi-Fi, the standard failed to gain a foothold in the market and was abandoned in 2003.

While 802.11b can transmit data up to 11 Mb/s for up to 150 feet, Bluetooth was developed for short-range communication at a data rate of 1 Mb/s. Bluetooth technology is built into devices, and Bluetooth-enhanced devices can communicate with each other and create an ad hoc network. The technology works with and enhances other networking technologies. For example, students with Bluetooth enabled laptop computers and Bluetooth enabled cellular phones in their backpacks can surf the Web while sitting in the park because the computer links to a phone that can connect to an Internet service provider (Frequently asked, n.d.).

Usually, a home network will involve not just one of the technologies discussed above, but several. It is not unusual for a home network to be configured for HPNA, Wi-Fi, and even traditional Ethernet. Table 21.1 compares each of the home networking technologies discussed in this section.

Table 21.1
Comparison of Home Networking Technologies

	How it Works	Specifications & Standards	Transmission Rate	Reliability	Cost	Security
Conventional Ethernet	Uses Cat 5 wiring with a server and hub to direct Traffic.	IEEE 802.3xx IEEE 802.5	10 Mb/s to 1 Gb/s	High	High	Secure
HomePNA	Uses existing phone lines and OFDM.	HPNA 1.0 HPNA 2.0 HPNA 3.0	1.0, up to 1 Mb/s 2.0, 10 Mb/s 3.0, 128 Mb/s	High	Low	High
IEEE 802.11a Wi-Fi	Wireless. Uses electro-magnetic radio signals to transmit between access point and users.	IEEE 802.11a 5 GHz	Up to 54 Mb/s	High	Moderate	High to Moderate
IEEE 802.11b Wi-Fi	Wireless. Uses electro-magnetic radio signals to transmit between access point and users.	IEEE 802.11b 2.4 GHz	Up to 11 Mb/s	High	Low	High to Low
IEEE 802.11g Wi-Fi	Wireless. Uses electro-magnetic radio signals to transmit between access point and users.	IEEE 802.11g 2.4 GHz	Up to 54 Mb/s	High	Moderate	High to Low
WiMAX	Wireless. Uses electro-magnetic radio signals to transmit between access point and users.	IEEE 802.16 Fixed IEEE 802.20 Mobile	Up to 75 Mb/s	High	High	Not yet determined
Bluetooth	Wireless.	Bluetooth SIG 4.2 GHz	1 Mb/s	High to Moderate	Low	High to Moderate
Power Line	Uses existing power lines in home.	HomePlug v1.0 HomePlug AV	Up to 14 Mb/s	Moderate	Moderate	High

Source: Meadows & Grant

Residential Gateways

The residential gateway, also known as a cable/DSL router, is what makes the home network infinitely more useful. This is the device that allows users on a home network to share access to their broadband connection. As broadband connections become more common, the one "pipe" coming into the home will most probably carry numerous services such as the Internet, phone, and entertainment. A residential gateway seamlessly connects the home network to a broadband network so all network devices in the home can be used at the same time.

The current definition of a residential gateway has its beginnings in a white paper developed by the RG Group, a consortium of companies and research groups interested in the residential gateway concept. The RG Group determined that the residential gateway is "a single, intelligent, standardized, and flexible network interface unit that receives communication signals from various external networks and delivers the signals to specific consumer devices through in-home networks" (Li, 1998).

Residential gateways can be categorized as complete, home-network-only, and simple.

A *complete residential gateway* operates independently of a personal computer and contains a modem and networking software. This gateway can intelligently route incoming signals from the broadband connections to specific devices on the home network. Set-top box and broadband-centric are two categories of complete residential gateways. A broadband-centric residential gateway incorporates an independent digital modem such as a DSL modem with IP management and integrated HomePNA ports. Set-top box residential gateways use integrated IP management and routing with the processing power of the box. Complete residential gateways also include software to protect the home network, including a firewall, diagnostics, and security log.

Home network only residential gateways interface with existing DSL or cable modems in the home. These route incoming signals to specific devices on the home network, and typically contain the same types of software to protect the home network found in complete residential gateways.

Simple residential gateways are limited to routing and connectivity between properly configured devices. Also known as "dumb" residential gateways, these have limited processing power and applications and only limited security for the home network.

Working Together: The Home Network and Residential Gateway

A home network controlled by a residential gateway or central router allows multiple users to access a broadband connection at the same time. Household members do not have to compete for access to the Internet, printer, scanner, or even the telephone. The home network allows for shared access to printers and peripherals. Using appropriate software, household members can keep a common schedule, e-mail reminders to each other, share files, etc. The residential gateway or router allows multiple computers to access the Internet at the same time by giving each computer a "virtual" IP address, with the household only needing one external IP address. The residential gateway routes different signals to appropriate devices in the home. For example, Web pages are sent to the specific computer requesting them at the same time that entertainment signals in the form of radio, video, and games can be routed to a stereo, television, or digital video recorder, (such as a TiVo unit), attached to the network.

Home networks and residential gateways are key to what industry pundits are calling the "smart home." Although the refrigerator that tells us we are out of milk may seem a bit over the top, utility management, security, and enhanced telephone services are just a few of the potential applications of this technology. Before these applications can be implemented, however, two developments are necessary. First, appropriate devices for each application (appliance controls, security cameras, telephones) have to be configured to connect to one or more of the different home networking topologies (wireless, HPNA, or power line). Next, software, including user interfaces and control modules, needs to be created and installed. It is easy to conceive of being able to go to a Web page to adjust the air conditioner, turn on the lights, or monitor the security system, but these types of services will not be widely available until consumers have proven that they are willing to pay for them.

Recent Developments

There has been significant change in the home networking landscape since 2002. New wireless, phone line, and power line standards have been introduced; others have failed; and the Wi-Fi network is fast becoming a regular part of the networked home and the public space.

In phone line networking, HPNA 3.0, introduced in June 2003, has transmission rates of 128 Mb/s and can reach 240 Mb/s with optional extensions (HomePNA, 2003), but the industry has been slow to release HPNA 3.0 equipment. Power line networking is beginning to gain some attention. Certified HomePlug networking products are being manufactured by established companies including Linksys, Belkin, and Siemens. HomePlug presented a new standard HomePlug AV in 2003, and products using HomePlug AV were demonstrated at the Consumers Electronics Show in January 2004. The Alliance says "HomePlug AV will enable distribution of data and multi-stream entertainment throughout the home, including high-definition television (HDTV) and standard definition television (SDTV), and is designed to co-exist with the current HomePlug technology" (HomePlug Powerline Alliance, 2004). The alliance forecasts that HomePlug AV products will be available by the end of 2004. While phone line and power line networking have been improving, HomeRF did not take off. Unfortunately for HomeRF, the competition from Wi-Fi was too great. The HomeRF working group disbanded in January 2003 (About HomeRF, n.d.).

Several developments in wireless networking are changing the way people access the Internet. First, Wi-Fi is now the term used to incorporate the 802.11 family of standards including 802.11a, 802.11b, and 802.11g (Wi-Fi Alliance, n.d.-a). Ratified in 2003, 802.11g is a high-speed wireless networking technology with transmission rates up to 54 Mb/s. The technology is backward-compatible with 802.11b and operates in the same 2.4 GHz spectrum. Even before it was ratified, 802.11g products were shipped from Apple (Airport Extreme), Linksys, and Belkin (Griffith, 2003).

Because wireless networks use so much of their available bandwidth for coordination among the devices on the network, it is difficult to compare the rated speeds of these networks with the rated speeds of wired networks. For example, 802.11b is rated at 11 Mb/s, but the actual throughput (the amount of data that can be effectively transmitted) is only about 6 Mb/s. Similarly, 802.11g's rated speed of 54 Mb/s yields a data throughput of only about 25 Mb/s (Atheros Communications, n.d.).

The issue of wireless network speed is even more complicated when two networks with the same frequencies are used together. Both 802.11b and 802.11g use the same frequency band, the 2.4 GHz band. 802.11g was designed to be compatible with its slower counterpart, but there is a cost—the data throughput for an 802.11g network slows to less than 15 Mb/s when an active 802.11b connection uses the same frequencies (Atheros Communications, n.d.).

The newest version of 802.11 is 802.11n. This technology should have transmission rates up to 108 Mb/s. The IEEE 802.11n Wi-Fi Group issued a request for proposals in January 2004. The group is at odds with industry groups such as the Wi-Fi Alliance over the speed at which the standard is ratified. With 802.11g, products were shipped before the standard was ratified; this too may happen with 802.11n because the demand for bandwidth is so compelling to manufacturers. The concern is that the standards process is compromised when industry demands the process move quickly (Wireless Watch, 2004).

Whether it is 802.11b or its other versions, Wi-Fi is no longer confined to the home or office, with Wi-Fi hotspots available in a variety of locations. Wi-Fi hotspots were expected to generate significant revenues. This has not happened. In-Stat/MDR estimates that U.S. Wi-Fi providers will generate $28 million in 2004, the same amount that Verizon Wireless makes in 12 hours (Thurm, et al., 2004). The problems with Wi-Fi hotspots include high charges to connect (for example, a Wi-Fi connection at the Las Vegas Convention Center in April 2004 was $24.95 per day!), lack of coverage, availability of some free connections, and alternatives such as wireless phones. Wi-Fi access is most often offered for a fee—per hour, day, or month—that applies to a particular network. So if you pay a monthly fee for T-Mobile's service, then you might have access in a Starbucks but not in Central Park in New York City, which has hotspots provided by Verizon. Wi-Fi providers are now looking to roaming agreements to give their subscribers more access to the service (Berman & Drucker, 2003). In addition, a growing number of hotels and businesses are offering free Wi-Fi to attract and keep business. Why pay a subscription fee from $6.00 an hour to $35.00 per month, when you can get the same service for free with a little planning (Thurm, et al., 2004)?

Internationally, Wi-Fi availability has been increasing in parts of Asia, but has been less successful in Europe. Industry analysts point out that price is a major factor in Wi-Fi's success. Users in South Korea can have access to more than 12,000 hotspots from KT Corporation for $13.00 a month, and if subscribers already have the company's residential broadband service, the price is $1 a month (Thurm, et al., 2004). In Europe, the typical price is much higher, and the attitude of providers is that Wi-Fi is for the business traveler who can afford it.

Security is an issue with any network, and Wi-Fi uses two types of encryption; WEP (Wired Equivalent Privacy) and WPA (Wi-Fi Protected Access). WEP has security flaws and is easily hacked. WPA fixes those flaws in WEP and uses a 128-bit encryption (Bradley, n.d.). There are two versions: WPA-Personal that uses a password and WPA-Enterprise that uses a server to verify network users (Wi-Fi Alliance, n.d.-b).

Competing with Wi-Fi are wireless telephony providers. Advanced wireless telephony networks offer high bandwidth, and users with these advanced phones can connect to the Internet anywhere they can get a signal. In the United Kingdom, Vodafone Group PLC sells cards that plug into a laptop to allow wireless Web surfing (Thurm, et al., 2004).

In the United States, Verizon Wireless, a joint venture of Vodophone Group PLC and Verizon Communications, is implementing a new wireless network called EV-DO. EV-DO stands for Evolution Data Optimized and offers transmission speeds averaging 300 Kb/s to 500 Kb/s. EV-DO works with the cellular network, so it is subject to dead spots, but it can be used in a moving vehicle. Verizon is investing $1 billion over two years to build out the network and hopes to reach 80 million people by the end of 2004 (Drucker, 2004). As of April 2004, Verizon's EV-DO service, called BroadBand Access, was available in Washington (DC) and San Diego for $80 per month. Lucent Technologies, Inc. has developed a "Wi-Fi on the Go" that works in conjunction with EV-DO. The technology would allow users with Wi-Fi enabled laptops to access the EV-DO network (Sheng, 2004).

Will EV-DO be the next step in wireless networks? Another technology being developed is WiMAX. WiMAX is also known as IEEE 802.16 and IEEE 802.20. With a range of up to 30 miles (compared to Wi-Fi's 30 meters), WiMAX IEEE 802.16 is a fixed wireless service operating in the 2 GHz to 11 GHz or the 10 GHz to 66 GHz frequencies. The service is designed to compete with other broadband services providers such as DSL and cable modems. IEEE 802.20 is the WiMAX standard for wireless data services to mobile users. This technology would compete with advanced wireless telephony networks such as EV-DO and EDGE (WiMAX explained, 2004). Intel has committed to manufacturing chips for WiMAX equipment. The technology has transmission speeds of up to 75 Mb/s, and the Yankee Group predicts there will be 12,000 WiMAX hotspots by the end of 2004 (WiMAX: the (next), 2004).

Current Status

The market for home networking depends on the penetration of broadband. ComScore Networks reported that U.S. broadband penetration reached 36% by the fourth quarter of 2003 (ComScore Networks, 2004). The number of multiple computer households also continues to grow. With that growth comes the adoption of home networks. IDC forecasts that 83% of multiple computer homes in the United States will have a home network by 2005 (Microsoft Developer's Network, n.d.).

The growth in broadband access continues to fuel growth in the residential gateway market. In 2003, 1.5 million units were shipped, with 2Wire leading the market with 43% share (Wolf, 2004). One reason for this growth is that broadband providers are offering residential gateways to new subscribers. For example, new subscribers to SBC's Yahoo! DSL service have the option to get a home networking or wireless networking kit that comes with a 2Wire HomePortal Residential Gateway for $50. IDC predicts that the market for gateway devices will reach $1.1 billion by 2005, and In-Stat/MDR predicts revenues of $3.9 billion (Microsoft Developer's Network, n.d.).

The overall market growth for Wi-Fi is a function of increases in both the number of hotspots and the penetration of laptop computers with wireless cards. Although no authoritative count exists, one estimate put the number of worldwide hotspots at 50,000 as of the end of 2003, with the United States having more than any other country (Wireless Web, 2004). Although the same source predicted that the number of hotspots would double in 2004, such a prediction probably understates the appeal of these wireless hotspots to businesses.

For example, hotels, restaurants, bars, bookstores, and airline terminals are among the businesses that are installing wireless networks to take advantage of consumers who have wireless connectivity built into their laptop computers. Many of these locations, including hotels and most airline terminals, see the wireless hotspot as a new revenue opportunity, while others, especially restaurants, offer wireless Internet access for free as a means of attracting customers and getting them to spend more time.

As of the end of 2003, one source estimated that there were 9,732 hotspots in the United States (U.S. hotspot, 2004). T-Mobile is the leading Wi-Fi hotspot provider with 4,546 hotspots as of April 2004, including locations at Starbucks, Borders Books & Music, and Kinko's (T-Mobile, n.d.). Unlimited monthly subscriptions start at $29.99, and the service can be purchased for $6.00 per hour.

Factors to Watch

The most dramatic change occurring in home networking is related to the devices that are being connected. Early home networks connected only computers. The introduction of residential gateways, with their ability to assign a unique IP address and route information to and from virtually any type of device that could be networked, has enabled the development of a new generation of devices that take advantage of this connectivity.

For example, some companies are producing set-top boxes that allow you to use your television to display any data stored on any networked computer in your home, including downloaded movies, MP3 files, and digital photographs. These devices can even allow you to surf the Internet on your television, although the quality of the TV display is usually inferior to most computer displays. Perhaps the most important aspect of the set-top box is the way it leverages the existing home network and the storage and processing power of computers located throughout the home to provide a new set of capabilities at a very low cost (typically $200 or less).

Similarly, the most advanced digital video recorders (discussed in Chapter 16) have Ethernet connections that allow these devices to download program guides over the Internet, eliminating the need for a dial-up modem and telephone connection. In addition to reducing the cost of providing the program guide to the DVR, these connections also enable users to share television programs that they have recorded with other users of the same device located across town or across the country. Users of this feature do not even have to record a program themselves in order to watch it later, as long as they can find someone else who has recorded the program. As useful as this feature might be for users, expect television producers to attempt to limit its use because of copyright considerations.

The latest generation of video game consoles (see Chapter 12) includes network connections that allow a user to play a game with anyone in the world that has network access. This capability can be expected to trigger a new generation of video games designed to take advantage of the Web.

Home networks can also be used to support Internet telephony. As discussed in Chapter 19, Internet connections for long-distance telephone calls enable the user to bypass the local phone network and its associated access charges. The home network is a logical last link in IP telephony, as it enables a person to make a phone call from any location in the home that has a computer or an IP-enabled telephone set.

Not all devices that take advantage of home networking have succeeded. The refrigerator that can sense that there is no milk and order more is an example of the impractical uses that have been proposed for home networks. "Internet radios" that received "broadcasts" over the Internet rather than through the airwaves were sold by the thousands in the late 1990s, but have, for the most part, disappeared from the market. Experiments with these and other applications will ultimately yield a new generation of devices utilizing an Internet connection that cannot be predicted today. (By comparison, would Thomas Edison have predicted the range of electrical appliances we connect to our power lines today, from curling irons to electric lawn mowers and vacuum cleaners?)

No issue is more important to an owner of a home network than security. Home networks are vulnerable to two types of intrusion. The first is through the broadband connection to the Internet. The same Internet connection with a unique IP address that allows a user access to any information on the Web also allows any other computer on the Web to attempt to access information on any computer. Home networks can be protected from this type of intrusion with the installation of "firewall" software on the residential gateway and on individual computers in the household.

The second type of security threat is related to the proliferation of wireless networks. A wireless base station creates a wireless "cloud" that can be seen by other wireless computers up to 150 feet away. A neighbor or someone sitting in a car outside a home can as easily connect to an unprotected wireless network in the home as anyone living there. Almost all wireless base stations come with encryption tools discussed earlier (WEP and WPA) that help protect the wireless network from both intrusion (unauthorized use of the network) and snooping (monitoring information flowing across a network), but not all users choose to activate this level of security.

Corporate information technology personnel are especially concerned about their employees protecting their home networks. The same employee whose communication might be fully secure at the office could take a laptop computer with wireless connection home, and every message sent and received by that computer could be monitored from outside the home. Most of us do not need to worry about snooping, but individuals working with sensitive information (bankers, government workers, and even teachers) need to know that an unprotected wireless network allows anyone to view private information flowing through that network.

Another issue to watch is who actually owns the home network. Early home networks used equipment that was bought and paid for by the resident, so the ownership of the network clearly belonged to that person. Newer business models treat the home network as a service, with a subscriber paying a monthly fee for the equipment and any needed technical support. Although the "service" model costs a great deal more in the long run, it has proven popular, as most neophyte home network owners require a significant amount of technical support in setting up their networks and making all the computer-related devices in the home work together. In the long run, however, the industry will provide "plug-and-play" systems that should require a minimum of technical support; at that time, the cost of owning equipment should be so much lower than renting it that few people will opt for the service model.

The final issue to watch is the speed of home networks. While most new business networks use ultra-fast Gigabit networking (one billion bits per second), the fastest home networks run at a tenth of that speed (100 Mb/s), and most Wi-Fi networks are even slower (10 Mb/s). Expect the introduction of new home network standards and equipment over the next 10 years that will dramatically increase the speed available throughout the home, in the process creating a market for replacement of most of the home networking equipment being purchased today.

Home networks are at the same stage today that electrical power was about 100 years ago. Service availability was increasing, and a limited number of devices (computers and light bulbs, respectively) could be connected. The most important innovations to watch are those that will use the home network to provide services and convenience that have yet to be imagined.

Bibliography

Atheros Communications. (n.d.). *802.11 wireless LAN performance*. Retrieved May 3, 2004 from http://www.atheros.com/pt/atheros_range_whitepaper.pdf.

About HomeRF. (n.d.). *Palowireless*. Retrieved April 27, 2004 from http://www.palowireless.com/homerf/about.asp.

Berman, D., & Drucker, J. (2003, November). Wi-Fi industry bets "roaming" will lure users. *The Wall Street Journal*. Retrieved March 26, 2004 from http://www.wsj.com/article/0,,SB106807596834502700,00.html?mod=article-outset-box.

Bradley, T. (n.d.). Wireless network security for the home. *About.com*. Retrieved March 26, 2004 from http://netsecurity.about.com/cs/wireless/a/aa112203_p.htm.

ComScore Networks. (2004). *Broadband usage poised to eclipse narrowband in largest U.S. markets*. Retrieved April 28, 2004 from http://www.comscore.com/press/release.asp?press=439.

Demand for broadband growing slowly, survey shows. (April 23, 2002). *Telecom Direct*. Retrieved April 23, 2002 from http://www.telecomdirect.pwcglobal.com/telecom/direct:TIH/Telecom_Buzz/Internet/InternetArt::/Article/reuters042302c.

Drucker, J. (2004, March). EV-DO technology may be next big thing. *The Wall Street Journal*. Retrieved March 26, 2004 from http://www.wsj.com/article/0,,SB107956799858955857 0,00html?mod=article-outset-box.

Frequently asked questions. (n.d.). *Bluetooth*. Retrieved April 24, 2002 from http://www.bluetooth.com/util/faq1.asp.

Gardner, S. (2001, October). *Home networking 101*. Presented at the Broadband Home Conference. San Jose, CA.

Gardner, S., Markwalter, B., & Yonge, L. (2002). HomePlug standard brings networking to the home. *CommsDesign*. Retrieved March 26, 2002 from http://www.commsdesign.com/main/200/12/0012feat5.htm.

Griffith, E. (2003, June 11). 802.11g: It's official. *Internetnews.com*. Retrieved March 26, 2004 from http://www.internetnews.com/wireless/article.php/2220701.

Handley, L. (2002). Powerline home networking to get new life with arrival of HomePlug Standard. *In-Stat/MDR*. Retrieved January 30, 2002 from http://www.instat.com.

HomePlug Powerline Alliance. (n.d.). *Coming soon to your TV: HomePlug AV*. Retrieved April 27, 2004 from http://www.homeplug.com/powerline/experience_page2.html.

HomePlug Powerline Alliance. (2004). *Homeplug Powerline Alliance demonstrates Homeplug AV technology at Consumer Electronics Show*. Retrieved April 27, 2004 from http://www.homeplug.org/news/press010704.html.

HomePNA. (2003). Home networking reaches 128 Mb/s and beyond with HomePNA 3.0. *Press release*. Retrieved March 26, 2004 from http://www.homepna.org/news/pressr.asp?releaseID=18.

In-Stat/MDR. (2003). *Survey finds broadband's on the mind of residential builders and developers*. Retrieved April 28, 2004 from http://www.instat.com/press.asp?ID=582&sku=IN0301083RC.

Li, H. (1998). Evolution of the residential-gateway concept and standards. *Parks Associates*. Retrieved April 22, 2000 from http://www.parksassociates.com/media/jhcable.htm.

Microsoft Developer's Network (MSDN). (n.d.). *Gateway industry trends*. Retrieved April 28, 2004 from http://msdn.microsoft.com/emdedded/devplat/gateways/trends/default.aspx.

Sheng, E. (2004, March 24). Lucent eyes mobile Wi-Fi for trains and automobiles. *The Wall Street Journal*. Retrieved March 26, 2004 from http://online.wsj.com/article/0,SB108014279996164166-search,00.html?collection=autowire%2F03day%vql_string=wi%2Dfi%3Cin%E28article%2Dbody%29.

T-Mobile. (n.d.). *Hotspot U.S. location map*. Retrieved April 28, 2004 from http//locations.hotspot.t-mobile.com/.

The Irvine Company. (n.d.). *Irvine Ranch*. Retrieved on April 25, 2002 from http://www.irvineranch.com.

Thurm, S., Pringle, D., & Ramstad, E. (2004, March 18). Chill hits Wi-Fi "hotspots." *The Wall Street Journal.* Retrieved March 26, 2004 from http://online.wsj.com/article/0,,SB107956780465958560-search, 00.html?collection=autowire%2F30day&vql_string=wi%2Dfi%3Cin%3E%28article%2Dbody%29.

U.S. hotspot figures for end-December 2003. (2004, March 16). *Total Telecom.* Retrieved April 28, 2004 from http://www.totaltele.com/view.asp?articleid=105997&pub=tt&categoryid=961.

Wi-Fi Alliance. (n.d.-a). *Wi-Fi overview.* Retrieved March 26, 2004 from http://www.wi-fi.org/OpenSection/why_Wi-Fi.asp?TID=2.

Wi-Fi Alliance. (n.d.-b). *Wi-Fi security.* Retrieved April 28, 2004 from http://www.weca.net/OpenSection/secure.asp?TID=2.

Wireless Watch. (2004, January 16). Will pressure to speed up 802.lln wreck standards process? *The Register.* Retrieved April 27, 2004 from http://www.theregister.co.uk/2004/01/16/will_pressure_to_speed_up/.

Wireless Web reaches out. (2004, January 5). *BBC Online.* Retrieved May 2, 2004 from http://news.bbc.co.uk/2/hi/technology/3341257.stm.

WiMAX explained. (2004, January 13). *Red Herring.* Retrieved April 28, 2004 from http://www.redherring.com/Article.aspx?f=articles/2004/01/94717807-e740-4271-809d-8e55cb60bbe2/94717807-e740-4271-809d-8e55cb60bbe2.xml.

WiMAX: The (next) great wireless hope. (2004, January 12). *Red Herring.* Retrieved April 28, 2004 from http://www.redherring.com/Article.aspx?f=articles/2004/01/cc2e6786-7ecd-41f7-a1b0-1793308641fa/cc2e6786-7ecd-41f7-a1b0-1793308641fa.xml.

Wolf, M. (2004, March 29). Home network market update. *Network World Fusion.* Retrieved April 28, 2004 at http://www.nwfusion.com/net.worker/columnists/2004/0329wolf.html.

22

Satellite Communications

Carolyn A. Lin, Ph.D.*

Artificial satellites have been serving the world's population in nearly every aspect of its cultural, economic, political, scientific, and military life during the past four decades. Since the successful launch of the very first communication satellite, Intelsat-I in 1965, the satellite was envisioned as the ultimate vehicle for linking the world together into a global village. As we begin the 21st century, that vision is being augmented as companion communication technologies—such as the Internet; digital voice, data, and video transmission; and networking systems—are evolving rapidly.

In a multimedia, multichannel communication environment, where wired and wireless communication technologies compete against each other for a finite pool of consumers, satellite technology has been able to maintain its favorable position in the marketplace as a broadband and global coverage communication delivery system. This is because satellite technology is uniquely suited for delivering a wide variety of communication signals to accomplish a great number of different communication tasks. There are three basic categories of non-military satellite services: fixed satellite service, mobile satellite systems, and scientific research satellites (commercial and noncommercial).

- Fixed satellite services handle hundreds of millions of voice, data, and video transmission tasks across all continents between fixed points on the earth's surface.

- Mobile satellite systems help connect remote regions, vehicles, ships, and aircraft to other parts of the world and/or other mobile or stationary communication units, in addition to serving as navigation systems.

---

* Professor & Coordinator, Multimedia Advertising Certificate Program and 2002 Distinguished Faculty Research Award Winner, Department of Communication, Cleveland State University (Cleveland, Ohio).

- Scientific research satellites provide us with meteorological information, land survey data (e.g., remote sensing), and other different scientific research applications such as earth science, marine science, and atmospheric research.

The satellite's functional versatility is imbedded within its technical components and its operational characteristics. Looking at the "anatomy" of a satellite, one discovers two modules (Miller, et al., 1993). First is the spacecraft bus or service module, which consists of five subsystems:

1) The *structural subsystem* provides the mechanical base structure, shields the satellite from extreme temperature changes and micro-meteorite damage, and controls the satellite's spin function.

2) The *telemetry subsystem* monitors the on-board equipment operations, transmits equipment operation data to the earth control station, and receives the earth control station's commands to perform equipment operation adjustments.

3) The *power subsystem* is comprised of solar panels and backup batteries that generate power when the satellite passes into the earth's shadow.

4) The *thermal control subsystem* helps protect electronic equipment from extreme temperatures due to intense sunlight or the lack of sun exposure on different sides of the satellite's body.

5) The *altitude and orbit control subsystem* is comprised of small rocket thrusters that keep the satellite in the correct orbital position and keep antennas pointing in the right directions.

The second major module is the communications payload, which is made up of transponders. A transponder is capable of:

- Receiving uplinked radio signals from earth satellite transmission stations (antennas).

- Amplifying received radio signals.

- Sorting the input signals and directing the output signals through input/output signal multiplexers to the proper downlink antennas for retransmission to earth satellite receiving stations (antennas).

The satellite's operational characteristics are literally "out of this world" and reach deep into outer space. Satellites are launched into orbit via a space shuttle (a reusable launch vehicle) or a rocket (a non-reusable launch vehicle). There are two basic types of orbits: geostationary and non-geostationary. The geostationary (or geosynchronous) orbit refers to a circular or elliptical orbit incline approximately 22,300 miles (36,000 km) above the earth's equator (Jansky & Jeruchim, 1983). Satellites that are launched into this type of orbit at that altitude typically travel around the earth at the same rate that the earth rotates on its axis. Hence, the satellite appears to be "stationary" in its orbital position and in the "line-of-sight" of an earth station, with varying orbital shapes, within a 24-hour period. Therefore, geostationary orbits are most useful for those communication needs that demand no interruption around the clock, such as telephone calls and television signals (see Figure 22.1).

Figure 22.1
Satellite Orbits

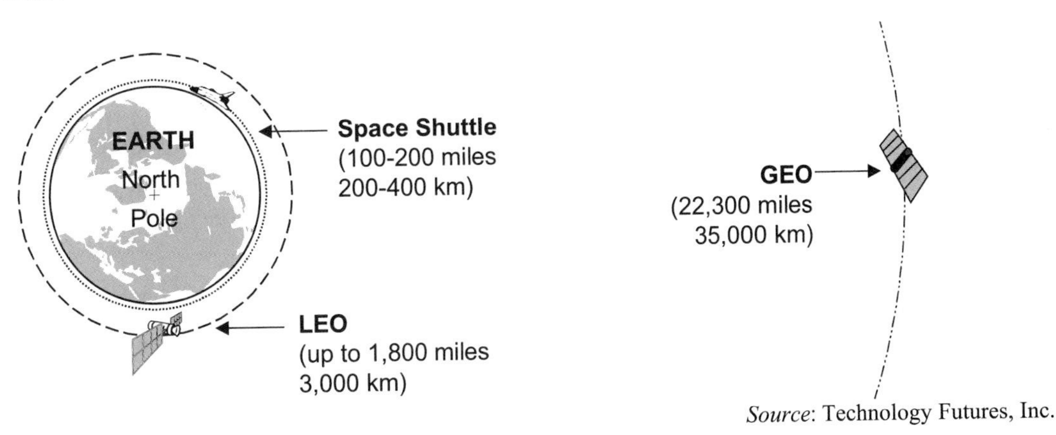

Source: Technology Futures, Inc.

Non-geostationary (or non-geosynchronous) orbits are typically located either above or below the typical altitude of a geostationary orbit. Satellites that are launched into a higher orbit travel at slower speeds than the earth's rotation rate; thus, they can move past the earth's horizon to appear for a limited time in the line-of-sight from an earth station. Those satellites that are launched into lower orbits (known as LEO, for low earth orbit) travel at higher speeds than the earth's rotation rate, so they race past the earth to appear more than once every 24 hours in the line-of-sight of an earth station. Non-geostationary orbits are utilized to serve those communication needs that do not require 24-hour input and output, such as scientific land survey data.

Background

During the early years of satellite technology development, the most notable events were the identification of the geostationary orbit by the English engineer Arthur C. Clarke in 1945 and the successful launch of the first artificial satellite, Sputnik, by the Soviet Union on October 4, 1957. On January 31, 1958, the first U.S. satellite, Explorer 1, was launched. In July 1958, Congress passed the National Aeronautics and Space Act, which established the National Aeronautics and Space Administration (NASA) as the civilian arm for U.S. space research and development for peaceful purposes. In August 1960, NASA launched its first communication satellite, Echo I—an inflatable metallic space balloon, or "passive satellite," designed only to "reflect" radio signals (like a mirror).

The next period of satellite technology development was marked by an effort to launch "active satellites" that could receive an uplink signal, amplify it, and retransmit it as a downlink signal to an earth station. In July 1962, AT&T successfully launched the first U.S. active satellite, Telstar I. In

February 1963, Hughes Aircraft launched the first U.S. geostationary satellite, Syncom, under a contract with NASA (Divine, 1993).

In tandem with these experimental developments came the creation of the Communication Satellite Corporation (Comsat), a privately-owned enterprise overseen by the U.S. government via the passage of the Communications Satellite Act of 1962. Meanwhile, an international satellite consortium, the International Telecommunications Satellite Organization (Intelsat), was formed on August 20, 1964. The 19 founding member nations—including the United States, 15 western European nations, Australia, Canada, and Japan—agreed to designate Comsat to manage the consortium (Hudson, 1990). The era of global communication satellite services began with the launch of Intelsat-I into geostationary orbit in 1965. Intelsat-I was capable of transmitting 240 simultaneous telephone calls.

Other important issues during this period involved satellite transmission frequency and orbital deployment allocation policy. Both issues were thrust into the management of the World Administrative Radio Conference (WARC) in 1963. WARC is a technical arm of the International Telecommunications Union (ITU), an international organization founded in 1865 to regulate radio spectrum use and allocation. Given that a geostationary satellite's "footprint" can cover more than one-third of the earth's surface, the ITU divides the world into three regions:

- Region 1 covers Europe, Africa, and the former Soviet Union.
- Region 2 encompasses the Americas.
- Region 3 spans Australia and Asia, including China and Japan.

The ITU also designated radio spectrum into different frequency bands, each of which is utilized for certain voice, data, and/or video communication services. Major categories of frequency bands include:

- L-band (0.5 to 1.7 GHz) is used for digital audio broadcast, personal communication services, global positioning systems, and non-geostationary and business communication services.
- C-band (4 to 6 GHz) is used for telephone signals, broadcast and cable TV signals, and business communication services.
- Ku-band (11 to 12/14 GHz) is used for direct broadcast TV, telephone signals, and business communication services.
- Ka-band (17 to 31 GHz) is used for direct broadcast satellite TV and business communication services.

With the passage of time, the satellite's status as the most efficient technology to connect the world seems firmly established. However, an array of other challenges remains concerning issues related to national sovereignty and the orbital resource as a shared and limited international commodity. At the 1979 WARC, intense confrontations on the issue of "efficient use" versus "equitable access" to the limited geostationary orbital positions by all ITU member nations took place, pitting western nations that owned-and-operated satellite services against the developing nations that did not. For the developed nations, equitable access to orbital positions suggests a waste of resources, as many developing nations lack the economic means, technical know-how, and practical needs to own and operate their own individual satellite services. The developed nations consider an "efficient use" (first-come, first-served) system to be more technically and economically feasible. By contrast, developing nations consider equitable access a must if they intend to protect their rights to launch satellites into orbital positions in the future, before developed nations exhaust the orbital slots. This dispute would carry over to the new millennium before it was fully resolved.

International competition for the share of satellite communication markets was not limited to the division between developed and developing nations. In response to U.S. domination of international satellite services, western European nations, for instance, established their own transnational version of NASA in 1964. It was renamed the European Space Agency in 1973, and was responsible for research and development of the Ariane satellites and launch programs. Modeled after the Intelsat system, Eutelsat was formed in 1977 to serve its western European member nations with the Ariane communication satellites launched by the European Space Agency. A parallel development was the establishment of the Intersputnik satellite consortium by the former Soviet Union, which served all of the eastern European Communist bloc, the former Soviet Union, Mongolia, and Cuba. Other developing nations owned-and-operated their own national satellite services during this early period as well, including Indonesia's Palapa, which was built and launched by the United States in 1976.

The success of Intelsat, initially as an international nonprofit organization providing satellite communication services, was repeated in another international satellite consortium—the international maritime satellite (Inmarsat) system—founded in 1979. Inmarsat now provides communication support for more than 5,000 ships and offshore drilling platforms using C-band and L-band frequencies. Inmarsat is also used to conduct land mobile communications for emergency relief work for such natural disasters as earthquakes and floods.

Recent Developments

During the 1980s, the satellite industry experienced steady growth in users, types of services offered, and launches of national and regional satellite systems. By the mid-1990s, satellite technology had matured to become an integral part of a global communication network in both technical and commercial respects. The following discussion will highlight a few of these important landmark events that set the stage for the future of satellite communication in 2000 and beyond.

Due to a new allocation of frequencies in the Ku-band and Ka-band, direct-to-home (DTH) satellite broadcasts or direct broadcast satellite (DBS) services became more economically and technically viable as higher-power satellites can broadcast from these frequencies to smaller-sized earth receiving stations (or dishes). This development helped stimulate the television-receive-only (TVRO) dish industry and ignite an era of DBS services around the world (see Chapter 4). The best success story of this particular type of satellite service can be found in Europe and Japan. For the Europeans, direct-to-home satellite services are deemed an economical means to both transport and share television programs among the various nations that are covered by the footprint of a single satellite, as cable television development has been slow due to lack of privatization in national television systems. The Japanese utilized this service to overcome poor television reception owing to mountainous terrain throughout the island nation.

As the economic benefits of satellite communication became apparent to those nations that owned and operated national satellite services during the 1970s, many countries followed suit in the 1980s and became active players. In particular, developing countries saw satellite services as a relatively cost-efficient means to achieve their indigenous economic, educational, and social development goals in the vastly-underdeveloped regions outside their selected urban centers. Following the example of Indonesia's Indosat (or Palapa) in 1976, the industry saw India's Insat in 1983, Brazil's Brasilsat in 1985, and Mexico's Satmex (or Morelos) in 1985 join the ranks of countries that became national satellite system owners. Later, the People's Republic of China burst onto the scene with two state-owned satellite services—Sinosat in 1994 and Chinasat in 1997.

To a lesser extent, regional satellite consortia were formed to pull together and share the financial resources as well as economic fruits of satellite development. For instance, 21 Arab nations established their own satellite communication system, Arabsat in 1985, to serve domestic and regional needs. Afsat, a joint-venture between Uganda, Kenya and Tanzania, began satellite service in the East-African region in 1994.

A parallel development occurred with large national satellite systems—either privately or state-owned, or a hybrid of private/state ownership—that target their services to different regions of the world. For example, launched in 1992, Spain's Hispasat provides Spanish-language services to the United States, Spain, Portugal, and all the Latin American countries. Asiasat (in 1989), the first privately operated satellite service in Asia, serves both China and Southeast Asia by carrying Chinese language television programs. Competing with Asiasat is the APT Group (in 1992)—a joint venture between the People's Republic of China and the private sector—that operates the APSTAR satellite system to serve Asia, Europe, and the United States. Similarly, U.S.-based satellite systems such as PanAmSat, Comsat, and Loral have also carved out an international market niche.

This growth in the number of satellite services available in the marketplace also provided the impetus for free market competition in both domestic and international satellite communications markets. The Federal Communications Commission's approval of five domestic private satellite systems to enter the international satellite service market in 1985 opened the floodgates for privatization. Subsequently, in 1996-1997, the Federal Communications Commission (FCC) permitted all U.S. satellite systems to offer both domestic and international satellite services and allowed private satellite networks authorized by the World Trade Organization (WTO) to provide their services to the U.S. market. With the domestic launch of PamAmSat and Orion (of Loral Space) satellite services in 1988 (Reese, 1990), Intelsat ended its historic near-monopoly existence.

Inmarsat was the first to become privatized on April 15, 1999, followed by Comsat in August 2000. Eutelsat became fully privatized in July 2001, and Intelsat followed suit two weeks later. These privatization developments were encouraged by the "Orbit Act" (the Open-market Reorganization for the Betterment of International Telecommunications Act) of 2000, passed by the U.S. Congress, which embraced full and open competition in the satellite communication marketplace and ordered the full privatization of intergovernment satellite organizations (NTIA, 2000).

Another aspect of this unstoppable growth trend in the satellite industry involves the expansion of satellite business services. The first satellite system launched for business services was Satellite Business Systems (SBS) in 1980. Its purpose was to provide corporations with high-speed transmission of conventional voice and data communications, as well as videoconferencing, to bypass the public switched telephone network. As earth uplink and downlink station equipment became more affordable, a number of corporate satellite communication networks were launched (e.g., Sears, Wal-Mart, K-Mart, and Ford). These firms leased satellite transponder time from SBS providers to perform such tasks as videoconferencing, data relay, inventory updates, and credit information verification (at the point-of-sale) using very small aperture terminals (VSATs)—earth stations (or receiving antennas) that range from 3-feet to 12-feet in diameter. In many areas, large and small satellite earth stations were clustered in "teleports," allowing a concentration of satellite services and a place to interconnect satellite uplinks and downlinks with terrestrial networks.

As the market has grown, digital transmission techniques have become the norm because they allow increased transmission speed and channel capacity. For instance, voice and video signals can be digitized and transmitted as compressed signals to maximize bandwidth efficiency. Internet via satellite is another fast-growing service area (including streaming media services) targeting corporate users by all major satellite services in the United States and Europe. Other digital transmission methods, such as time division multiple access (TDMA), demand assigned multiple access (DAMA), and code division multiple access (CDMA), can be used to achieve similar efficiency objectives. TDMA is a transmission method that "assigns" each individual earth station a specific "time slot" to uplink and downlink its signal. These individual time slots are arranged in sequential order for all earth stations involved (e.g., earth stations 1 through 50). Since such time allotment and sequential order is repeated over time, all stations will be able to complete their signal transmission in these repeated sequential time segments using the same frequency.

DAMA is an even more efficient or "intelligent" method. In addition to the ability to designate time slots for individual earth stations to transmit their signals in a sequential order, this method has the capability to make such assignments based on demand instead of an *a priori* arrangement. CDMA utilizes the spread spectrum transmission method to provide better signal security, as it alternates the frequency at which a signal is transmitted and hence allows for multiple signals to be transmitted at different frequencies at different times.

Yet another important technological advance during this period involved launching satellites with greatly-expanded payload capability and multiband antennas (e.g., C-band, Ku-band, and/or Ka-band receiving and transmitting antennas). These multiband antennas enable more versatile services for earth stations transmitting signals in different frequency bands for different communication purposes, as well as more precise polarized spot beam coverage by the satellite. A spot beam focuses a signal from a particular transponder on a small area of the satellite's footprint, increasing the strength of the signal in that part of the footprint and allowing the same frequency to be used for a different signal in another part of the footprint. Such refined precision in spot beam coverage allows for targeted satellite signals to be received by a large number of smaller and more economical earth receiving stations that aim to serve a wide range of individual and business sector users.

Current Status

At the World Administrative Radio Conference of 1995, the ITU's 189 member nations successfully passed a set of regulations to resolve the single-most politically contentious and technically challenging issue ever faced by the organization: balancing equitable access to and efficient use of geostationary orbital positions for all member nations. These regulations enabled those who have actual usage needs to obtain a designated volume of orbit/spectrum resources—on an efficient "first-come, first-served" basis—through international coordination, while guaranteeing all nations a pre-determined orbital position associated with free and equitable use of a certain amount of frequency spectrum for the future. Subsequently, during WARC '97, decisions involving frequency allocation for non-geostationary satellite services were resolved, as shown in Table 22.1.

Table 22.1
Frequency Allocation for Non-Geostationary Satellite Services

Service	Frequency Band	Frequency Allocation
Fixed Satellite Service	Ka-band	18.9/19.3 GHz and 28.7/29.1 GHz
Mobile Satellite Systems	Ka-band	19.3/19.7 GHz and 29.1/29.5 GHz
Low-Speed Mobile Satellite Systems	L-band	454/455 MHz

Source: C. A. Lin

Non-geostationary satellites are launched into low-earth orbits or medium-earth orbits (MEOs) to provide two-way business satellite services. An example of this type of service is Teledesic, which proposes to provide high-speed global voice, data, and video communication, as well as "Internet-in-the-sky" services. A similar, voice-only service is offered by Iridium (see Figure 22.2). Recent market emergence in cellular radio, personal communication services (PCS), and global positioning systems (GPS) represents another reason non-geostationary mobile satellite services are on the rise. Cellular radio and PCS both provide wireless mobile telephone services. When radio signals are being relayed via a satellite utilizing digital transmission technologies, these signals can reach a wide area network to accommodate global mobile communication needs (Mirabito, 1997). For instance, Globalstar's LEO satellites can transmit a phone signal originated from a wireless or fixed phone unit to a terrestrial gateway, which then retransmits the signal to existing fixed and cellular telephone or PCS networks around the world. GPS is a mobile satellite communications service that interfaces with geographic data stored on CD-ROMs, and was originally developed for military use. Today, GPS is also used as a navigational tool to pinpoint locations in remote regions and disaster areas, and on vehicles, ships, and aircraft for civilian applications.

The increasing interest in launching business satellite services in non-geostationary orbits resulted in an almost exponential market growth around the world toward the end of the 20th century. Due to the rapid expansion of the constellations of satellites launched by competing systems, several major players went through financial instabilities and were eventually reinvented under new management. For instance, ICO (Intermediate Circular Orbit) went into bankruptcy protection in August 1999

(Glasner, 1999) and was later integrated into Teledesic in May 2000 to keep its 12 MEO satellite service afloat (New life, 2000). Another service that emerged from bankruptcy in 2000 was the LEO satellite service provider Orbcomm, which had its first launch in December 1997 (Lloyd, 2001). Similarly, Iridium first launched its system in May 1997, declared bankruptcy in May 1999 (Iridium falls, 1999), and emerged from bankruptcy in December 2000 to run its 66 LEO satellite fleet (Iridium, 2001). Most recently, Globalstar—which launched its system in February 1998—maintains service to its clients with its 48-satellite LEO operation under a new corporate parent (Globalstar, 2004).

Figure 22.2
Iridium System Overview

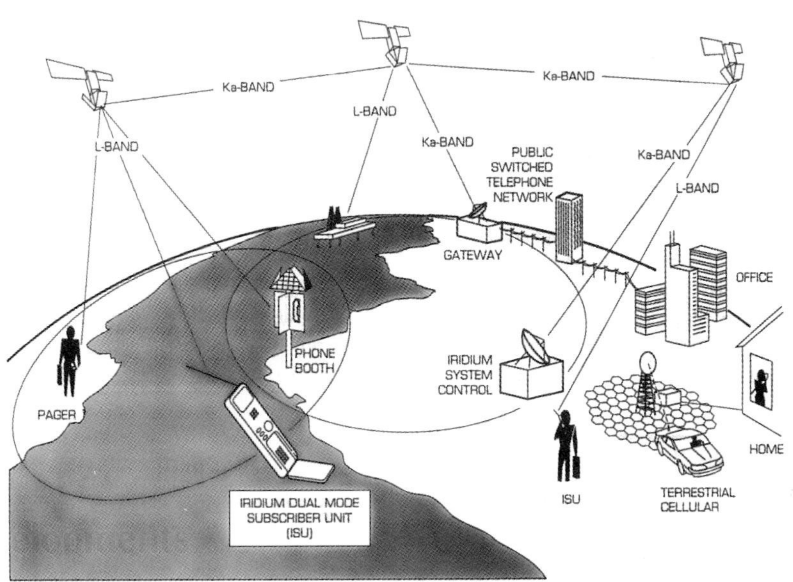

Source: Iridium, Inc.

To help accommodate the rapid development of mobile telecommunications, the recent WRC-2000 (World Radiocommunications Conference of 2000) reached the following major decisions (ITU, 2000):

1) Three additional frequency bands have been designated for international terrestrial mobile telecommunications: 806-960 MHz, 1710-1885 MHz, and 2500-2690 MHz.

2) To prevent frequency interference between geosynchronous and non-geosynchronous systems in order to provide high-quality communications services including telephony, television, and Internet applications, power limits were set for each type of satellite network services.

3) The rules for Ku-band (10-18 GHz) sharing between the two were defined.

4) Additional spectrum allocations were made for a new generation of radio-navigation-satellite services such as Europe's Galileo and Russia's GLONASS (global navigation satellite system) and the U.S. GPS.

Three other significant ITU resolutions were also reached by WRC-2003 (ITU, 2003a):

1) The frequency bands 27.5-28.35 GHz and 31-31.3 GHz have been designated for the use of high altitude platform stations (HAPS)—airships and aircraft—in the fixed service.

2) The guidelines for the implementation of high-density applications through the fixed-satellite service (HDFSS) in a number of different frequency bands were approved.

3) Earth stations on-board vessels (EVS) that transmit their signals in the frequency bands 5.925-6.425 MHz and 14-14.5 GHz will be permitted to use fixed-satellite service (FSS) networks to access a wide variety of telecommunications services including Internet services.

Factors to Watch

As the telecommunications industry continues to embrace the mobile communications modality, the satellite communications industry likewise enters a phase of intense competition between systems operated in non-geostationary orbits (i.e., low- or medium earth orbits) and fixed satellite services operated in geostationary orbits. The former represents a more nimble application of services ranging from telephony and Internet to telemedicine and distance education, while the latter remains the staple for large-scale voice, data, and video communications solutions.

The technologies that are driving the most new interests in this largely privatized competitive marketplace include very small aperture satellites, high-speed Internet services, and cellular network and mobile communications solutions. These technologies are all considered integral components of an integrated corporate communications network. The attraction of the VSAT-based solutions is economics. For instance, the VSAT service can now transmit more than 2,000 voice signals on a 36-MHz transponder by digitizing these signals using new digital signal processing (DSP) techniques (instead of the pulse code modulation technique). There are more than 500,000 VSAT terminals grounded in more than 120 countries worldwide today (Global VSAT Forum, 2004).

High-speed Internet services (including wireless DSL [digital subscriber line], Webcasting and streaming) and cellular network connections have now become a backbone for global business communication within and between corporations. Mobile communications services, including civilian applications (e.g., truck fleet tracking and disaster relief) and military applications employed by the U.S. Defense Department (e.g., Operation Iraqi Freedom) and the Homeland Security Department (e.g., nuclear generators or border entry-point surveillance), are also strong areas for industry growth (Noguchi, 2003).

Other emerging technologies that figure prominently in the wireless fixed communications service industry include high-density fixed satellite-service, which deploys a large number of high-power earth stations (Hayden, 2003), and high altitude platform stations, which use high-altitude aircraft to provide relay services similar to that provided by satellites, but at a lower cost and with a much smaller footprint (OFCOM, 2004). Experts (e.g., Pelton, 1998) have predicted that satellites launched into LEOs and MEOs will provide an efficient global telecommunications network. In conjunction with other mobile communications services, such as the experimental high altitude long endurance (HALE) aircraft, these systems are valuable because they "eliminate the 'last mile' problem faced by land-based carriers" (Reagan, 2002, p. 67) to provide wireless services that can reach even the most remote regions of the world.

The competition offered by emerging new satellite technologies and service solutions is driven by two main factors. The first pertains to the stable supply of satellite launch services and a longer shelf life for satellites. Reliable launch programs—both domestic and foreign—now include Boeing, Orbital, and Lockheed-Martin, as well as Long March (People's Republic of China), Ariane (European Space Agency), and Russia-based launch services. The second factor is characterized by the hastened trend toward deregulation and privatization for both national and international satellite services, in part due to the increased influence of the World Trade Organization and the European Union on both global and regional economic policies.

The ITU in the 21st century is cognizant of the rising trend of global trade liberalization and telecommunications industry privatization. It has begun to facilitate an intergovernmental process to take on the issues of the digital divide between developed and underdeveloped nations (Zeitoun, 2002), as well as the principles of universal access and competition for building a national telecommunications model (Perrone, 2002). The ITU's 2003 Telecommunications Development Conference (WTDC-02) released a digital access index (DAI) to measure an individual nation's access to information and communication technology (ICT). The two index items most relevant to the utilization of satellite technology solutions are Internet and mobile telephony access (ITU, 2003b).

For example, an innovative approach for creating Internet access through e-post services, delivered via VSAT services to individuals living in remote regions of developing nations, was partnered between the ITU and the Universal Postal Union (ITU, 2004a). As mobile telephony already reached 1.35 billion users in 2003, compared with the 1.2 billion fixed-line telephony users, broadband mobile communications services are developing more economically and expediently worldwide (ITU, 2004b). The ultimate objective of the ITU is to bring about equitable distribution of ICT development across all member nations.

Numerous predictions have fallen by the wayside over the past two decades regarding how high-speed fiber optics networks might overtake satellites to become the main providers of broadband services due to their technical reliability and economic efficiency. As wireless and mobile communications technologies continue to advance and their popularity increases, satellite services will outpace fiber optic networks as the primary providers of global telecommunications services. Next time you spot the International Space Station at night in "a steady white pinpoint of light moving slowly across the sky" (NASA, 2002), it will be a great reminder of the remarkable contributions that satellite services have brought us—an exciting era of space communication!

Bibliography

Divine, R. (1993). *The Sputnik challenge*. New York: Oxford Press.

Glasner, J. (1999, November 2). Craig McCaw's master plan. *Wired News*. Retrieved March 8, 2002 from http://www.wired.com/news/business/0,1367,32247,00.html.

Global VSAT Forum. (2004). *The VSAT Industry*. Retrieved March 9, 2004 from http://www.gvf.gov/vsat_indstury/trends.

Globalstar. (2004, March 11). *U.S. Federal Communications Commission approves Globalstar operating license transfer: Major condition fulfilled for acquisition by Thermo Capital partners*. Retrieved March 13, 2004 from http://www.globalstar.com/view_pr.jsp?id=355.

Hayden, T. (2003). *WRC-2003 Advisory Committee, Draft U.S. proposal on WRC-03 agenda item 1.25*. Retrieved March 9, 2004 from www.fcc.gov/ib/WRC-03/files/docs/advisory_comm/mtg6/WAC087.pdf.

Hudson, H. (1990). *Communication satellites: Their development and impact*. New York: The Free Press.

International Telecommunications Union. (2000, June 2). World Radicommunication Conference concludes on series of far-reaching agreements. *New release, press, and publication information*.

International Telecommunications Union. (2003a, July 4). *World Radiocommunication Conference concludes: Agreements define future of radiocommunication*. Retrieved July 10, 2003 from http://www.itu.int/newsarchives/press_releases/ 2003/19.html.

International Telecommunications Union. (2003b, December 4). *ITU World Telecommunication Development Report 2003*. Retrieved March 9, 2004 from http://www.itu.int/newsarchive/press_releases/2003/32.html.

International Telecommunications Union. (2004a, January 16). *ITU, India, and UPU partnership helps Bhutan bridge the digital divide: E-services link remote post offices with citizens*. Retrieved March 9, 2004 from http://www.itu.int/newsroom/press_releases/2004/NP01.html.

International Telecommunications Union. (2004b, March 16). *Shaping the future mobile information society: New initiatives workshop in the Republic of Korea attracts telecommunication stakeholders*. Retrieved March 20, 2004 from http://itu.int/newsromm/press-relaease/2004/NP03.html.

Iridium falls out of orbit. (1999, August 13). *Wired News Report*. Retrieved March 8, 2002 from http://www.wired.com/news/business/0,1367,21267,00.html.

Iridium Satellite LLC. (2001). The world's first global handset. *About Iridium*. Retrieved March 11, 2002 from http://www.iridium.com/corp/iri_corp_story.asp?storyid=4.

Jansky, D., & Jeruchim, M. (1983). *Communication satellites in the geostationary orbit*. Dedham, MA: Artech House.

Lloyd, W. (2001, May 1). Orbcomm FCC ECFS submissions. *Lloyd's satellite constellations*. Retrieved March 11, 2002 from http://www.ee.surrey.ac.uk/Personal/L.Wood/constellations/orbcomm.html.

Mirabito, M. (1997). Wireless technology and mobile communication. *The new communications technologies*. Boston: Focal Press.

Miller, M., Vucetic, B., & Berry, L. (1993). *Satellite communications: Mobile and fixed services*. Norwell, MA: Kluwer Academic Publishers.

NASA Human Space Flight. (2002, April 1). *Sighting opportunities: Viewing them from the ground*. Retrieved April 3, 2002 from http://spaceflight.nasa.gov/realdata/sightings/help.html.

National Telecommunications and Information Administration. (2000). *Open-market Reorganization for the Betterment of International Telecommunications (ORBIT) Act*. Pub. L. No. 106-180, U.S. Department of Commerce.

New life. (2000, May 17). *Wired News Report*. Retrieved March 8, 2002 from http://www.wired.com/news/business/0,1367,36408,00.html.

Noguchi, Y. (2003, April 17). With war, satellite industry is born again, *The Washington Post*. Retrieved April 18, 2003 from http://washingtonpost.com.

OFCOM, (2004). HAPS—*High altitude platform stations*. Federal Office of Communication. Retrieved March 10, 2004 from www.bakon.ch/imperia/md/content/English/funk/forchungundentwicklung/studien/HAPS/pdf.

Pelton, J. (1998). Telecommunications for the 21st century. *Scientific American, 278*, 80-85.

Perrone, L. (2002, March 26). Harmonizing idealism with solvency: The basic challenge of the modern telecommunications regulator. *ITU News Magazine*. Retrieved April 2, 2002 from http://www.itu.int/itunews/issue/2002/02/harmonizing.html.

Reagan, J. (2002). The difficult world of predicting telecommunications innovations: Factors affecting adoption. In C. A. Lin & D. J. Atkin (Eds.). *Communication technology and society: Audience adoption and uses*. Cresskill, NJ: Hampton Press.

Reese, D. (1990). *Satellite communications: The first quarter century of service*. New York: Wiley Interscience Publications.

Zeitoun, T. (2002, March 22). Transforming the digital divide into digital opportunities: ITU–D's challenge over the next four-year period. *ITU News Magazine*. Retrieved April 2, 2002 from http://www.itu.int/itunews/issue/2002/02/transforming.html.

23

Distance Learning Technologies

Kyle Nicholas, Ph.D.*

Distance learning is a combination of technologies, teaching practices, and industrial and institutional arrangements that some say are transforming the very nature of education. Distance learning is a "socio-technological system," and the implementation of new technologies highlights fundamental questions of pedagogy, global and regional challenges in distributing education, and the interpenetration of business practices into public educational institutions (Hughes, 2001).

The term "distance learning" in this chapter serves as a surrogate for a variety of related concepts, including distance education, distributed education, computer-assisted learning, and "open university." The open university model implies "free" secondary education, while distributed learning includes the use of technologies to provide on-campus education. While each of these terms has specific applications, their common denominators are digital instructional technologies and advanced telecommunications networks. These various forms also share a common history that illustrates the tight relationship between social necessity and technological evolution.

Background

Distance learning has evolved from its origins in British shorthand courses in 1832 into a multi-billion dollar set of systems with significant transnational implications. The first distance-learning program, as we would know it today, began in 1872 at the University of Chicago. William Harper, University of Chicago President at the time, believed that education via mail would overtake classroom studies in terms of students served. He began classes via mail in 1892.

---

* Assistant Professor, Old Dominion University (Norfolk, Virginia).

This is generally regarded as the first distance education program in the United States. Regulation of distance learning was a concern almost from the beginning. By 1915, learning programs via mail, or "correspondence courses," were popular enough that the National Continuing Education Association (later renamed the University Continuing Education Association) was formed at the University of Wisconsin to create national standards for quality and credit transfer. Subsequent accreditation associations were formed to meet challenges of new technologies and the for-profit educational sector. Universities began to receive radio station licenses in the United States in the early 1920s, and almost immediately began using their frequencies to distribute education. Early players included universities in Utah, Minnesota, and Pennsylvania. Although courses were popular, newer technologies eventually surpassed radio and, by the 1940s, only one such course remained (PBS, 2004).

In 1934, Iowa State University was the first to broadcast televised courses for credit, two decades before most Americans could view them. This practice forecast the current state of distance learning wherein universities tend to employ new technologies ahead of their diffusion curve. It is no coincidence that public universities, often situated in rural areas, tend to explore distance-learning alternatives first. State universities have public mandates for the democratization of education, and rural areas pose particular challenges to centralized institutions such as universities. This combination of public policy and geography has created a persistent irony, however. Although underserved rural areas are key to the development of distance-learning programs (at least at public universities), they are often the least able to take advantage of new technologies.

Although private television stations lobbied against them, distance learning programs got a boost in the early 1960s when the Federal Communications Commission (FCC) set aside television frequencies specifically for accredited programs (Zechowsky, 2004). Later in that decade, the British Open University system was founded; it used televised instruction and the mail to deliver textbooks and supplementary materials. The university now offers 360 degrees to more than 200,000 students annually (Open University, 2004).

The open university concept, combined with new communication technologies and increasing enrollment pressures, encouraged the development of educational consortia in the 1990s. The idea behind most of these consortia was to offer courses via the World Wide Web and to generate interest by leveraging university "brands." As of this writing, most have become portals to courses offered at individual universities. Most no longer offer degrees (for example, see the California Virtual Campus [2004] at http://www.cvc.edu/aboutus/), and some of the most prestigious are either defunct or have shifted educational activities to the private sector. For example, Universitas 21, a consortium of 17 universities, offers courses through Thomson, the global publishing company (Universitas 21, n.d.).

The other key development in the last decade has been the rise of private, specialized online universities, such as the University of Phoenix (www.universityofphoenix.com) and Jones International University (www.jonesinternational.edu). Rather than targeting rural and underserved populations, these organizations market advanced degrees to busy professionals. While universities continue to invest in computers at record rates, spending nearly $570 million in one quarter in 2003 (IT dollars, 2003), public educators who once saw distance learning technologies as the solution to myriad funding and population problems appear to be reconsidering their options as the real costs of technologies and educators become apparent (Kobulnicky et al., 2002).

Key Technologies

Distance learning employs a wide variety of communication modalities in almost endless combinations. However, depending on the mission of the educational institution and the socio-economic context of learners, a few key models are emerging. Most of these variations can be plotted on two key dimensions: synchronicity and interactivity (distributed education has the additional dimension of on/off campus). Synchronous courses are concerned with providing instruction in "real time," often with both a "live," on-campus classroom, as well as students in remote locations. Asynchronous technologies, including CD ROM, textbooks, and Web-based instructional designs, afford students great flexibility in scheduling instruction, but they generally suffer from low interactivity. Synchronous instruction provides the closest thing to traditional face-to-face instruction, but it generally demands greater resource commitments by both the institution and the student. The emphasis is on a "live" classroom experience and at least some level of simultaneous interaction between students and instructor or among students.

Distance instructional techniques also vary in the amount and direction of interactivity as well. At the high end of the spectrum, two-way video interaction—as in teleconferencing—best simulates the multidimensional interaction of the classroom. The potential of two-way video instruction has been enhanced by increased institutional investments in satellite distribution systems that can connect instructors to specially-designed remote classrooms. Classrooms equipped with two-way video allow students to "see and be seen" by instructors. This innovation provides a key additional benefit: it reduces the time and energy required to adapt traditional courses to the distance earning format. This is an important consideration, as anecdotal evidence suggests that faculty resistance to course transformation creates an obstacle to increased distance learning offerings. Some of this friction may be reduced by emerging digital content organization schemes, e.g., SCORM (Sharable Content Object Reference Model—a protocol for organizing course information), although those technologies carry their own disruptive seeds.

If two-way video is closest to "ideal" communication, it is also ideal in the sense that it is rare and expensive. Synchronous communication is more often achieved through less robust channels, such as one-way video with audio or text return. In this system, the instructor is seen and heard, but the student feedback channel is either synchronous audio via speaker at a remote site, or nearly synchronous text return via online chat software.

Most synchronous audio/video courses depend on broadcast television technologies, typically satellite transmission to a closed-circuit cable headend at a remote site. Once broadcast to the remote site, courses can be distributed to several classrooms or recorded for asynchronous courses in a variety of means. Courses streamed via the Internet provide an alternative to the broadcast model. These courses can be delivered to any networked site, either at a remote institution or directly to the student's home. Video streaming is generally delivered from a server at the origination site (school or university) via a high-bandwidth network to remote computers equipped with media software and speakers. Interactivity requires microphones and voice processing software at the student site. Originally expected to engender an explosion in distance learning (as well as on-demand video distribution and picture phone applications), streaming has been constrained by technical considerations and a general lack of consumer enthusiasm (see Rheingold, 2003 for the most effervescent predictions). For educators, the nature of streaming, with its severe signal compression and low resolution, reduces communication effectiveness and, in most cases, renders substantial interactivity in all but the smallest classes unfeasible.

Figure 23.1
Distance Learning Center, Old Dominion University

Source: Old Dominion University (Norfolk, Virginia)

Recent Developments & Current Status

In addition to the prohibitive costs of the initial investment in broadcast technologies, synchronous audio/video courses may not be appropriate for two key markets for distance learning. Rural and developing areas often have difficulty implementing the infrastructure required for video capture and distribution (although basic television transmission is generally available), and the high-bandwidth networks necessary for effective streaming may not be available. The on-time, on-site commitment demanded by synchronous modalities reduces student's scheduling flexibility, a key attribute for the working professionals considered to be a vital constituency of distance learning.

Asynchronous technologies reduce the immediacy of the communication experience, but offer tremendous flexibility to institutions, instructors, and students. In its most basic form, contemporary distance learning depends on network-based information technologies to maintain communication between instructors and students. E-mail remains a key communication medium both on campus and in global systems. Because of the ubiquity, flexibility, and low bandwidth requirements of e-mail, it is an efficient stand-alone channel that is increasingly integrated into more sophisticated courseware and course management technologies.

New technologies have created new legal challenges for distance learners. For years, copyright problems haunted distance learning programs as copyright holders sought royalties and redress for the distribution of materials via learning networks. In the United States, Congress adopted the Technology Education and Copyright Harmonization Act. The new law gives teachers and institutions new flexibility to distribute and store copyrighted materials and allows materials to be transmitted directly to home computers rather than only closed-circuit satellite campuses (TEACH Act, 2001).

This new distribution flexibility, along with increased availability of Web-based resources and the development of easy-to-use HTML editors, has helped accelerate the diffusion of Web-based course materials. Along with e-mail, course Web pages are essential ingredients in any distance learning program. Increasingly, these elements are incorporated into courseware packages, such as Blackboard, that provide self-contained Web-based course management and content distribution. These packages combine synchronous communication, e.g., chat channels, with a variety of asynchronous dimensions, including newsgroups, digital drop boxes, grade spreadsheets, and other forms of exchange. Web-based course management provides a rich, multifaceted communication channel. Course management is a combination of software and hardware; institutions are required to either purchase servers or lease space on proprietary servers associated with the software (Blackboard, 2004; WebCT, 2004).

These technologies are part of an evolving direction in education toward separating and systematizing the role of instructors into distinct, computer-assisted functions, some of which are produced by proprietary services. This disintegration of idiosyncratic knowledge creation, compilation, and sharing is not unique to distance learning. Distance learning, however, represents the vanguard of technological implementation in learning, and any informed discussion requires at least a brief mention of the broader issues associated with some of these technologies.

While early fears that media companies and other global corporations would displace traditional educational institutions in the Internet age appear unfounded, at least for now, there has been a significant adoption of corporate culture and technique in distance education (Cunningham, 1998; Quinn, 2003). Although the distance-learning "market" appears to have cooled since the early part of the decade, market principles and efforts to rationalize learning through the application of computers continue to drive much of the innovation in educational technologies. For many educators, the introduction of market-based efficiency and generic education "modules" goes against underlying principles of democratic education and, in particular, liberal arts traditions (Hughes, 2001). Others bemoan the slow pace of reform in public education and see corporatization of public education as inevitable or even preferable (Collis, 2003).

Beyond the adoption of business models to public education, private Internet-based operations are proliferating around the globe. These technologically-based educational corporations are different from traditional private institutions in that they are generally for-profit businesses that concentrate on only the most profitable aspects of education. For instance, while one can find hundreds of online MBA programs, dot-com universities that concentrate on the liberal arts are scarce.

The emergence of technical standards as part of the academic mission is rationalizing course creation and teaching practices in several dimensions. At MIT, the Open Knowledge Initiative (OKI) is creating free course management software based on open computing standards. "The result of this collaboration is an open and extensible architecture that specifies how the components of an educational software environment communicate with each other and with other enterprise systems" (OKI, 2004). The key to OKI, like many other initiatives, is the creation of education modules that can be interchanged via computer networks. The obsession with standardization in educational technologies and the rationalization of education informs private initiatives, such as the Global Alliance for Transnational Education (GATE), as well as public groups (GATE, n.d.). While GATE focuses on "best practices," the IMS Global Learning Consortium seeks to globalize standards for transnational education delivery systems (IMS, 2004).

The U.S. Department of Defense, one of the largest users of distance training programs, sponsors the Advanced Academic Distributed Learning Co-Lab, whose partners include some of the largest U.S. universities. This is one of several projects using metatags, identifiers in computer code, to categorize and catalog "content objects." The SCORM project works by fragmenting and recombining educational materials (SCORM, 2004). For instance, a professor's lecture can be rendered into so many "content objects" by dissecting and cataloging the notes, slides, quotes, and reference material. Each little chunk of material is assigned a metatag that identifies its subject matter, date, place of origin, and digital material existence (slide, music clip, Word document, etc.). These chunks can then be searched via the Internet and recombined by others in endless variations. In other words, teachers (and possibly students) picking and choosing from other courses will create courses of the future.

In distance learning, this recombination increasingly is performed in an international arena (Ali, et al., 1997). For now, the rhetoric of globalized education outstrips reality, but a powerful set of players is exploring the field. The United Nations lists distance learning initiatives as a key priority in development efforts, and the World Bank has developed a series of policy statements encouraging distance learning in the developing world. Along with educational effects, the international lender envisions the acceleration of a political agenda via distance learning: "Distance education is a means of meeting changing government priorities more rapidly and flexibly than conventional institutions may be able to" (World Bank, 1999).

Factors to Watch

Distance learning epitomizes the cyclical relationship between technological advancement and educational evolution. As new communication technologies are developed, individuals need increased knowledge and training. Subsequently, educational institutions, feeling the challenge of increased demand, turn to technologies to leverage resources, leading to new innovations in technologies and new requirements for populations. Although recent technical developments cast education as a product, or a means to an end, a close inspection of this relationship reveals distance learning to be part of a complex social process that must evolve to meet new challenges in a global arena.

Bibliography

Ali, M., Shamsher, A., Haque, E., & Rumble, G. (1997). The Bangladesh Open University: Mission and promise. *Open Learning, 12* (2), 12-17.

Blackboard, Inc. (n.d.). *Product information*. Retrieved March 25, 2004 from http://blackboard.com/products/index.htm.

California Virtual Campus. (2004). *Welcome to the California Virtual Campus*. Retrieved April 20, 2004 from http://www.cvc.edu/aboutus/.

Collis, D. (2003). New business models for higher education. In D. J. Collis (Ed.). *The future of the city of intellect: The changing American university*. Palo Alto, CA: Stanford University Press, pp. 181-202.

Cunningham, S., Tapsall, S., Ryan, Y., Stedman, L., Bagdon, K., & Flew, T. (1998). *New media and borderless education: A review of the convergence between global media networks and higher education provision.* A report to the Department of Employment, Education, Training and Youth Affairs, Commonwealth of Australia.

GATE. (n.d.). *Global Alliance for Transnational Education.* Retrieved March 25, 2004 from http://www.edugate.org/.

Hughes, T. (2001). Through a glass darkly: Anticipating the future of technology enabled education. *Educause Review, 36* (4), 1-16. Retrieved March 4, 2004 from http://www.educause.edu/ir/library/pdf/erm0140.pdf.

IMS Global Learning Consortium, Inc. (n.d.). *About IMS.* Retrieved March 25, 2004 from http://www.imsglobal.org/aboutims.cfm.

IT dollars: Higher education increases spending on hardware, information technology in the news. (2003). *Educause Review, 38* (6), 4–8.

Kobulnicky, P., Rudy, J., & the Educause Current Issues Committee. (2002). Third annual Educause survey identifies current IT issues. *Educause Review, 2* (1), 8-21.

Open Knowledge Initiative. (n.d.). Retrieved March 25, 2004 from http://web.mit.edu/oki/.

Open University. (n.d.). About_the Open University. Retrieved March 25, 2004 from http://www3.open.ac.uk/media/factsheets/Information_about_The Open University/Background Information.pdf.

Public Broadcasting System. (2004). *A brief history of distance learning: Timeline.* Retrieved February 24, 2004 from http://www.pbs.org/als/dlweek/history/.

Quinn, J. (2003). Services and technology: Revolutionizing economics, business, and education. *Educause.* Retrieved April 29, 2004 from http://www.educause.edu/ir/library/pdf/ffp0106s.pdf.

Rheingold, H. (2003). *Smart mobs.* Cambridge, MA: Perseus Publishing.

SCORM (Sharable Content Object Reference Method). (n.d.). *SCORM overview.* Retrieved March 25, 2004 from http://www.adlnet.org/index.cfm?fuseaction=scormabt.

Technology Education and Copyright Harmonization (TEACH) Act. (2001). Retrieved March 25, 2004 from http://www.copyright.iupui.edu/teach_summary.htm or http://www.arl.org/info/frn/copy/TEACH.html.

Univeristas 21. (n.d.). *Information.* Retrieved March 25, 2004 from http://www.universitas21.com/u21global.htm#Information.

WebCT. (n.d.) *WebCT software.* Retrieved March 25, 2004 from http://webct.com/products.

World Bank. (n.d.). *Global Distance Education Net mandate and mission statements.* Retrieved March 25, 2004 from http://www1.worldbank.org/disted/Policy/Institutional/mandate.html.

Zechowski, S. (n.d.). *Television in education.* Retrieved March 25, 2004 from http://www.museum.tv/archives/etv/E/htmlE/educationalt/educationalt.htm.

24

Teleconferencing

Michael R. Ogden, Ph.D.*

Teleconferencing has always promised increased productivity and efficiency, improved communications, enhanced business opportunities, and reduced travel expenses. But, "we humans are a 'touchy-feely' species, who, in general, prefer travel and face-to-face encounters over the more impersonal [teleconference] experience" (Kuehn, 2002). According to the Travel Industry Association, business travel rose 14% between 1994 and 1999, with nearly half of the 44 million travelers in 1998 attending a meeting, trade show, or convention. Forecasts predicted a continued increase of up to 3% per year well into the early 21st century (Seaberry, 1999).

Following the events of September 11, 2001, however, many executives accustomed to frequent cross-country trips began grappling with growing concerns over their business travel. As evidence, the National Business Travel Association (NBTA) found companies cutting travel plans by as much as 60% (Bamnet, 2001), and airline ticket sales in the United States fell 20% to 40% in the months immediately after the attack (McLuhan, 2002). Worldwide, many telecommunications companies also reported seeing their videoconferencing usage increase by as much as 85% in the first six weeks following September 11, while audioconferencing was up almost 30% (Perera, 2001). However, as Elliot Gold, publisher of the influential telecommunications news and analysis Webzine, *Electronic Telespan*, has been quick to point out, "9-11 made teleconferencing a 10-year or, more precisely, a 20-year overnight success. The fact is that usage of teleconferencing was growing quite well before September 11, 2001. Conference calls alone [grew] at an average rate of 39% per year for the five years before 2001 … [and the] growth rate for 2001 was 42.5%" (Gold, 2004).

* Associate Professor, Film & Video Studies, Department of Communication, Central Washington University (Ellensburg, Washington) and Affiliate Graduate Faculty, Telecommunications and Information Resource Management (TIRM) Graduate Certificate Program, University of Hawaii, Manoa (Honolulu, Hawaii).

"Lucy Johnson, senior VP of daytime television for CBS, used to fly from LA to New York every two months to meet with the East Coast staff. Now, they meet via videoconferences, phone conference calls, and a plethora of e-mails" (Ryssdal, 2001). Even as more recent NBTA surveys found evidence of a rebound in travel by the end of 2001 (Bamnet, 2001), most companies were already cutting travel budgets to varying degrees prior to September 11, replacing expensive business travel with less expensive electronic conferencing options. Quaker Chemical Corporation typically spent "...upward of $30,000 to fly participants to a single [conference] site. [V]ideoconferencing and teleconferencing costs are minuscule by comparison—about $2,000, ... $50 an hour for video hookups and $10 an hour for telephone connections" (Chabrow, 2001).

Successful Meetings, a magazine serving the needs of meeting planners across all industries, observed in their "State of the Industry Report" that, increasingly, face-to-face meetings were being replaced by teleconferencing (State of the industry, 2003). When surveying their constituency, 58.5% of meeting planners reported that teleconferencing was substituting for some meetings with the majority (49.1%) stating that videoconferencing was the preferred substitute, Webcasting was a close second (44%), and satellite broadcasts and other means of conducting meetings electronically came in a distant third (State of the industry, 2003). A Yankee Group survey of small and home-based businesses indicated that nearly 68% of these businesses have broadband access (Dukart, 2002), leading industry observers to foresee increased demand for teleconferencing—video, audio, and computer-based solutions—as companies look for ways to connect with employees and customers while remaining on *terra firma* (Swanson, 2001).

Definitions of what exactly constitutes "teleconferencing," however, tend to differ across the industry. Attempts to find a consensus in defining terms is not entirely an academic exercise. It has been the experience of many industry professionals that disputes over the feasibility of teleconferencing are often perpetuated by the fact that few of the negotiating parties know exactly what the other party believes a teleconference is (Hausman, 1991).

In the broadest since, teleconferencing can be defined as "small group communication through an electronic medium" (Johansen, et al., 1979, p. 1). Similarly, *Electronic Telespan* defines teleconferencing as "an electronic meeting that allows three or more people to meet, be it across time zones or across office cubicles" (Gold, 2004). However, these very broad definitions have not proven very useful in defining the widely varying ways of conducting "electronic meetings." Jan Sellards, then president of the International Teleconferencing Association, defined the term in 1987 as, "the meeting between two or more locations and two or more people in those remote locations, where they have a need to share information. This does not necessarily mean a big multimedia event. The simplest form of teleconferencing is an audioconference" (Hausman, 1991, p. 246). Some practitioners prefer to call conferences using video "videoconferences," those employing mainly audio "audioconferences," and those using a range of computer-based technologies either "computer-conferences" or "Web-conferences"—oftentimes arranged on a continuum to distinguish those technologies that facilitate the "most natural" type of meetings from those that are "least natural" (see Figure 24.1). More typically, the term "teleconferencing" is used as a shorthand term to represent an array of technologies and services ranging from a three-way telephone conversation to full-motion color television to highly interactive, multipoint Web-based electronic meeting "spaces"—each varying in complexity, expense, and sense of immediacy.

Figure 24.1
Teleconferencing Continuum

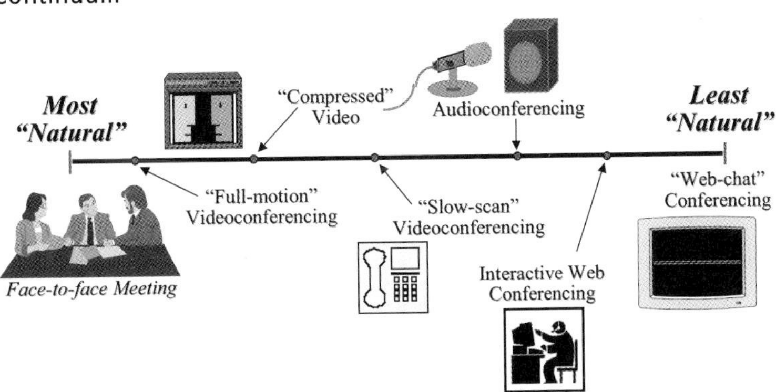

Source: M. Ogden

Recently, "integrated collaboration" has emerged as the preferred term used to emphasize the functionality of teleconferencing without dwelling excessively on the technology. Whereas most definitions of teleconferencing focus on the use of audio/video communications to facilitate meetings without the burden of travel, "integrated collaboration is a *process* that allows two or more users to interact with audio, video, and/or data streams in both real-time and non-real time communications modes across packet and circuit switched networks" (Davis, 1999, p. 5, emphasis added). According to Andrew Davis, Managing Partner at Wainhouse Consulting Group, teleconferencing is primarily about meetings, but "integrated collaboration is about meetings, corporate communications, sales, training, and enhanced customer services" (Davis, 1999, p. 5).

As what constitutes teleconferencing continues to evolve, users look for easy-to-use, standards-based solutions that can integrate across all their communication needs. Whether referred to as "integrated collaboration," "collaboration solutions" (Glenwright, 2003), "conferencing and collaboration" (Polycom, 2003), or any of the other process-oriented, use-based terms, multimedia communications are fuelling a paradigm shift, as geographically-dispersed work teams become the rule rather than the exception and as price reductions and connectivity improvements make the various technologies of teleconferencing more dynamic and available to a wider constituency.

While seeking a meaningful definition for "teleconferencing," it is also important to distinguish between the two different "modes" of teleconferencing: *broadcast* and *conversational*. In broadcast teleconferences, the intention is to reach a large and dispersed audience with the same message at the same time, while providing a limited opportunity for the audience to interact with the originator. Typically, this takes the form of a "one-to-many" video broadcast with interaction facilitated via audience telephone "call-ins." These events (e.g., "state of the organization" addresses, employee relations conferences, new product kickoffs, etc.) require expensive transmission equipment, careful planning, sophisticated production techniques, and smooth coordination among the sites to be successful (Stowe, 1992).

On the other hand, conversational teleconferences usually link only a few sites together in an "each-to-all" configuration with a limited number of individuals per site. Although prior arrangement is usually necessary, such meetings are often more spontaneous, relatively inexpensive, and simple to conduct. They seldom require anything more complicated than exchanging e-mail messages, synchronizing meeting times (especially if across time zones), and/or making a phone call to a colleague. Conversational teleconferencing supports numerous kinds of business activities from management and administrative meetings (e.g., project, budget, staff, etc.), to marketing, sales, finance, and human resources (Stowe, 1992).

Another way to categorize teleconferencing applications is according to whether the facilities are used *in-house* as part of the normal routine, or *ad hoc*, only occasionally as the need arises. Once the exclusive domain of Fortune 500 firms, in-house or institutional teleconferencing—dedicated, room-based, and/or roll-about audio- and/or videoconferencing systems—is becoming more commonplace across a wide spectrum of organizations (e.g., schools, churches, hospitals, etc.). For example, Quaker Chemical offices have maintained PictureTel Corporation (now owned by Polycom) videoconferencing systems for years (Chabrow, 2001). Alyeska Pipeline Service Company, headquartered in Anchorage with business units in Fairbanks and Valdez and with seven active pump stations, uses their dedicated, in-house videoconferencing network to battle isolation, save travel costs, and improve training for its geographically-dispersed and weather-challenged workforce (Rosen, 2001).

Ad hoc teleconferencing, more commonly referred to as "conferencing on demand," can run the gamut from large-scale, one-off videoconferences, to much smaller-scale, desktop conferencing. Many of the major hotel chains (e.g., Holiday Inn, Hilton, Marriott, Sheraton, and Hyatt) first began offering teleconferencing services in the 1980s to attract corporations and professional associations (Singleton, 1983), and several business copy centers such as Kinko's have developed their own teleconferencing networks tailored for occasional use by small businesses and/or individuals. For the much smaller-scale solutions, there is no need to schedule special rooms or equipment, "just push the conference button on your touchtone phone, switch on the … camera on top of your PC, and you're ready to meet with others in your organization or outside it" (Weiland, 1996, p. 61).

As a case in point, take the Los Angeles law firm of Paul, Hastings, Janofsky & Walker LLP, which had scheduled a mid-September, face-to-face meeting in New York City for all of the managers from its seven U.S. offices. Because of uncertainty after the events of September 11, no one could travel. So the firm set up an online meeting through Latitude Communication Inc.'s MeetingPlace that provided an integrated voice and Web conference solution with only one day's notice. "The meeting wasn't perfect. Some managers were in hotel rooms and had to choose between participating through [an audio] conference or through the Web, and others had slow modem connections. However, the six-hour meeting took place" (Chabrow, 2001).

For most people, the key benefit of teleconferencing is that it eliminates the need for travel and enhances communications because many people can share information directly and simultaneously. Businesses are implementing teleconferencing strategies to reduce non-telecommunications costs, while improving productivity. Educational institutions are using teleconferencing for cost-effective distance learning programs, while government agencies are using it for crisis management, as well as daily information exchange (Muller, 1998). Because more people can participate in teleconferences than can affordably travel to face-to-face meetings, more input into problem-solving as well as new channels of communication can be opened, resulting in broader support for tough decisions.

On the downside, broader participation can also bring to the surface underlying conflicts within an organization, aggravating differences instead of cultivating unity. Because of limitations unique to each of the respective teleconferencing options, electronic meetings tend to be more orderly and focused. But there is a thin line between orderly agendas and narrow, repressive ones. Finally, those who ascribe to the "integrated collaboration" model of teleconferencing believe that the long-term growth of the industry will contribute real human value only to the extent that it improves the productivity of its users beyond the conservation of time and material resources expended in travel. Given the ease and flexibility of today's teleconferencing options, the most difficult challenge will be to hold more meaningful and productive meetings, instead of just more meetings.

Background

The basic technology for teleconferencing has existed for years. In its simplest form, teleconferencing has been around since the invention of the telephone—allowing people for the first time to converse in "real time," even though they were separated physically (Singleton, 1983). Today, we understand teleconferencing to be more involved than a simple telephone conversation. Some contend that teleconferencing has its roots in comic strips and science fiction. Certainly, comic strips such as *Buck Rogers*, popular in the early 1900s, made it easy to imagine people of the future communicating with projected images, or *Dick Tracy* calling instant meetings through a communication device on his wrist. The videophone in Stanley Kubrick's *2001: A Space Odyssey* furthered the notion of interactive video communications as a common feature of our near future. Likewise, *Star Trek*'s holodeck popularized the notion of fully-immersive virtual environments for diversion or business as accepted conventions of our far future.

The teleconferencing gear that invaded science fiction was pioneered by AT&T in the 1950s and 1960s. These early efforts were impressive for their time, but they were also considered little more than novelties and failed to catch on with consumers (Noll, 1992). Today, these visions have been brought to life in three alternatives to face-to-face meetings easily classified by their broad medium of application: audio, video, and computer-mediated teleconferencing.

Audio

Audioconferencing relies only on the spoken word, with occasional extra capacity for faxing documents or "slow-scan" image transmissions (see discussion under *Video* below). Historically, audioconferencing began with the familiar "conference call," generally set up by an operator working with the local telephone company (Stowe, 1992). Today, an audioconference can be implemented in a variety of ways. For a basic telephone conference involving a limited number of participants between two sites, a telephone set with either a three-way calling feature on the line or a conferencing feature supported by the PBX or key system is all that is required. If participants are expected to be in the same room at each location, a speakerphone can be used. Adding additional sites to an audioconference requires an electronic device called a *bridge* to provide the connection (Singleton, 1983)—simply plugging several telephone lines together will not yield satisfactory results. Audio bridges can accommodate a number of different types of local and wide area network (LAN and WAN) interfaces and are capable of linking several hundred participants in a single call, while simultaneously balancing the volume levels (allowing everyone to hear each other as though they were talking one-on-one) and reducing noise (echo, feedback, clipping, dropout, attenuation, and artifacts) (Muller, 1998).

A bridged audioconference can be implemented in several ways: dial-out, prearranged, or meet-me. In the dial-out mode, the operator (or conference originator) places a call to each participant at a prearranged number and then connects each one into the conference. Prearranged audioconferences are dialed automatically, or users dial a predefined code from a touch-tone telephone. In either case, the information needed to set up the conference is stored in a scheduler controlling the bridge. In the meet-me mode, participants are required to call into a bridge at a prearranged time to begin the conference. Regardless of the mode of implementation, the more participants brought into an audioconference through the bridge, the more free extensions are required (Muller, 1998).

Video

Since AT&T demonstrated the Picturephone at the 1964 World's Fair, corporate America has had an on-again, off-again fascination with video communications (Borthick, 2002). The assumption behind videoconferencing, usually unquestioned, has been that, the closer the medium can come to simulating face-to-face communications, the better (Johansen, et al., 1979). Engineers have thus struggled to make video images more lifelike in size and quality. As a result, full-motion videoconferencing offers glamour, but has yet to catch on as a routine business tool. Also, videoconferencing has traditionally had a technical complexity that tended to limit its scale and scope, despite periodic engineering breakthroughs, steady price/performance improvements, and gradual market growth.

Today, the technology has progressed to the point where videoconferencing has become crucial to the new, post-September 11 business reality (Kontzer 2001). Recent data reports show that there are only a few large players that continue to dominate the videoconferencing hardware market. In 2001, Polycom (after acquiring PictureTel) continued to dominate the market with 61% market share (down from 65%), with Tandberg (16%) and Sony (11%) showing solid gains since 2000. Whereas VTel slipped to 3%, the most noticeable change came at the expense of smaller companies that saw their collective market share reduced by almost half (to 9%) between 2000 and 2001 (Big boys, 2002).

Videoconferencing systems are frequently grouped into two main types of end-point systems: group systems and personal systems. The growing popularity of videoconferencing can be confirmed by the fact that nearly 100,000 new group videoconferencing systems are shipped each year, with more than 400,000 group videoconferencing systems in regular use today around the world, according to *Electronic TeleSpan* (Gold, 2004). In addition, there is growing demand for desktop or personal videoconferencing systems, with close to 40,000 such systems installed each year. Between three-quarters of a million and one million PC-video systems are purchased each year by consumers—often used to keep in touch with relatives (Gold, 2004). Prices for room-based group videoconferencing systems such as Polycom's ViewStation FX—designed for boardrooms, large conference rooms, or classrooms (Polycom, 2002)—can range from $15,000 to over $100,000, depending on options and/or equipment add-ons.

Roll-about group videoconferencing systems—once considered viable solutions for organizations that did not have extensive need for dedicated videoconferencing facilities—are on the wane as the range of options available for set-top and personal videoconferencing systems increase in popularity. Typical roll-about systems consist of a television monitor, a video codec, a single camera, and microphone (the last three items are usually packaged as an integrated unit) on a mobile cart. They range in price from $4,000 to $9,000 depending on options. Newer systems such as Polycom's iPower line of videoconferencing devices (a PictureTel heritage product)—although not technically a "roll-about" solution—are filling the niche for small conference rooms, executive suites, and professional office environments that were previously the reserve of roll-about systems (Polycom, 2002). Set-top videoconferencing systems, such as Sprint's turnkey IP-based service (Rendleman, 2001) or Polycom's ViaVideo, are becoming increasingly popular because they allow organizations to leverage existing assets. Videoconferencing becomes just another application running on the user's computer desktop. Fully-equipped desktop videoconferencing systems are available in the $599 to $5,000 per unit range, depending on options (Gold, 2004).

The videophone concept popularized in the 1968 film *2001: A Space Odyssey* has never really taken off, but that has not prevented people from trying to make it work. Sorensen Media is the latest company to enter the market with a consumer-based broadband videophone solution being made available to the OEM (original equipment manufacturer) market (Sorensen, 2004). According to the Sorensen Website, "with a broadband Internet connection, [standard] television, a remote control and the Sorensen VP-100 videophone device, users [will be] able to make and receive video calls" quickly and easily (Sorensen, 2004). Whereas, no OEM partners have been identified as of mid-2004, Sorensen is confident that with the device's H.323 (Session Initiation Protocol or SIP) compliance, G.723.1 and G.711 audio support, and standard television video display support, the videophone will find a niche market by the third quarter of 2004 when units are expected to become commercially available (Sorensen, 2004).

On the other hand, slow-scan or freeze-frame videophones, once popular in the 1980s for more impulsive, one-on-one communications between sites similarly equipped, have all but disappeared from the business videoconferencing radar screen. Slow-scan videophone units include a small screen (typically black and white), a built-in camera, video codec, audio system, and a keypad. The handset lets the unit work as an ordinary telephone as well as a videoconferencing system. Mitsubishi's VisiTel and Luna Picturephone systems are typical examples of consumer-grade freeze-frame videophones. By using special lenses, participants can send each other close-up images of themselves, photographs, graphics, or even drawings. The transmission takes about five seconds; during this time, neither party can talk as the device requires the full telephone bandwidth to send the picture. Commercial-quality slow-scan devices, such as the transceivers from Colorado Video, greatly increase the size and improve the resolution of the image, while allowing for two-way voice conversation to continue while the image is being sent (Stowe, 1992).

Videoconferencing between more than two locations requires a multipoint control unit (MCU). An MCU is a switch that acts as a video "bridge" connecting the signals among all locations and enabling participants to see one another, converse, and/or view the same graphics (Muller, 1998). The MCU also provides the means to control the videoconference in terms of who is seeing what at any given time. Most MCUs include voice-activated switching, presentation or lecture mode, and moderator control (Muller, 1998). Likewise, because of the large amount of information contained in an uncompressed full-motion video signal (about 90 million bits per second), two-way videoconferencing that even remotely approaches broadcast television quality would require incredible bandwidth at equally great cost. Therefore, all but the most expensive full-motion videoconferencing options utilize video compression/decompression technology to take advantage of the fact that not all of the 90 million bits of a video signal are really necessary to reconstruct a "watchable" image. In fact, the majority of the information in a typical videoconferencing image is redundant: most of the image remains exactly the same except for the speaker's head movements and occasional gestures.

Video compression techniques take advantage of this and other factors to greatly reduce the number of bits that must be transmitted, and thus reduce the bandwidth requirements (Stowe, 1992). A popular transmission rate for higher-end videoconferencing technology is about 1.5 Mb/s (Megabits per second). This is only about one-sixtieth of the original information in the video image, but it produces acceptable pictures for most purposes. More important, this rate corresponds to the telephone T1 rate so that, with the proper equipment, such videoconferences can be transmitted by telephone circuits in and among most major cities (Stowe, 1992). Other videoconferencing systems have been developed that run at varying rates up to 384 Kb/s (kilobits per second), while smaller roll-about and/or desktop systems typically run at 128 Kb/s. However, at lower data rates, some design tradeoffs are made in the relationship between quality per frame and frames per second. "Some vendors have engineered their products to maintain a constant frame rate by sacrificing clarity when there is a high motion component to the image. This compromise presents a problem, since there are no established units of measurement for how clear an image appears, or whether the video is smooth or choppy" (Finger, 1998).

Computer-Mediated

Murray Turoff developed one of the first computer conferencing systems in 1971 for the Office of Emergency Preparedness (EMISARI, a management information system) to deal with the wage-price freeze (Lucky, 1991). Since then, the capacity of two or more personal computers to interconnect, send, receive, store, and display digitized imagery has expanded rapidly—thanks in large part to rapid developments in computer hardware and the spread of the Internet. At the most basic level, computer conferencing is the written form of a conference call. Participants in a computer conference could communicate via a simple conferencing application such as Internet Relay Chat (IRC) or AOL's Instant Messenger, both of which provide a simple text-based chat function.

Another form of computer conferencing takes advantage of the ability to "time-shift" a presentation via Webcasting. In a Webcast, the originator tries to anticipate the viewer's questions and concerns during the recording of a presentation, then makes the finished product available to anyone who wants to download it at a more convenient time—sort of an asynchronous broadcast on the Web. To make a computer-based meeting more interactive, however, most users want to do either real-time audio or video (or both), while sharing documents or a common workspace. Enter Web conferencing, perhaps one of the fastest growing sectors in teleconferencing with projected cumulative average growth rates of 35% over the next seven years and a market expected to reach $2 billion by 2008 (O'Keefe, 2002). Web conferencing allows participants to use either software or a service to show presentations or work on the same program or application in real time simultaneously, all while being linked on a shared telephone line (Lafferty, 2002). Not only is Web conferencing affordable—running on a typical office computer with little or no added equipment beside a Webcam—it also "...offers considerable functional advantages over typical on-site meetings with features such as interactive Q-and-A sessions, real-time collaboration, and the ability to digitally record and archive the event for playback from a company Web site" (O'Keefe, 2002).

There is no denying that Web conferencing is an area of high interest to conferencing users. The technology has seen explosive growth because it faces fewer technological barriers than videoconferencing, has better "presence" than audioconferencing, and delivers a unified conferencing platform that integrates voice and video bridging capabilities with the ability to share and view data collaboratively (Heyworth, 2004). Still, Curtis O'Keefe, president and co-founder of Communiqué Conferencing, cautions that "Web conferencing technology is not a panacea and is not meant to replace all business travel. However, it can certainly supplement travel in many cases" (O'Keefe, 2002).

Obviously, choosing the right teleconferencing technology for the right purpose is the subject of great concern for many business executives. Social psychologists have pointed out that "in terms of the immediacy that they can afford, media can be ordered from the most immediate to the least: face-to-face, [videoconferencing], picturephone, telephone, [below this, synchronous and asynchronous computerized conferencing] … the choice of media in regard to intimacy should be related to the nature of the task, with the least immediate or intimate mode preferable for unpleasant tasks" (Hiltz & Turoff, 1993, p. 118). This would suggest that, for the less intimate task, the most immediate medium (face-to-face) would lead to favorable outcomes. In this same vein, a slightly more intimate task would require a medium of intermediate immediacy (videoconferencing), while those tasks that are highly intimate, perhaps embarrassing, personal, or charged with potential conflict, would benefit from using a medium of lowest immediacy (audioconferencing or even computer conferencing).

Interestingly, though, Hiltz and Turoff (1993)—reporting the results of a 1977 study—indicated that "numerous carefully conducted experiments … found all vocal media to be very similar in effectiveness [including face-to-face, audio, and video]. However, … people *perceived* audio to be less satisfactory than video," and both to be less satisfactory than face-to-face (emphasis added, p. 121). Little recent experimentation has been done in this area, and it appears that there is a great deal of room for further research.

Recent Developments

The combination of the September 11 terror attacks and the 2001 economic downturn markedly curtailed business travel. However, according to Carol Devine, NBTA President and CEO, whereas "[t]he economy and business travel are closely related … the challenges of the past few years have taught us to be cautious, [even as] economic trends and travel purchase indicators bode well for an increase in business travel—with the possibility of significant volume increases at the end of the year and into 2005" (NBTA, 2004). Opinions vary, but past circumstances have led many corporate managers to take a long, hard look at teleconferencing and collaboration tools (Krapf, 2002).

In the aftermath of September 11, stocks for companies in the teleconferencing and Web businesses fared better than most. Stock prices for companies such as Polycom, ACT Teleconferencing, and WebEx Communications surged as analysts noted many companies were turning to Web, voice, and video communications to conduct conferences instead of traveling to meetings. Indeed, *Successful Meetings* observed that the average number of face-to-face meetings that planners handle dropped significantly since 2000, from 47 per year to just 20 (State of the industry, 2003), primarily due to the increasing use of teleconferencing solutions.

Additionally, businesses that provided teleconferencing services saw their product usage increase by as much as 50% (Schaffler & Wolfe, 2001), as a newly reinforced aversion to business travel, along with advances in teleconferencing technologies and services, began to bring teleconferencing capabilities to the masses at reasonable costs (Lafferty 2002). Leading vendors say the teleconferencing industry was growing steadily at about 30% prior to the events of September 11. Since then, the industry's growth has reached 40%, and all indications point to a steady rise (Vinas, 2002). Jay

Williams, senior vice president and chief technology officer for Concours Group global consultants in Kingwood (Texas), believes that "the sole focus of collaborative technologies [today] is not to replicate the live experience of a meeting; that's been a fallacy that we are going to be able to recreate our casual meeting environment online. We are going to create something new, and we are going to create something that works" (Vinas, 2002, p. 30). Of course, as David Woolley, President of Thinkofit, an online communication consulting firm, points out, "[t]his can only happen if everyone adopts the same tool, or if the various tools become sufficiently standardized that they can talk to each other, allowing each participant in a conference to use whatever tool and user interface they're most comfortable with" (Good, 2003b).

Many business people still maintain that face-to-face meetings will always offer the best level of interactivity. But they are also beefing up their Web sites for actual end users "...using animation for demonstrations of products as well as using streaming video for testimonials" (Vinas, 2002, p. 32). The key point in the use of teleconferencing is ease of use and ease of access. Daniel Shefer of InterWise believes that, in order for conferencing and collaboration tools (Web conferencing specifically) to gain widespread adoption, they will have to become as transparent and easy to use as the telephone—perhaps even integrating with the telephone into one suite (Good, 2003a).

The other obstacles to ubiquitous adoption are cultural—many people are uncomfortable speaking to a camera and very uncomfortable knowing they are being recorded. Paul Saffo, a director of the Institute for the Future in Menlo Park (California), believes this is normal. "It takes about 30 years for new technolog[ies] to be absorbed into the mainstream. Products for videoconferencing [and by extension, Web conferencing] ... invented 15 to 20 years ago, are just now beginning to come into their own" (Weiland, 1996, p. 63). By Saffo's standard, U.S. society is probably entering its second stage of acceptance of teleconferencing technology—perhaps accelerated by the events of September 11—and, while this transformation is underway, the industry may seem in a constant state of flux and somewhat confusing. The most confusing, yet promising of the newest teleconferencing developments are not in the hardware or software, but in the international standards that make it all work.

Standards

Certain international standards for teleconferencing have been worked out by the International Telecommunications Union (ITU). These standards have set the foundation for network transmission technologies and vendor interoperability in the teleconferencing industry, especially videoconferencing. Prior to the establishment of these standards, vendors employed a range of proprietary algorithms and packaged hardware and software so that systems from different vendors could not communicate with each other. By establishing worldwide teleconferencing standards, the ITU helps ensure that teleconferencing technologies can "talk" to each other regardless of brand. Likewise, as new innovations are developed, the new systems remain backward-compatible with existing installed systems. The most important standards for the implementation and adoption of teleconferencing are under the umbrella standard of H.320 that defines the operating modes and transmission speeds for videoconferencing system codecs, including the procedures for call setup, call teardown, and conference control (Muller, 1998). All videoconferencing codecs and MCUs that comply with H.320 are interoperable with those of different manufacturers. Other important standards for the implementation and adoption of teleconferencing are:

- *H.322*—LAN-based videoconferencing with guaranteed bandwidth (ITU, 1996).

- *H.323*—LAN-based videoconferencing with non-guaranteed bandwidth such as LAN and WAN networks using packet switched Internet protocol (IP) (ITU, 2000). This standard supports the move by many desktop teleconferencing systems to sophisticated bridging systems and IP multicasting software that reduces desktop clutter and frees up valuable bandwidth on the network. Microsoft's NetMeeting, Polycom's WebOffice, and one of the biggest names in Web conferencing, WebEx, all utilize H.323 IP videoconferencing specifications to deliver their services to the desktop. Likewise, many popular Web browsers have H.323-compliant videoconferencing capabilities embedded into their latest versions (Lafferty, 2002).

- *H.324*—Developed to facilitate low bit-rate desktop teleconferencing via standard telephone lines (ITU, 1998). Many computer manufacturers are now including teleconferencing capabilities as part of their standard personal computer package to take advantage of the growing home teleconferencing and videophone markets brought about by the growth in telecommuting.

It is important to note that the ITU audio coding recommendations for teleconferencing for all H.3xx videoconferencing recommendations (except H.324) mandate the G.711 recommendation for audio. G.711 is the oldest compression algorithm and codes toll-quality (3 KHz analog bandwidth) audio into 48 Kb/s, 56 Kb/s, or 64 Kb/s. Mandatory for H.324, but optional for all other H.3xx recommendations, is G723.1 which codes toll-quality audio into 5.3 Kb/s or 6.3 Kb/s (Paul, 2000). In addition to these important video and audio recommendations, the ITU has also generated recommendations for multiplexing, control, multipoint services, security, and communications interface (see http://www.itu.int).

According to David Brown of Worldcom, "[i]t has only been [within] the last 10 years that the standards have settled in" now that almost all conferencing equipment sold today works over the H.320 standard (Perera, 2001). The next generation of videoconferencing technology, the IP-based H.323 standard for conferencing and collaboration via LANs or the Internet, is catching on in a big way leading Brown to predict that, "[i]n two or three years, you'll just dial an IP address instead of a phone number to start a videoconference" (Perera, 2001). Indeed, H.323 is widely regarded as the eventual standard for real-time communications. "It is cheaper to provide, it is nearly ubiquitous in most developed areas, and the technology is flexible and can support different types of applications in a converged, integrated environment, delivering new levels of functionality and manageability" (Heyworth, 2004). As these standards become more widely implemented, and as more bandwidth is added to the Internet, the quality of Internet videoconferencing will likely improve dramatically.

Current Status

It is ironic that some of the simplest ideas can be put into practice only when a very complex level of development has been reached in a related field. As technologies have expanded, morphed, and matured, teleconferencing options have transformed and expanded as well. As discussed earlier, teleconferencing is no longer a simple conference call, nor is it exclusively the domain of complex two-way video hook-ups. Roopam Jain, a strategic analysis for Frost & Sullivan, noted that videoconferencing has reached a new level of accessibility despite still having issues with quality, reliability, and overall cost (Lafferty, 2002). Still, Jain is optimistic, "videoconferencing systems worldwide are expected to grow from $819.9 million in 2001, to $1.55 billion by 2005. At the same time, ... videoconferencing services in the United States are expected to grow from $1.68 billion in 2001 to $2.54 billion by 2005" (Lafferty, 2002).

As the cost of codecs and other end-user equipment has fallen, making videoconferencing much more attractive than ever before, network access has grown to represent by far the largest portion of operational costs. The greatest expense for most companies using today's videoconferencing systems "is the transport rate or cost-per-minute, not monthly access, installation, or one-time initial equipment investment. This cost-per-minute varies based on carrier and on the type of terminating and originating access" (Earon, 1998). Prices are, however, often lower than commonly perceived: "So-called full-bandwidth, ISDN-based calls [384 Kb/s] cost between $0.50 to $1.20 per minute, and IP-based calls can be virtually free, depending on the type and design of the corporate network and application of the video call" (Wisehart, 2002, p. 8). In their editorial for January 2004, *Electronic Telespan* predicts that, as service bureaus introduce "lite" Web conferencing (Web conferencing products purchased as part of voice conferencing services clients buy from conference call providers), the prices will drop to $0.05 a minute above voice and toll—ultimately, by 2006, the price will fall to $0.01 or less a minute (I can see, 2004).

Because of such expectations in the market, there is a gradual shift toward IP networks. "Devices including ISDN [Integrated Services Digital Network] switches and H.323/H.324 gateways provide dial plan support, allow users to take advantage of competitive tariffs from multiple carriers, and enable users to access the least expensive type of bandwidth as is needed" (Earon, 1998). According to Jain, IP will ultimately bring cost efficiencies and network reliability that ISDN has not been able to deliver and will eventually bring videoconferencing into the mainstream (Lafferty, 2002). In the business environment, IP communications has already taken a strong hold in regions such as North America and Asia Pacific (Heyworth, 2004). Anticipating the coming broadband/IP wave, Polycom, one of the most recognized names in videoconferencing, has begun to equip all of their ViewStations with Ethernet jacks and has even brought out their own line of IP-only conferencing systems (Krapf, 2002). To date, Polycom claims that approximately 75% of customer inquiries for videoconferencing products are related to IP (Heyworth, 2004).

Paul Berberian, co-founder, president, and CEO of audio and data conferencing provider Raindance Communications, notes that, "for every boardroom-type videoconference, there are thousands of audio conferences" (Borthick, 2002). Audioconferencing over IP networks using voice over IP (VoIP) has also garnered some attention lately as Internet telephony software makes it possible for users to engage in long-distance conversations between virtually any location in the world without regard for per-minute usage charges. In most cases, all that is needed is an Internet-connected computer equipped with telephony software, a sound card, microphone, and speakers or a headset (Muller, 1998). However, because the voice of each party is compressed and packetized, there may be significant delays that result in noticeable gaps in a person's speech. In the future, some industry analysts contend VoIP's use of SIP signaling protocols could provide increased reliability and enable new audioconferencing features such as breakout sessions, and sub-conferencing sessions without losing connectivity (Borthick, 2002).

Lewis Ward, a senior analyst with Collaborative Strategies, LLC, a San Francisco-based management consulting firm, predicts that, by 2005, more than 79 million people and more than 94,000 companies and organizations will be using Web-based conferencing systems worldwide (Ward, 2002). This is good news for companies such as WebEx, which describes itself as "an interactive communications infrastructure provider" (WebEx, 2002), with a global network consisting of 10 data centers around the world and privately leased bandwidth (Lafferty, 2002). On top of this network rides a software platform that allows people to connect computer-to-computer to share files and applications in real time (Figure 24.2). According to Subrah Iyar, CEO of WebEx Communications, the company enjoyed dramatic growth every quarter during its first three years of operation, resulting in an over 300% increase in revenue and usage rising by nearly 50% (Schaffler & Wolfe, 2001). WebEx now claims to be the largest provider of Web conferencing services in Europe, Asia, and North America and is viewed as the primary competitive service to Polycom's recently rolled-out WebOffice application (Krapf, 2002).

Figure 24.2
WebEx Screen Shots

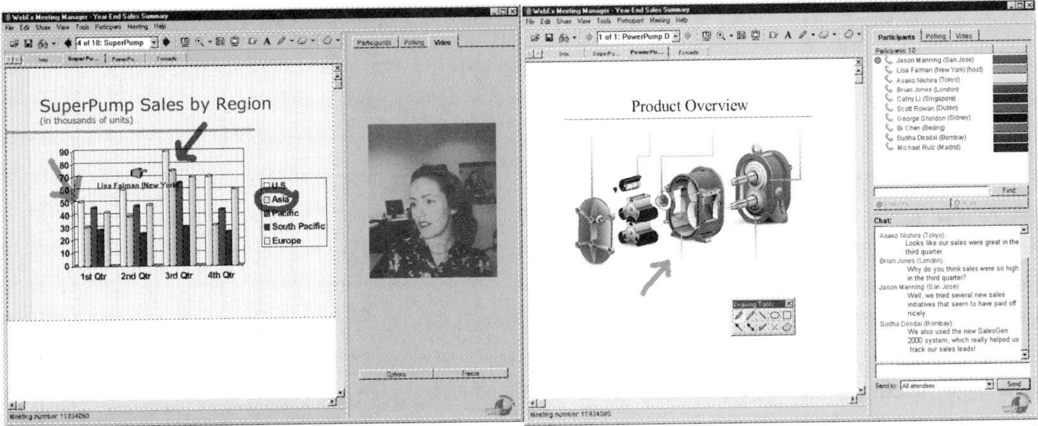

Source: WebEx

Other companies have also joined the Web conferencing bandwagon. Raindance, a Colorado-based Web conferencing service, offers "reservationless" conferencing through its Web Conferencing Pro service (Lafferty, 2002). Recently, Raindance introduced their new "Meeting Edition" software that offers to integrate Web, audio, and desktop videoconferencing (via Web cameras) all from one user interface (Raindance announces, 2004). Raindance's Meeting Edition software also has a feature that uses an automated attendant to call all meeting participants, eliminating dial-in, ID and PIN number hassles, and gives the speaker a private window in the interface to see their upcoming slides and notes while keeping the content hidden from the audience (Raindance 2004).

Microsoft entered the nascent Web conferencing market with the roll-out of NetMeeting 3.0 software back in 1999 as part of the Windows Operating System. According to some product reviewers, NetMeeting 3.0 was (and still is) a great Web conferencing application—it is free, and it offers Windows users H.323 compatibility, whiteboard function, text chats, audio, video, and application and file sharing (Microsoft NetMeeting, 1999). Unfortunately for Microsoft, in December 2003, after a U.S. District Court in Richmond (Virginia) found Microsoft had violated patents with its NetMeeting software, Microsoft agreed to pay SPX and its Imagexpo unit $60 million for the right to use the whiteboard technology covered under patent #5,206,934 (Festa, 2003). Whereas NetMeeting is still available as a free download, effort has more recently been put into promoting Microsoft Office Live Meeting. Live Meeting is Microsoft's new Web conferencing service that can be integrated in a corporate Intranet environment with extensive branding capabilities or facilitate interactive meetings via the Internet in real time at a moment's notice with everyone participating from their desktops (Microsoft, 2004).

A relative newcomer to the Web conferencing market, Macromedia's latest release of Breeze offers a number of advantages that competitive products, such as WebEx, Raindance, and Genysys Conferencing, do not: rapid time to deployment, Macromedia's Flash technology, and "a large and growing base of millions of designers and developers that are fanatically loyal to Macromedia," according to Yankee Group analyst Paul Ritter (Montalbano, 2004). Macromedia Breeze was originally designed to allow Flash-based rich media content to be created from PowerPoint presentations for delivery over the Web. As a software suite, Macromedia's Breeze includes modules for building online presentations, creating and managing Web-delivered courseware, and adding live discussions among participants (Macromedia, 2004). Some of the features of Breeze include a whiteboard, polling, and application and desktop sharing. The whiteboard feature can even be used on top of video. Presentations can be both live and on-demand, with the on-demand version indexed and searchable (Meserve, 2004).

Factors to Watch

As the teleconferencing and real-time, integrated collaboration industries go through fast—and at times, monumental—changes, the ability to understand the macro process taking place, as well as the future direction(s) they will follow, is difficult to assess. "The evolution of collaboration systems and the ability of Web conferencing development companies to truly understand and integrate an effective approach to collaboration … has to be fundamentally different from what [is currently being offered now]" (Good, 2004a). Over the past few years, a lot of consolidation has occurred in the multimedia conferencing market. "The announced integration between Yahoo IM and WebEx and the similar integration between LiveMeeting and MS Messenger" is just the start (Good, 2004a).

Growing industry interest in "rich media communications and real-time conferencing" has been borne out by Collaborative Strategies senior analyst Lewis Ward's research, showing that the fastest-growing type of teleconferences are those in which conferees can talk with each other while viewing the same Web-based data (Borthick, 2002; Ward, 2002). Ward predicts that, by 2005, "the ongoing globalization and decentralization of corporations, coupled with the wider availability of easy-to-use conferencing and integration tools will keep the three elements of the real-time collaboration market—audio, video, and data/Web conferencing—growing at a healthy rate, both separately and in combination" (Ward, 2002, p. 53). The results of a recent survey of *VideoSystems* subscribers conducted by Primedia Business magazines and Media, Inc. indicate that, by 2005, most respondents expect 42% of their work to be done via Internet/streaming technologies (How will, 2002). Virtual communication and collaboration will arguably be the lifeblood of business in the 21st century.

Still, industry keynoters rarely leave full-motion video off their list of future "killer" applications. Primarily, this is because many users of today's videoconferencing technology frequently state that, "While the experience wasn't as good as meeting in person, … the technology was so strong [they] could see gestures, facial expressions, everything" (Goodridge, 2001). Others contend that videoconferencing is better for conducting meetings that need to produce decisions, whereas Web conferencing is the preferred tool for a working session or an informal presentation to a group of people (Porter, 2002). Videoconferencing would thus seem to be a better substitute for face-to-face meetings than audio-only conferencing or those mediated by computer-based technology. However, many argue that videoconferences have a different "flavor" than face-to-face meetings; they are typically more formal, there is little opportunity for side conversations, and the technology itself is fairly intrusive (Lucky, 1991). Furthermore, videoconference participants have difficulty making eye contact due to camera placement limitations and often complain of a lack of shared workspace for collaborative brainstorming (Ditlea, 2000).

Lately, the most attractive thing about videoconferencing is its high-bandwidth requirement—a good use for overbuilt, underutilized fiber networks and a reason to believe in broadband access (Borthick, 2002). Although it does not seem likely that face-to-face meetings will be replaced by electronic togetherness, in the near future, new kinds of computer-mediated interactions among people are nevertheless likely to be facilitated where participants can share an aural as well as visual virtual meeting space (Lucky, 1991).

What is the next step in the pursuit of a more realistic virtual meeting? Some would say the answer is "tele-immersion." As one of the principal applications anticipated for Internet 2—tomorrow's much faster, next-generation Internet—"tele-immersion visually replicates, in real time and in three dimensions, slabs of space surrounding remote participants in a cybermeeting." The result is a "shared, simulated environment that makes it appear as if everyone is in the same room" (Ditlea, 2000, p. 28). Demonstrated for the first time in May 2000, researchers at Advanced Network and Services in Armonk (New York), the University of Pennsylvania in Philadelphia, and the University of North Carolina at Chapel Hill tested a three-dimensional virtual meeting room where participants could see and interact with their life-sized colleagues viewed through "windows," when each participant was actually physically located hundreds of miles away from the others (Ditlea, 2000).

Although described as being somewhat crude, requiring users to wear awkward goggles and head tracking devices (not to mention requiring nine separate video cameras per participating location), the test provided vindication for two years of collaborative research between the participants and afforded a brief glimpse of what might lie beyond today's videoconference room. Jaron Lanier, chief scientist for the project and one of the primary scientists who helped invent and popularize virtual reality (VR) in the 1980s and 1990s, believes the test demonstrated viewpoint-independent, real-time scene sensing, and reconstruction (Ditlea, 2000). Despite suffering from video glitches, Lanier states that the ultimate tele-immersion experience is meant to be seamless (Ditlea, 2000). As discussed in Chapter 15, scientists continue to work on autostereo screens (for three-dimensional views without bulky headgear) and advances in haptics (for tactile simulations).

Smart VR has taken teleconferencing one step further with the introduction of SmartVerse, a low-bandwidth, many-to-many conferencing system using VoIP technology (Smart VR, 2004). SmartVerse features several technology components to make communications inside 3D space effective including voice conferencing, text chat, a shared whiteboard, and user avatars. Voice sound streams are 3D spatialized and appear to come from the speaker's avatar, which can be commanded by users to perform a range of preset motion sequences such as waving or bowing among other emotional expressions, including keeping eye contact (Smart VR, 2004). The stated goal is to make interaction in 3D space come as close to real-life interaction as possible.

Another attempt at bridging the gap between videoconferencing and the "holodeck" of *StarTrek: The Next Generation* is the Virtual Conference Room from TeleSuite that strives to create a videoconferencing environment that is more true to life (Regenold, 2003). TeleSuite attempts to accomplish this with a 16 × 4-foot display that fills the field of view of the human eye and accommodates up to 20 conference participants, thus creating a virtual meeting space that is closer to the real thing. The Virtual Conference Room package includes a panoramic screen, a wide-angle camera, projectors, lighting and audio components, and the necessary videoconferencing and network software all for about $80,000 per room (Regenold, 2003). As broadband access continues to increase, fully-immersive virtual conferencing may become a reality, but today, it is still awkward and "kludgy" at best. Still, if we go back just three or four years, who would have imagined that virtual events—real-time integrated collaboration via audio, video, and Web conferencing—would be a business reality today?

Bibliography

Bamnet, S. (2001, December). Rethinking telecom architectures. *Business Communications Review*. Retrieved March 24, 2002 from http://www.bcr.com/bcrmag/2001/12/p16.asp.

Big boys dominate videoconferencing. (2002, April). *VideoSystems, 28* (4), 12.

Borthick, S. (2002, March). Video: Nice but not necessary? *Business Communications Review*. Retrieved March 24, 2002 from http://www.bcr.com/bcrmag/2002/03/p10.asp.

Chabrow, E. (2001, November 5). Technology brings far-flung colleagues together. *InformationWeek.com*. Retrieved March 21, 2002 from http://www.informationweek.com/story/IWK20011102S0014.

Davis, A. (1999, June). *Integrated collaboration: Driving business efficiency into the next millennium*. Tempe, AZ: Forward Concepts.

Ditlea, S. (2000, September/October). Meeting the future: Tele-immersion makes virtual conferencing more real. *Technology Review, 103* (5), 28.

Dukart, J. (2002, January). The broadband age cometh. *Utility Business*, 27. Retrieved January 18, 2002 from http://industryclick.com/magazine.asp?magazineid=11&siteid=30 &releaseid=9786.

Earon, S. (1998, November). The economics of deploying videoconferencing. *Business Communications Review*. Retrieved March 24, 2002 from http://www.bcr.com/bcrmag/1998/11/p43.asp.

Festa, P. (2003, December 26). Microsoft settles in whiteboard patent dustup. *C/Net News.com*. Retrieved April 1, 2004 from http://zdnet.com.com/2100-1104-5133588.html.

Finger, R. (1998, June). Measuring quality in videoconferencing systems. *Business Communications Review*. Retrieved March 24, 2002 from http://www.bcr.com/bcrmag/1998/06/p51.asp.

Glenwright, T. (2003, January). Digital dialogue. *PM Network, 17* (1), 47, 49.

Gold, S. (Ed.) (2004). *Telespan's definitive buyer's guide to teleconferencing*. Retrieved April 12, 2004 from http://www.telespan.com/buyersguide/index.html.

Good, R. (2003a, September 23). *The future of Web conferencing: Good interviews Daniel Shefer*. Retrieved April 1, 2004 from http://www.masternewmedia.org/2003/09/23/the_future_of_web_ conferencing_good_interviews_daniel_shefer.htm.

Good, R. (2003b, October 24). *The future of Web conferencing: Good interviews David Woolley*. Retrieved April 1, 2004 from http://www.masternewmedia.org/2003/12/19/the_future_of_web_ conferencing _good_interviews_david_wooley.htm

Goodridge, E. (2001, October 22). Virtual meetings yield real results. *InformationWeek.com*. Retrieved March 21, 2002 from http://www.informationweek.com/story/IWK20011018S0082.

Hausman, C. (1991). *Institutional video: Planning, budgeting, production, and evaluation.* Belmont, CA: Wadsworth Publishing Company.

Heyworth, T. (2004, January). Businesses switch on to videoconferencing. *Telecommunications Online* [International Edition]. Retrieved April 1, 2004 from http://www.telecommagazine.com/default.asp?journalid=2&func=articles&page=0401i18&year=2004&month=1.

Hiltz, S., & Turoff, M. (1993). *The network nation: Human communication via computer.* Cambridge, MA: The MIT Press.

How will you display your work? (2002, April). *VideoSystems, 28* (4), 13.

I can see for miles, and miles, and miles: *Telespan's* predictions for 2004 (2004, January 19). *Electronic Telespan. 24* (2). Retrieved April 12, 2004 from http://www.telespan.com/editorial.html.

International Telecommunications Union. (1996, March). *Visual telephone systems and terminal equipment for local area networks which provide a guaranteed quality of service (H.322).* Retrieved April 14, 2002 from http://www.itu.int/rec/recommendation.asp? type=folders&lang=e&parent=T-REC-h.322.

International Telecommunications Union. (1998, February). *Terminal for low bit-rate multimedia communications (H.324).* Retrieved April 14, 2002 from http://www.itu.int/rec/recommendation.asp? type=folders&lang=e &parent=T-REC-h.324.

International Telecommunications Union. (2000, November). *Packet-based multimedia communications systems (H.323).* Retrieved April 14, 2002 from http://www.itu.int/rec/recommendation.asp? type=folders&lang=e &parent=T-REC-h.323.

Johansen, R., Vallee, J., & Spangler, K. (1979). *Electronic meetings: Technical alternatives and social choices.* Reading, MA: Addison-Wesley.

Kontzer, T. (2001, October 12). Video-to-desktop set to emerge as killer app. *InformationWeek.* Retrieved March 21, 2002 from http://www.informationweek.com/story/IWK20011012S0032.

Krapf, E. (2002, January). Web conferencing from Polycom. *Business Communications Review.* Retrieved March 24, 2002 from http://www.bcr.com/bcrmag/2002/01/p62.asp.

Kuehn, R. (2002, January). 2002: Year of the conundrum. *Business Communications Review.* Retrieved March 24, 2002 from http://www.bcr.com/ bcrmag/2002/01/p66.asp.

Lafferty, M. (2002, February). Convergence: Teleconferencing takes-off! *Communications Engineering & Design.* Retrieved February 8, 2002 from http://www.cedmagazine.com/ced/2002/0202/02a.htm.

Lucky, R. (1991). In a very short time. In D. Leebaert (Ed.), *Technology 2001: The future of computing and communications.* Cambridge, MA: The MIT Press.

Macromedia. (2004). *Corporate Website.* Retrieved April 1, 2004 from http://www.macromedia.com/software/breeze/.

McLuhan, R. (2002, January 17). Webcasts bolster access to events—Conferences are turning to technology to widen their reach. *Marketing,* 27.

Meserve, J. (2004, February). Making presentations a Breeze. *Network World.* Retrieved April 1, 2004 from http://napps.nwfusion.com/weblogs/multimedia/archives/004107.html.

Microsoft. (2004). *Corporate Web site.* Retrieved April 1, 2004 from http://www.microsoft.com/office/livemeeting/.

Microsoft NetMeeting 3.0 Preview. (1999, May). *Internet telephony.* Retrieved April 1, 2004 from http://www.tmcnet.com/articles/itmag/0599/0599labs1.htm.

Montalbano, E. (2004, February). Macromedia: It's a Breeze. *CRN.* Retrieved April 1, 2004 from http://crn.channelsupersearch.com/news/crn/47802.asp.

Muller, N. (1998). *Desktop encyclopedia of telecommunications.* New York: McGraw-Hill.

National Business Travel Association. (2004, January 30). Positive economic trends seen fueling business travel upturn. *Press release.* Retrieved March 20, 2004 from http://www.nbta.org/newsroom/ pr_1_30_04.cfm.

Noll, A. (1992). Anatomy of a failure: Picturephone revisited. *Telecommunications Policy,* 307-317.

O'Keefe, C. (2002, April 26). The potential of Web conferencing is virtually here. *Washington Business Journal.* Retrieved April 1, 2004 from http://washington.bizjournals.com/washington/stories/2002/04/29/focus6.html.

Paul, G. (2000, February). An overview of the ITU videoconferencing and collaboration standards. *The Edge Perspectives, 1* (1). Retrieved April 1, 2004 from http://www.mitre.org/news/edge_perspectives/february_00/paul.html.

Perera, R. (2001, October). Teleconferencing demand up since September 11. *ComputerWorld Hong Kong*. Retrieved April 23, 2004 from http://www.idg.com.hk/cw/printstory.asp?aid=20011030008.

Polycom. (2003). *Polycom guide to conferencing and collaboration*. Pleasanton, CA: Polycom.

Polycom. (2002). *Corporate Web site*. Retrieved April 22, 2002 from http://www.polycom.com/naindex.html.

Porter, S. (2002, April). The new conferencing: A meeting of minds … and media. *VideoSystems, 28* (4), 43-50.

Raindance. (2004). *Corporate Web site*. Retrieved March 3, 2004 from http://www.raindance.com.

Raindance announces Meeting Edition. (2004, March 3). *Presentations Industry Update*. Received via email distribution.

Regenold, S. (2003, May). Virtual reality trends. *Presentations.com*. Retrieved June 23, 2003 from http://www.presentations.com/presentations/trends/article_display.jsp?vnu_content_id=1911037.

Rendleman, J. (2001, June 25). Sprint offers IP-based videoconferencing. *InformationWeek, 843*, 91.

Rosen, E. (2001, April 16). Extreme videoconferencing. *InformationWeek*. Retrieved March 21, 2002 from http://www.informationweek.com/833/oovideo.htm.

Ryssdal, K. (Anchor), & Gray, C. (Reporter). (2001, October 15). Some executives using teleconferencing instead of flying across country for meetings [radio program]. *Marketplace Morning Report,* Minnesota Public Radio (6:50 A.M. ET), Syndicated.

Schaffler, R., & Wolfe, C. (2001, September 21). Teleconferencing, Web-based meetings on the rise. *CNNfn Market Call* (09:30 A.M. EST), Transcript #092108cb.105, Federal Document Clearing House, Inc.

Seaberry, J. (1999, December 6). Business travel continues to increase. *The Dallas Morning News*. Retrieved February 26, 2002 from http://www.dallasnews.com/.

Singleton, L. (1983). *Telecommunications in the information age*. Cambridge, MA: Ballinger.

Smart VR. (2004). *Corporate Web site*. Retrieved April 24, 2004 from http://www.smartvr.com.

Sorensen (2004). *Sorensen VP-100 videophone*. Retrieved April 24, 2004 from http://www.sorenson.com/solutions/videophone_more_info.php.

State of the industry report. (2003, January). *Successful Meetings*. Retrieved March 20, 2004 from http://www.successmtgs.com/successmtgs/reports_analysis/reports_archive.jsp.

Stowe, R. (1992). Teleconferencing. In A. Richardson (Ed.). *Corporate and organizational video*. New York: McGraw-Hill.

Swanson, S. (2001, September 27). Travel fears fuel Web conferencing. *InformationWeek.com*. Retrieved March 21, 2002 from http://www.informationweek.com/story/IWK20010927S0017.

Vinas, T. (2002, February). Meetings makeover. *Industry Week, 251* (2), 29-35.

Ward, L. (2002, March). The rise of rich media and real-time conferencing. *Business Communications Review*. Retrieved April 13, 2002 from http://www.bcr.com/bcrmag/2002/03/p53.asp.

WebEx. (2002). *Corporate Website*. Retrieved April 14, 2002 from http://www.webex.com.

Weiland, R. (1996). 2001: A meetings odyssey. In, R. Kling (Ed.). *Computerization and controversy: Value conflicts and social choices*, 2nd edition. San Diego, CA: Academic Press.

Wisehart, C. (2002, April). Can you see me now? *VideoSystems, 28* (4), 8.

25

Wireless Telephony

Philip J. Auter, Ph.D. & Tyrone Adams, Ph.D.*

If you walk around almost any urban or suburban area and observe carefully, you may notice two related trends—almost everyone from 12 to 75 has a cell phone and seems to be using it constantly. Meanwhile, payphones are disappearing (Bunkley, 2004).

U.S. society has become dependant upon cellular technology and personal communication services (PCS), primarily for instantaneous phone conversations, but also for instant messaging, gaming, taking and sending photos, and a host of other uses. Phones have evolved from large and clunky mobile car phones in the 1970s and 1980s to slim, sleek, multifunctional devices.

Background

The history of the cellular telephone begins with the history of mobile radio in the 1920s. The first land mobile systems were used by public safety agencies, primarily police departments. The earliest system was tested in Detroit beginning in 1921. World War II demonstrated the superiority of two-way FM transmission—only U.S. forces used significant numbers of FM battlefield systems—which proved easier to use and more difficult to jam than two-way AM systems. Surplus military radio equipment, for example the Motorola "Handie-Talkie," entered civilian life as taxi dispatch radios, particularly in New York (SRI, 1998).

---

* Phil Auter is an Assistant Professor, Department of Communication, University of Louisiana at Lafayette. Ty Adams is an Associate Professor and Director of Graduate Studies, University of Louisiana at Lafayette (Lafayette, Louisiana).

In 1947, experts marveled at the idea of a cellular phone, a new concept that could improve the coverage of existing mobile phones. Engineers realized that by using a small range of service areas (cells) with frequency reuse, they could increase the traffic capacity of mobile phones substantially (Bellis, 2004a). The technology of the day, however, was not as advanced as theory. The concept of a cellular telephone was still new, and the Federal Communications Commission (FCC) did not support the service at the time.

Lars Magnus Ericsson, founder of Ericsson in 1876, operated the first car telephone at the turn of the 20th century (Privateline, n.d.). The first major step in wireless communication came in 1912 with the passage of the *Radio Act of 1912* by the U.S. government. This law required stations and operators to apply for licenses in order to get on the air. Bell Laboratories is said to have invented the first version of a mobile, two-way, voice-based radio telephone in 1924 (Privateline, n.d.). On March 1, 1948, the first fully automatic radiotelephone service began operating in Richmond (Indiana), eliminating the requirement for a human operator to place calls. Some researchers claim the Swedish Telecommunications Administration's S. Lauhren designed the world's first automatic mobile telephone system, with a Stockholm trial in 1951 (Privateline, n.d.).

The late 1940s were important years for mobile radio. AT&T introduced the first commercial land mobile radio telephone system in St. Louis in 1946. However, the service was limited by a lack of communication channels (frequencies), and the system was cumbersome to use, with "push to talk" features and manual connections via human operator. Nonetheless, 25 U.S. cities had mobile service by year's end (SRI, n.d.).

Later that year, the FCC decided to make available separate radio frequencies for mobile calling. However, the commission only allowed 23 cell phone conversations in any given calling area (Cell phones, n.d.). During the 1950s, the FCC declined to allocate significant frequencies for mobile radio (cellular phones). This resulted in limited research and development in the field. However, scientists and engineers at Bell Labs continued limited investigation into the cellular concept and published several internal papers on the topic (SRI, n.d.).

Two decades passed with limited growth in the area of cellular telephony. Public radiophone testing began in 1977 with a trial run of service in Chicago. Bell signed up 2,000 customers for this initial run. This was soon followed by trial service in Washington, DC and Baltimore by American Radio Telephone Service, Inc. (ARTS) (a progenitor of CellularOne), in partnership with Motorola (Cell phones, n.d.; SRI, n.d.). In 1979, the first commercial cellular system was installed by NTT in Tokyo. From there, the concept of cellular technology took off. In 1983, both the Illinois Bell and ARTS trial services began regular commercial operation. Technology was, to say the least, both rudimentary and expensive. Motorola shipped the first commercial portable cellular telephone in 1984 with the suggested retail price of $3,000 to $4,000. Initial systems were big, bulky bag phones and equally large car phones. Competition first entered the arena in 1984 in Washington, DC. Services were popular with businesses and upper-income individuals. Basic analog cellular services grew tremendously in the 1980s. By 1988, cellular systems in Los Angeles and New York had become overloaded by high demand.

In 1990, there were an estimated five million cellular subscribers. By 1995, the number quintupled to over 25 million subscribers and, by 1997, it was estimated to be over 50 million subscribers. According to the Cellular Telecommunications & Internet Association (CTIA), a trade organization, the number of U.S. cellular subscribers was nearly 117 million by 2002 (Robinson, 2002). This number jumped to 155 million by the end of 2003 and has been projected to climb to over 245 million by 2007 (Matteo, 2004).

One of the reasons cell phones were adopted so quickly was because all mobile telephones are interconnected with the public switched telephone network (PSTN). Because of that connection, all wireless phone adopters can immediately reach or be reached by anyone with a phone. Figure 25.1 shows cellular telephone network architecture. Users make a call, and the cell phone tower relays the signal to the mobile telephone switching office (MTSO). If the call is going to another wireless customer, it is routed to the recipient's cell phone through the corresponding tower. If the call is to someone on a landline phone, the signal is routed from the MTSO to the PSTN through an access trunk.

Figure 25.1
Cellular Telephone Network Architecture

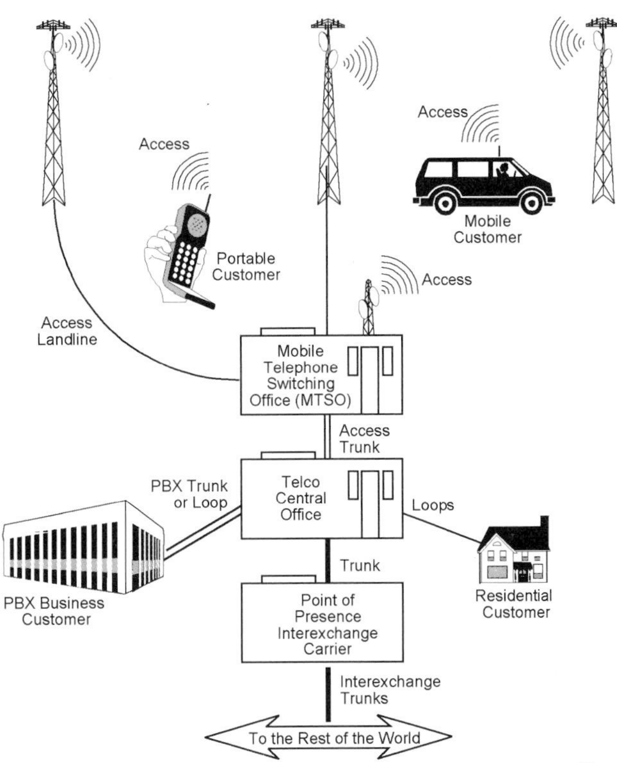

Source: Technology Futures, Inc.

Although cell phones are thought of as an added convenience by many people, they have also been labeled a nuisance by many others. Legislators in many states are trying to ban the use of cell phones while driving. For example, in 2001, New York banned talking on hand-held phones while driving (Robinson, 2002).

The cellular telecommunications industry is facing other battles, as well. The Food and Drug Administration (FDA) has issued warnings of health risks possibly linked to excessive use of cell phones, ranging from headaches to brain tumors. Testing is still being conducted, and no conclusive evidence of health risks from cellular telephone transmissions has been found.

April 3, 2003 marked the 30th anniversary of the first cellular phone call made by Martin Cooper, now chairman, chief executive officer, and co-founder of ArrayComm, Inc. He placed the call on April 3, 1973 when he was general manager of Motorola's Communications Systems Division. Who did he call? He called one of his rival's at AT&T's Bell Labs from the streets of New York City (Bellis, 2004b).

Recent Developments

Actions and events that took place since the beginning of 2003 will drastically affect the future development of cell phone service, technological advances, and costs to the consumer. On February 17, 2004, AT&T Wireless agreed to be acquired by Cingular, to the tune of $41 billion (Cingular agrees, 2004). The combined entity will have $32 billion per year revenue, subscribers in 49 states, and 97% of top market service. Other companies such as Sprint PCS are predicted to "ramp up" to handle the competition. A merger of Verizon and Sprint's wireless divisions could be a possible outcome (Shafer, 2004). Cingular's action happens in a period when telecom stocks are performing well on the market. Consumer groups believe that the deal will negatively impact customers—some say mergers create poorer service; others say less competition equals higher prices. Industry insiders, of course, say otherwise.

Interestingly, business deals in early 2004 have taken on an international flavor. In a reaction to the (at the time) proposed AT&T/Cingular merger, Japan's largest cellular outfit, NTT DoCoMo, made overtures to negotiate with Cingular (they had been a sizable stakeholder in AT&T since 2000) (Williams, 2004).

Another important development is mobile number portability (MNP). This hot-button issue garnered much controversy and discussion in the press from late 2003 through mid-2004. MNP was mandated in the United States beginning on November 24, 2003. The decision was quite expensive and resulted in over 7,000 complaints as of April 2004 (Kimball, 2004). The FCC ruling stipulates that companies in various areas must allow consumers to retain their phone number if they decide to switch from hard to wireless lines (or vice versa), and that they can also take their numbers from one wireless carrier to another (CTIA, n.d.). Attorneys for Verizon and a trade association tried to halt the measure, saying the FCC was overstepping its bounds (Court won't halt, 2003). Verizon eventually bowed out of the fight, which may have been a factor in pushing the decision through.

The decision produced no immediate rush of customer switching, as was predicted. Just over one million took advantage. As one observer said, only "those in pain" ported. Early adopters experienced technical difficulties. The supposed three-hour switching procedure often was slow, or did not work properly. AT&T Wireless lost the most subscribers to competitors.

In the months surrounding the portability decision, predictions surfaced as to the long-term effects on both price and service. In a thorough listing of the possible effects of portability, San Diego wireless consulting firm inCode warned that telemarketers could easily infiltrate switched cell numbers because of the way the portable technology works (inCode releases, 2003).

Cellular providers are also implementing a litany of technical advances. Wireless data transmission may be the most important. Transmission of data (anything other than voice) constitutes only 3% of cellular industry revenue as of mid-2004 (of an annual $81 billion) (Trewyn, 2004). Data includes photos, motion graphics, text, games, and the World Wide Web.

New developments are also occurring on the hardware side. A significant problem with wireless phones is battery life. Direct-methanol fuel cells are a point of discussion among technologists (Ashok, 2003). At present, the technology boasts three times the life of conventional power supplies (PolyFuel delivers, 2004). In a minor but interesting note, Celldar—a passive technology that is part cellular and part radar—is emerging, and is discussed mainly in terms of military applications (Port, 2003). In development since 1997, the technology could be used to track how signals from cell phone base stations interact with objects such as cars, trucks, or planes. Minor but interesting hardware entries include a Bluetooth wireless headset that connects Bluetooth-enabled cell phones to a wireless hands-free environment (Cardo launches, n.d.).

Current Status

Despite the economic slowdown after the "dot com crash," cell phone use continues to grow, particularly among young adults. Competition and legislation are driving the cost of service down, even as technological advancements increase the availability of a wide variety of enhanced services such as color Web browsing, e-mail, and file transfer. Cell phone penetration in the United States continues to grow, even though it lags behind Europe and Japan. The most recent report by Scarborough Research shows that national average penetration in 2003 was 66%, a growth rate of over 30% since 1999. Many larger cities such as Houston, Miami, and Atlanta had broken 70% penetration (McFarland & Mongrain, 2003). From culture to culture, it seems that cell phone "have-nots" tend to be older, more likely female, and with lower education and household income than "haves" (Leung & Wei, 1999).

Time spent using cell phones appears to be on the rise, but research has pointed out the disparity between self-report measures of cell phone use and real-time observance of usage levels (Cohen & Lemish, 2003). Although there were predictions of a small (9%) adoption rate among adults overall, the growth trend seems to be in the adolescent (12 to 17) and younger adult (18 to 24) segments of the market (Charny, 2002; McFarland & Mongrain, 2002; McVicker, 2001; Wrolstad, 2002). The youth market—currently with a penetration rate of only 25%, but growing fast—is considered a "gold mine" of opportunities to increase flagging industry profits (Charny, 2002; McVicker, 2002).

U.S. mobile phone subscribers predominantly fall into one of two categories: 800 MHz cellular and 1900 MHz PCS. The primary digital technology in the United States is CDMA (code division multiple access), although GSM (global system for mobile communication) is popular in other regions of the world and gaining here. AT&T Wireless uses GSM. Worldwide, both technologies are replacing analog. This is happening a bit more slowly in the United States where a large analog cell phone base arose before the development of digital cellular technologies (Robbins & Turner, 2002). While the trend toward digital service continues to rise, there is a significant analog infrastructure that is both costly to replace and, interestingly, utilizes a system that requires fewer towers. While providers scramble to upgrade their networks, companies such as Analog Devices work to reduce infrastructure costs while improving signal service (Analog Devices launches, 2004).

Enhanced 911 services are being implemented that provide emergency services with the caller's location. This technology uses a variant of the global positioning system (GPS) to identify a caller's location, which is a significant advance from the previous system that could only give the location of the tower closest to the person making a cellular phone call. One concern is that the service essentially makes cell phone homing devices that could allow the government and others to track someone's every move (Robbins & Turner, 2002).

Even as mobile phone penetration has increased in the United States, user demographics have changed. Women now represent the majority of U.S. cell phone users. The mean age and income level of cell phone users has also declined over the years (McFarland, 2002; Robbins & Turner, 2002), contrary to research done earlier in the adoption cycle of the cell phone and in other cultures (e.g., Leung & Wei, 1999). The primary reason for use appears to have shifted from business to personal communication (McFarland, 2002; Robbins & Turner, 2002). Growth and penetration trends seem to lean toward younger potential customers. People most likely to fail to adopt cell phones now or in the future tend to be much older, have a lower education, and no children (Leung & Wei, 1999; Robbins & Turner, 2002). These trends become more pronounced when looking at potential users of advanced cell phone technologies—such as text messaging and Internet access—that go beyond the confines of conventional voice communication (Robbins & Turner, 2002).

Prepaid phone cards have made it even easier for adolescents and young adults to obtain cell phone service despite limited funding and essentially no line of credit. The purchase of these cards—often the responsibility of the adolescent, not his or her parent—has increased the accessibility of cell phones to this demographic, while alleviating the concerns of parents that their children will use excessive amount of minutes, resulting in an extremely high bill at the end of the month (Ling & Yttri, 2002). Limited adolescent finances appear to have also resulted in their making strategic communication choices such as waiting until "nighttime" minutes (often unlimited) start or sending text messages that do not count against daytime minutes.

Although usage overall is certainly up, research is showing that much use is concentrated on interpersonal voice communication. A recent study of college-aged cell phone users shows that predominant usage patterns centered around features such as caller ID, voice messaging, call screening, and voice communication. These early adopters of new technologies and features had little interest in games, cameras, and other high-tech features (Auter, 2004).

Factors to Watch

Cell phone penetration in the United States should continue to increase dramatically as companies target younger and younger users as well as traditional technology laggards. In the second quarter of 2002, mobile worldwide phone sales jumped .8%. A 2003 report by In-Stat/MDR predicted that, despite the sluggishness at the time of writing, there will be more than 931 million new subscribers over the next five years. This is echoed by forecasts from the Semiconductor Industry Association—

they reported in January 2004 that cell phone sales would increase by 10% and cameras by 14% (Davis, 2003).

The hot-button issues and gadgets of the next wave center around 3G technology. 3G networks have a standard, IMT-2000 (International Mobile Communications 2000 Initiative) issued by the ITU in 1999. Table 25.1 reviews the important capabilities of 3G systems including data transmission rates up to 2 Mb/s. As of January 2004, it was reported that wireless carries are working overtime to "upgrade their networks" to meet 3G standards, which consist of five operating modes, including three based on CDMA technology.

The migration from 2G digital standards (CDMA, GSM) to 3G standards is difficult. There is an intermediate step, referred to as 2.5G. These services are digital like 2G, but they employ packet switched data transmission. GSM networks use GRPS for 2.5G and will go to EDGE and W-CDMA (wideband CDMA) for 3G. CDMA networks migrate to cdmaOne for 2.5G and cdma2000 or EV-DO for 3G. Figure 23.2 reviews the migration to 3G.

Table 25.1
3G System Capabilities

Capability to support circuit and packet data at high bit rates:

- 144 kilobits/second or higher in high mobility (vehicular) traffic
- 384 kilobits/second for pedestrian traffic
- 2 Megabits/second or higher for indoor traffic

Interoperability and roaming

Common billing/user profiles:

- Sharing of usage/rate information between service providers
- Standardized call detail recording
- Standardized user profiles

Capability to determine geographic position of mobiles and report it to both the network and the mobile terminal

Support of multimedia services/capabilities:

- Fixed and variable rate bit traffic
- Bandwidth on demand
- Asymmetric data rates in the forward and reverse links
- Multimedia mail store and forward
- Broadband access up to 2 Megabits/second

Source: FCC

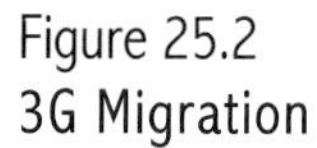

Figure 25.2
3G Migration

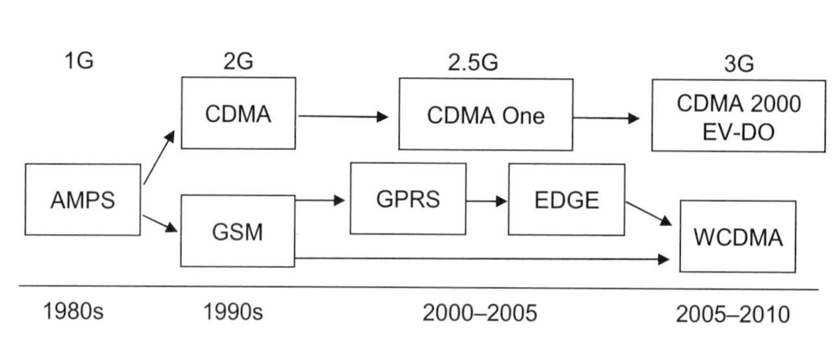

Source: J. Meadows

These upgrades are packet-based and are at the center of advanced wireless networks. CDMA is used by 71 million people in the United States; GSM is used by 22 million. The new standard, EV-DO (evolution-data optimized), is planned for national introduction (as of January 27, 2004) by Verizon in late 2004. It is actually a data only system; it requires a "dedicated slice of spectrum," but is exceedingly fast at 300 Kb/s to 600 Kb/s. Sprint has picked a similar technology, EV-DV, which handles data and voice. The EVs up the ante on quality and quantity. GRPS (General Radio Packet Service) has been in use by AT&T wireless since 2002 and provides speeds up to 144 Kb/s.

The advent and proliferation of these wireless standards facilitates the convergence of multiple applications and technologies into a single unit.

There are many newfangled applications and hardware that could be converged including camera phones with up to 640 pixels, video streaming, tracking devices, GPS, e-mail attached animation, and complex role-playing games. Indeed, six million camera phones were sold in the United States in 2003, and cameras are predicted to be standard on all phones in 2005. Three-dimensional content—primarily games—is also on the horizon (ATI Technologies Inc., 2004). In its ultimate expression, users could interface with home appliances with the wireless phone as a remote control. Companies such as BlueTie, Incorporated are working on ways to allow cell phone users to secure real-time access to business data (BlueTie Inc., 2004).

Not everyone is banking on convergence though. Some are leery of too much technology shoved into one tiny package, and a recent survey found that many consumers want more practical applications such as services that provide step-by-step instructions.

Bibliography

Analog Devices launches new products for cell phone towers. (2004, March 18). *The Business Journal.* Retrieved May 2, 2004 from http://www.bizjournals.com/triad/stories/2004/03/15/daily31.html?jst=s_cn_hl.

Ashok, B. (2003, October 1). Can micro fuel cells supplant batteries? *Power Electronics Technology, 29.* Retrieved April 17, 2004 from http://powerelectronics.com/mag/power_micro_fuel_cells_2/.

Auter, P. (2004, April). *College student gratifications from cell phone usage.* Poster session presented at the meeting of the Broadcast Education Association, Las Vegas, NV.

ATI Technologies, Inc. (2004, March 12). ATI and SK Telecom to cooperate on next generation 3D content service platform. *Press release—2004.* Retrieved May 2, 2004 from http://www.ati.com/companyinfo/press/2004/4736.html.

Bellis, M. (2004a). Martin Cooper—History of cell phone. *About.com.* Retrieved April, 17 2004 from http://inventors.about.com/cs/inventorsalphabet/a/martin_cooper.htm.

Bellis, M. (2004b). Selling the cell phone. *About.com.* Retrieved April, 17 2004 from http://inventors.about.com/library/weekly/aa070899.htm.

BlueTie, Inc. (2004). BlueTie delivers real-time business data with release of BlueTie Mobile. *Press releases.* Retrieved May 2, 2004 from http://www.bluetie.com/press/pr_article.asp?id=&res=&releaseId=435.

Bunkley, N. (2004, March 14). Pay phones a dying breed in cellular age. *The Detroit News.* Retrieved April 18, 2004 from http://www.detnews.com/2004/technology/0403/16/c01-90877.htm.

Cardo launches Bluetooth adapter: Converts standard cell phones into wireless BT devices to work with the Allways Bluetooth headset. (n.d.). *PR Newswire.* Retrieved May 2, 2004 from http://www.prnewswire.com/cgi-bin/stories.pl?ACCT=109&STORY=/www/story/04-26-2004/0002159572&EDATE=.

Cell phones: History of. (n.d.). *Cell phone world,* Chap. 3. Retrieved April 17, 2004 from http://www.eden.rutgers.edu/~cang/history/public.html.

Charny, B. (2002, June 20). Virgin, Sprint create teen venture. *C/Net News.* Retrieved April 17, 2004 from http://news.com.com/2100-1033-937859.html.

Cellular Telecommunications & Internet Association. (n.d.). Court backs telecom competition, upholds FCC order for intermodal number portability. *CTIA News.* Retrieved on May 2, 2004, from http://www.wow-com.com/articles.cfm?ID=1354.

Cingular agrees to buy AT&T Wireless. (2004, February 17). *Silicon Valley.* Retrieved May 2, 2004, from http://www.siliconvalley.com/mld/siliconvalley/7972175.htm.

Cohen, A., & Lemish, D. (2003). Real-time and recall measures of mobile phone use: Some methodological concerns and empirical applications. *New Media & Society, 5* (2), 167-183.

Court won't halt phone number switching. (2003, November 21). *USA Today.* Retrieved on May 2, 2004 from http://www.usatoday.com/money/industries/telecom/2003-11-21-portability-suit_x.htm.

Davis, J. (2003, December 31). Consumer products leading new boom. *Electronic News.* Retrieved April 17, 2004 from http://www.reed-electronics.com/electronicnews/article/CA372503?spacedesc=news.

inCode releases top 10 predictions for wireless carriers in 2004. (2003, November 18). *PR Newswire.* Retrieved May 2, 2004 from http://www.findarticles.com/cf_dls/m4PRN/2003_Nov_18/110268160/p1/article.jhtml.

Kimball, R. (2004, April 29). Wireless portability complaints: Approximately 7,040 consumer complaints since porting began on November 24. *FCC News.* Retrieved on May 2, 2004, from http://hraunfoss.fcc.gov/edocs_public/attachmatch/DOC-246604A1.doc.

Leung, L., & Wei, R. (1999). Who are the mobile phone have-nots? Influences and consequences. *New Media & Society, 1* (2), 209-226.

Ling, R. & Yttri, B. (2002). Hyper-coordination via mobile phones in Norway. In J. E. Kats & M. A. Aakhus (Eds.). *Perpetual contact: Mobile communication, private talk, public performance.* Cambridge, UK: Cambridge University Press, pp. 139-169.

Matteo, S. (2004, March 17). Tech research services forecasts U.S. cellular/wireless subscribers to increase to over 245 million by year-end 2007. *PR Web: The Free Wire Service.* Retrieved May 2, 2004 from http://www.prweb.com/releases/2004/3/prweb111510.php.

McFarland, D., & Mongrain, A. (2003, October 14). Atlanta, GA; Detroit, MI; and Austin, TX ring the loudest when it comes to cell phone ownership. *Scarborough Research, Inc.* Retrieved May 2, 2004 from http://www.scarborough.com/scarb2002/press/pr_cell.htm.

McVicker, D. (2001, May 14). Teen angels? Young Americans hold the keys to the kingdom, as far as wireless providers are concerned. *Teledotcom.* Retrieved April 16, 2004 from http://www.teledotcom.com/article/TEL20010511S0010/.

Port, O. (2003, October 20). Super-radon, done dirt cheap. *Business Week.* Retrieved May 2, 2004 from http://www.businessweek.com/magazine/content/03_42/b3854113.htm.

PolyFuel delivers breakthrough fuel cell membrane for portable fuel cell systems. (2004, January 19). *PolyFuel press releases.* Retrieved May 2, 2004 from http://www.polyfuel.com/press_pr_011904.shtml.

Privateline.com. (n.d.). *Mobile telephone history.* Retrieved April 16, 2004 from http://www.privateline.com/PCS/history.htm.

Robbins, K., & Turner, M. (2002). United States: Popular, pragmatic and problematic. In J. E. Kats & M. A. Aakhus (Eds.). *Perpetual contact: Mobile communication, private talk, public performance.* Cambridge, UK: Cambridge University Press, pp. 80-93.

Robinson, L. (2002, February 6). Cell phone users: 21st century pariahs. *USA Today.* Retrieved April 16, 2004 from http://www.usatoday.com/tech/techreviews/2001-06-19-phone-pariahs-reut.htm.

Shafer, A. (2004, February 23). Analysts speculate on Sprint's merger prospects. *USA Today.* Retrieved May 2, 2004 from http://www.usatoday.com/tech/techinvestor/techcorporatenews/2004-02-23-pcs-merge_x.htm.

SRI International. (1998, October 22). The cellular telephone. In *The science and technology policy program,* Chapter 4. Retrieved April 17, 2004 from http://www.sri.com/policy/stp/techin2/chp4.html.

Trewyn, P. (2004, April 26). U.S. Cellular spends $100 million on upgrade; others follow suit. *The Business Journal of Milwaukee.* Retrieved on May 2, 2004 from http://milwaukee.bizjournals.com/milwaukee/stories/2004/04/26/story1.html?t=printable.

Williams, M. (2004, March 1). DoCoMo considers Cingular future. *ComputerWeekly.* Retrieved May 2, 2004 from http://www.computerweekly.com/Article128773.htm.

Wrolstad, J. (2002, August 20). Study: Youth market critical for wireless carriers. *Wireless News Factor.* Retrieved April 15, 2004 from http://www.wirelessnewsfactor.com/perl/story/19082.html

V
Conclusions

26

Media Convergence

August E. Grant, Ph.D.*

Among all of the trends explored throughout this text, one of the most important is the emerging trend toward media convergence. At least four types of convergence have the potential to alter the communication technology landscape over the next decade. The purpose of this chapter is to explore these individual types of convergence to generate an understanding of the larger role that media convergence is playing in the adoption and evolution of these technologies.

The four trends include:

- *Technological convergence*, caused primarily by the transition of virtually all communication technologies from analog to digital signals that can be manipulated, stored, and duplicated using computer technology.
- *Organizational convergence*, including cross-ownership of media and cooperation among media organizations to facilitate the delivery of media content across a variety of technology platforms.
- *Convergent journalism*, by which a single news organization gathers information that is then reported across print, broadcast, and online media.
- *Media use convergence*, in which consumers simultaneously use multiple media.

* Associate Professor, College of Mass Communications and Information Studies, University of South Carolina (Columbia, South Carolina)

The following sections discuss each of these four in relation to the Umbrella Perspective on Communication Technologies introduced in Chapter 1.

Technological Convergence

There was a time when a "radio" technology referred almost exclusively to broadcast communication and wired technology referred to telephony. Over time, however, engineers and entrepreneurs have adapted technology from one medium to another, such that virtually any communication medium can be used to carry virtually any type of message.

In the process, virtually all communication technologies have adopted digital technologies that reduce messages to streams of "zeroes" and "ones." Because these digital streams are all designed to be stored, processed, and transmitted using one or more types of computer, the technologies have necessarily become more similar to each other. The next logical step is adapting equipment, algorithms, etc. from one medium to another, resulting in an even greater impetus for media convergence.

The result of this digital convergence is that virtually any digital communication technology can be used to transmit or receive virtually any type of digital information. Perhaps the best example is the latest generation of cellular telephones, which not only are designed to send and receive voices, but can also send and receive pictures, video, and data (including e-mail and Web access). The computer processor and memory that is the heart of these telephones also enables them to function as small computers or PDAs (personal digital assistants), with some of these phones having more computing power and memory than the first manned spacecraft sent to the moon just three decades ago.

Just as a cellular telephone can be designed to perform the functions of a radio, PDA, camera, or voice recorder, so can virtually any other device. High-definition television sets also include massive amounts of processing capability that could be used for other purposes. Digital video recorders are simple computers with massive memory that are already used for storage and display of photos and music as well as television shows; these devices could easily be adapted to record any other type of data, as well as record images from other sources such as security systems.

With technology no longer serving as the primary barrier between technologies, the focus has to shift to users and their wants and needs. Inventive entrepreneurs (including you?) may find new fortunes in discovering ways to utilize the capabilities of technological convergence to meet existing or future needs of users. Along the way, though, there are likely to be many more failures than successes. As discussed in Chapter 10, the combination computer/television set has never succeeded despite being introduced numerous times over the past two decades, primarily because of differences in the way people use computers and watch television. On the other hand, the inventors of the cellular telephone never envisioned their product being used for photography, yet cameras are now a standard feature on high-end cellular phones.

The most important enabling technology may not be digital technology itself, but broadband access that allows virtually any device connected through a local network to a broadband network to receive information from and send information to virtually any other computer on the Internet. The most important existing barrier to the utilization of broadband networks for applications beyond Web surfing, e-mail, and the like is the lack of ubiquitous networking in the home. This situation parallels the early stage of diffusion of electricity to the home, where a single electric line was run to the home to serve a small number of lights and appliances. Just as 100 years of innovation in the uses of electricity have resulted in a wide range of electrical appliances and the installation of multiple electrical outlets in nearly every room in a house, broadband access could also be expected to proliferate in the home, enabling a powerful new array of appliances that use digital technology to take advantage of that connection.

Given the capabilities of today's technology, it is easy to see how direct broadcast satellite can also be used to deliver Internet access, how telephone lines can deliver video, and how cable television can provide telephone service. As Chapter 1's Umbrella Perspective suggests, however, factors at the individual user, organizational, and system levels encompass a wide range of enabling and limiting factors. In short, just because it is technically possible for a particular type of media convergence to exist does not mean that there are not significant barriers remaining at other levels of the umbrella. Because this chapter is focused on analyzing convergence, the following sections discuss a few of the enabling factors at these other levels.

Organizational Convergence

Organizational convergence encompasses a number of ways that organizations can work together, ranging from cross-ownership (where one organization owns multiple media) to cooperation (where two or more organizations work together to deliver their messages). The global trend in media ownership since 1990 has been one toward consolidation, with smaller media companies being purchased by larger ones that, in turn, become even larger, with more market power.

Almost every chapter in this book has some discussion of organizational convergence. In cable television, Comcast's purchase of AT&T Broadband (discussed in Chapter 3) made Comcast the largest cable operator in the United States, capable of exercising disproportionate power over the cable channels from which it gets its service. News Corporation's purchase of DirecTV (discussed in Chapter 4) adds both horizontal and vertical integration to that multinational corporation. Hewlett-Packard's merger with Compaq (discussed in Chapter 10) created a computer giant that is poised to challenge Dell as the number one computer manufacturer in the world. Cingular's proposed purchase of AT&T Wireless (discussed in Chapter 24) may significantly reduce the number of choices of cellular telephone providers in the United States. (The list could go on....)

There is no place this trend became clearer than in the creation of this book. Between the time the initial drafts of each chapter were written in early 2004 and the time the final text was set in May 2004, chapter authors contributed almost no changes in the hardware discussions in each chapter. However, the area of greatest change from the first drafts to the final text was in new developments regarding the organizational infrastructure of specific technologies. Of all of the areas of the umbrella perspective explored, no area enjoys greater change in the short term than the organizational infrastructure, and the trend of these changes is toward organizational convergence.

There is continuing debate regarding the desirability of organizational convergence. Those speaking in favor of convergence cite economic efficiencies and the power of large media companies to bring more resources to smaller media and smaller markets. They also point to the proliferation of new media including the Internet, satellite television, and 500+ channel cable systems as forces that are reducing their market share, indicating that they must grow in order to keep pace with these emerging media. Opponents of convergence, on the other hand, decry the loss of multiple editorial voices in the media and the resulting concentration of media influence in the hands of a relatively small number of companies and individuals.

It follows that the greatest changes in the technologies discussed in this book between the time the text of the book was finalized in May 2004 and the time you are reading this sentence is in the organizational infrastructure for these technologies. Accordingly, your best update to these "updates" is an examination of changes in the organization infrastructure of the technology you are studying.

Convergent Journalism

Within schools of journalism, no area of media convergence has received more attention over the past few years than convergent journalism. The fact that a single news organization can leverage its resources to gather information delivered across a variety of media, rather than through a single medium, promises a revolution in the practice of journalism.

The impetus for convergent journalism is the World Wide Web. Companies that risked massive expenditures to develop online news services, as well as those that held back to see how the medium would develop, quickly saw that practical business models to support delivery of news over the Internet were difficult to come by. No news organization wanted to be left behind, however, as the public expected delivery of up-to-date news through the Web. The answer for both newspapers and television stations was the creation of Web sites that leveraged content gathered for the traditional media to make up the bulk of online content. With these Web sites as a common distribution medium for news across newsrooms, it was not long before companies started to look at the practicality of gathering news for delivery across multiple media.

As of mid-2004, the United States is well behind the rest of the world in the practice of convergent journalism (Quinn, 2003). The primary reason for the lag is a set of regulations that discourages cross-ownership of media. Although, in 2003, the Federal Communications Commission proposed relaxing media ownership rules to allow cross-ownership of television stations and newspapers in a single market, public outcry over the potential for this cross-ownership to reduce the number of "voices" in the media led to a decision to temporarily put these rule changes on hold.

In the meantime, media in most of the rest of the world have no such prohibition, and savvy owners have aggressively pursued strategies designed to realize synergies from the convergence of multiple media in a single operation. The primary driver behind these convergent media forces is economic, as media owners seek to get more content out of as few people as possible (Quinn, 2003).

Although FCC regulations bar cross-ownership of television stations and newspapers in the same market in the United States, a few such operations that were in place before the rules were adopted were "grandfathered," providing a setting for field experiments in media convergence. The most cited example is Media General's marriage of *The Tampa Tribune*, WFLA-TV, and Tampa Bay Online in a single building in Tampa (Florida). Scholars who have studied the Media General operation have concluded that it heralds the future of journalism (Garrison & Dupagne, 2003), and it illustrates new challenges in managing media (Killebrew, 2002).

More than two dozen universities are also experimenting with converged programs that are designed to train journalists to work across media (Duhé & Tanner, 2003). There is no standard curriculum across these programs. Some programs combine print, broadcast, and online journalism in the introductory courses, while others do not attempt the convergence until the final courses in the sequence.

One common trait shared by these convergent media programs is the recognition that a converged newsroom creates new journalistic roles. These new roles are a not a function of combining different types of journalism, but rather a function of the new opportunities inherent when a team of journalists is gathering content and delivering news across multiple media. One set of these roles has been identified by Ifra's Kerry Northrup, founder of Ifra Newsplex at the University of South Carolina.

Four specific roles have been identified by the Newsplex team. The "Newsflow Manager" oversees all stories in progress across all media, allocating the appropriate resources to individual stories and then directing the stories to each of the output media in the converged newsroom. The "Storybuilder" supervises all aspects of an individual story, coordinating the reporters, photographers, and other personnel assigned to a story in the gathering of information and the distribution of the stories produced across media. The "News Resourcer," an information specialist who is a resource to all of the journalists and editors in a newsroom, provides information from archives, databases, the Internet, and other secondary sources to assist in the production of stories. Finally, the "Multiskilled Journalist" is a reporter trained to gather information and write stories for each of the output media in the converged newsroom (Grant, 2004).

The most important thing to understand about these new roles is that they do not necessarily reflect individuals or specific positions in a newsroom. Rather, each of the four represents a new set of responsibilities and activities in a newsroom. In convergent journalism training, individuals may be assigned to each role, but, in newsrooms, the roles may overlap across individuals or may be split, with two or more people combining to serve the role.

The study of convergent journalism has yielded a great deal of insight into the cultural and language differences across the media that must be overcome in a converged newsroom (Silcock & Keith, 2002), new management styles that must be implemented (Killebrew, 2002), and descriptive data comparing converged operations across small and large markets (DeMars, 2003).

Individual Users

Consumers are at the forefront of media convergence as well, utilizing multiple media simultaneously. Recent research on media multitasking indicates that consumers spend an average of 10.5 hours a day consuming media content, with one-quarter of that time spent using two or more media simultaneously, demonstrating the increasing trend toward using more than one medium at the same time (Papper & Homes, 2003).

This multitasking has important implications for media professionals in at least two respects. First, the attention level to each medium decreases when multiple media are used simultaneously, creating new challenges for advertisers. Second, this trend presents an interesting opportunity for entrepreneurs seeking new ways to combine media to better serve individual audience members.

Conclusions

The most important lesson from this chapter is that changes in communication technology are as likely to emerge from the convergence of technologies, organizations, and uses as they are from the development of new technologies. In this respect, communication technologies are no different from any other technologies, with innovation emerging from the merger of disparate systems as well as from development of new systems.

The study of communication technologies must therefore include an analysis of the manner in which these technologies, organizations, and uses combine to create new opportunities. In the process, we are certain to see short-term revolutions in the structure of media and long-term revolutions in the structure of society itself.

Bibliography

DeMars, T. (2003, November). *What's working and what's not: A survey of print and broadcast news convergence.* Paper presented to Media Use in a Changing Media Environment, Columbia, SC.

Duhé, S., & Tanner, A. (2003, November). *Convergence education: A nationwide look.* Paper presented to Media Use in a Changing Media Environment, Columbia, SC.

Garrison, B., & Dupagne, M. (2003, November). *A case study of media convergence at Media General's News Center in Tampa, Florida.* Paper presented to Media Use in a Changing Media Environment, Columbia, SC.

Grant, A. E. (2004). New roles in converged newsrooms. *The Convergence Newsletter, 1* (9). Retrieved May 7, 2004 from http://www.jour.sc.edu/news/convergence/issue9.html.

Killebrew, K. (2002, November). *Distributive and content models issues in convergence: Defining aspects of "new media" in journalism's newest venture.* Paper presented to The Dynamics of Convergent Media, Columbia, SC.

Quinn, S. (2003, November). *Lessons from the edge: Convergence outside the United States.* Paper presented to Media Use in a Changing Media Environment, Columbia, SC.

Papper, B., & Holmes, M. (2003, November). *Middletown media studies: Three investigations of media use patterns.* Paper presented to Media Use in a Changing Media Environment, Columbia, South Carolina.

Silcock, B. W., & Keith, S. (2002, November). *Translating the Tower of Babel: Issues of language and culture in converged newsrooms: A pilot study.* Paper presented to The Dynamics of Convergent Media, Columbia, SC.

27

Conclusion

Jennifer H. Meadows, Ph.D.*

The 9th edition of the *Communication Technology Update* is filled with updated information about important communication technologies. While it is essential to consider each chapter on its own, it is also useful to review the entire book to see important factors that reach across all chapters. Some of these factors, such as digital-to-analog conversion, have been in progress for a decade or more, while others have emerged since 2000.

The conversion from analog-to-digital technology continues in all areas explored in the previous chapters. Digital television adoption continues to be slow, but there is hope. Consumer sales of digital sets are at an all time high. Many of the owners of these digital television sets, though, do not receive their digital television from over-the-air broadcasts. Instead, they receive high-definition television through cable or direct broadcast satellite service. That makes sense because most Americans receive television programming through cable or DBS. What is frustrating for many of these digital-set-owning consumers is the lack of available HDTV programming. Oftentimes, the cable or DBS system does not carry the DTV/HDTV signal of local broadcasters. Consumers could get the digital signal over the air, but many of the sets sold in the United States are HDTV monitors without digital tuners. In many cases, the consumer is not likely to purchase and set up a digital tuner. Imagine the frustration of buying a new, widescreen HDTV set for the Super Bowl and then not being able to see the game in HDTV because your cable system does not carry the channel.

Broadcast radio is just starting its conversion to digital, with the slow adoption of Ibiquity's IBOC "HD radio" technology. Satellite radio is faring slightly better than expected with more than two million subscribers. The question for digital audio broadcasting is whether people are willing to

* Associate Professor, Department of Communication Design, California State University, Chico (Chico, California).

buy new radios when analog AM and FM will still work. There needs to be a significant benefit for people to upgrade. Will sound quality and data information be enough? Or will the appeal of commercial-free music be the driving force behind consumer adoption of satellite radio, eventually forcing terrestrial broadcasters to adapt to this new type of competition (something radio has done every few decades since its inception almost a century ago).

Other technologies have completed the conversion from analog to digital including audio, home video, and wireless telephony. It will be interesting to see how the "traditional" mass media of television and radio fare in the next few years, as they are likely to be among the last analog, electronic communication technologies.

Another factor that reaches across chapters is increasing bandwidth or transmission rates. Does it seem like things are moving faster? That is because they are. Many different forms of data transmission technologies have faster speeds. Home networking technologies, such as HomePNA, have gone from 1 Mb/s to 128 Mb/s in just six years. Wireless phone networks have gone from 1 Mb/s at best to 300 Kb/s to 500 Kb/s with EV-DO. More homes have broadband access with DSL and cable modems. Wi-Fi has gone from 11 Mb/s with IEEE 802.11b to 54 Mb/s with IEEE 802.11g. WiMAX and 802.11n have speeds up to 70 Mb/s and 108 Mb/s, respectively.

With increased bandwidth comes software that takes advantage of that bandwidth. Video streaming has become not only easier and more efficient, but more pleasurable to watch as well. Services such as TiVo's Home Media Option allow users to program their digital video recorder remotely. Replay DVR users can share programming over a broadband connection. Distance learning software such as Blackboard and WebCT include more multimedia components to enhance learning. The cell phone is now a multifunctional device so users can take and send pictures, play games, download music, and surf the Web. Digital cable customers in many markets enjoy video on demand and subscription VOD services over advanced cable networks. Overall, networks are becoming faster.

Much like the increase in bandwidth, the speed of computer processors continues to grow each year, reaffirming Moore's Law. Along with that increased speed comes less expensive memory. These developments impact a wide range of technologies, from digital audio to distance learning. In addition, solid-state memory devices are quickly replacing memory devices that use moving parts, such as CD drives and hard drives. Just look at two popular technologies: the portable MP3 player and the flash memory keydrive. When they go for a jog, music lovers no longer need to deal with CDs that skip. Computer users can quickly swap files with a tiny drive that plugs into a USB port and holds 254 MB of information.

Wireless networking has grown in popularity since 2000. Wireless capabilities are being incorporated into more technologies and are being extended for others. An increasing number of laptop computers come with wireless cards. To allow laptop users wireless Internet access, the number of Wi-Fi hotspots continues to grow. The Bluetooth short-range wireless standard is even taking off, with Bluetooth-enabled wireless phones and computer accessories such as keyboards being adopted. Wireless phone use continues to grow worldwide. New wireless data transmission technologies are under development that will bring unheard of transmission rates. These technologies, such as WiMAX, can often be deployed for lower cost than traditional wired networks. Wireless technologies have brought remote communities around the world telephone and Internet access that would have never come had the service been limited to wired networks.

Finally, when looking at many of the chapters, it appears that there have not been major changes in hardware and software. In many cases, the major changes in technology have occurred in policy and regulation. Cable, DBS, and radio are under scrutiny by the Federal Communications Commission for profanity and obscenity, sparked in part by the infamous "wardrobe malfunction" at the 2004 Super Bowl and Bono's brief expletive in his acceptance speech at the 2003 Grammy Awards. This crackdown may lead to changes in program scheduling.

Even more important than regulatory changes are changes in the organizational infrastructure of the communication industries. Changes in ownership limits have led to increasing consolidation in radio and television. Consolidation has also become a major force in other industries, particularly telephony. Cingular's purchase of AT&T Wireless is perhaps the best example of consolidation in the months preceding publication of this book. The landline telephone industry is one of the most concentrated of any industry, with SBC and Verizon now the telephone companies for most of the nation. Landline phone service is seeing increasing competition from wireless phone service, and VoIP promises low-cost competition to both plain-old telephone service and wireless telephony. Telephone customers can now take their numbers from their landline phones to wireless phones, and can take wireless phone numbers to other carriers. Finally, no discussion would be complete without the Internet. The U.S. Senate voted in May 2004 to extend the tax-free status of Internet access another four years. These are just a few of the policy changes that have occurred since 2000.

As of May 2004, the technology sector appears to be heading slowly toward an economic recovery as the nation as a whole recovers. Internet businesses such as eBay, Amazon, and Yahoo are showing profits, and Google launched its IPO in May 2004. Computer sales are up, and consumers are spending money on digital entertainment,

In the midst of all these changes, a few constants emerge. People still need to communicate. For example, while POTS may be losing customers, wireless telephony is gaining ground. Mass media is still an important sector in communication technology. According to the principle of relative constancy, the percentage of disposable income people spend on mass media remains relatively constant, but the categories of spending within mass media vary. So, while spending on music may be down, people are spending more on video games and streaming media. Similarly, the number of people watching broadcast television may be down, but cable and DBS viewership is up. The Internet, of course, has become an increasingly important source of information, entertainment, and commerce.

One of the most exciting—and frustrating—things about communication technology is that it changes quickly. In fact, by the time this book is published, there will have been significant developments in every area discussed. Please visit http://www.tfi.com/ctu/redev, the Communication Technology Web site, at to read updates on the technologies discussed in the book. To follow continuing developments in technology, you can also look at the bibliographies of particular chapters. The authors have used a variety of reliable sources to gather information about communication technologies.

Glossary

2-wire line. The set of two copper wires used to connect a telephone customer with a switching office, loosely wrapped around each other to minimize interference from other twisted pairs in the same bundle. Synonymous with twisted pair.

802.11. An IEEE specification for 1 Mb/s and 2 Mb/s wireless LANs.

802.11a. An IEEE specification for 54 Mb/s high rate wireless LANs in the 5 GHz band that uses orthogonal frequency division multiplexing.

802.11b. An IEEE specification for 11 Mb/s high rate wireless LANs in the 2.4 GHz band that uses direct sequence spread spectrum technology. (Also known as 802.11HR—for high rate.)

802.11g. An IEEE specification for 54 Mb/s high rate wireless LANs in the 2.4 GHz band that is backward-compatible with 802.11b, which uses the same band.

802.16. An IEEE specification for fixed and mobile wireless connectivity offering speeds up to 120 Mb/s using the 10-66 GHz frequency range. See also Wi-MAX.

802.3. An IEEE specification for SCMA/CD based Ethernet networks.

802.5. An IEEE specification for token ring networks.

A

Adaptive transform acoustic coding (ATRAC). A method of digital compression of audio signals used in the MD (MiniDisc) format. ATRAC ignores sounds out of the range of human hearing to eliminate about 80% of the data in a digital audio signal.

Addressability. The ability of a cable system to individually control its set-top boxes, allowing the cable operator to enable or disable reception of channels for individual customers instantaneously from the headend.

ADSL (*asymmetrical digital subscriber line*). A system of compression and transmission that allows broadband signals up to 6 Mb/s to be carried over twisted pair copper wire for relatively short distances.

Advanced television (ATV). Television technologies that offer improvement in existing television systems.

Agent. Any being in a virtual environment.

Algorithm. A specific formula used to modify a signal. For example, the key to a digital compression system is the algorithm that eliminates redundancy.

AM (*amplitude modulation*). A method of superimposing a signal on a carrier wave in which the strength (amplitude) of the carrier wave is continuously varied. AM radio and the video portion of NTSC TV signals use amplitude modulation.

American National Standards Institute (ANSI). An official body within the United States delegated with the responsibility of defining standards.

American Standard Code for Information Interchange (ASCII). Assigns specific letters, numbers, and control codes to the 256 different combinations of 0s and 1s in a byte.

Analog. A continuously varying signal or wave. As with all waves, analog waves are susceptible to interference, which can change the character of the wave.

ANSI. See *American National Standards Institute*.

ASCII. See *American Standard Code for Information Interchange*.

Aspect ratio. In visual media, the ratio of the screen width to height. Ordinary television has an aspect ratio of 4:3, while high-definition television is "wider" with an aspect ratio of 16:9 (or 5.33:3).

Asynchronous. Occurring at different times. For example, electronic mail is asynchronous communication because it does not require the sender and receiver to be connected at the same time.

ATM (*asynchronous transfer mode*). A method of data transport whereby fixed length packets are sent over a switched network. Speeds of up to 2 Gb/s can be achieved, making it suitable for carrying voice, video, and data.

Audio on demand. A type of media that delivers sound programs in their entirety whenever a listener requests the delivery.

Augmented reality. The superimposition of virtual objects on physical reality. For example, a technician could use augmented reality to display a 3D image of a schematic diagram on a piece of equipment to facilitate repairs.

Avatar. An animated character representing a person in a virtual environment.

B

Backbone. The part of a communications network that handles the major traffic using the highest-speed, and often longest, paths in the network.

Bandwidth. A measure of capacity of communications media. Greater bandwidth allows communication of more information in a given period of time.

Bit. A single unit of data, either a one or a zero, used in digital data communications. When discussing digital data, a small "b" refers to bits, and a capital "B" refers to bytes.

Bluetooth. A specification for short-range wireless technology created by a consortium of computer and communication companies that make up the Bluetooth Special Interest Group.

Bridge. A type of switch used in telephone and other networks that connects three or more users simultaneously.

Broadband. An adjective used to describe large-capacity networks that are able to carry several services at the same time, such as data, voice, and video.

Buy rate. The percentage of subscribers purchasing a pay-per-view program divided by the total number of subscribers who can receive the program.

Byte. A compilation of bits, seven bits in accordance with ASCII standards and eight bits in accordance with EBCDIC standards.

C

Carrier. An electromagnetic wave or alternating current modulated to carry signals in radio, telephonic, or telegraphic transmission.

CATV (community antenna television). One of the first names for local cable television service, derived from the common antenna used to serve all subscribers.

C-band. Low-frequency (1 GHz to 10 GHz) microwave communication. Used for both terrestrial and satellite communication. C-band satellites use relatively low power and require relatively large receiving dishes.

CCD (*charge coupled device*). A solid-state camera pickup device that converts an optical image into an electrical signal.

CCITT (*International Telegraph and Telephone Consultative Committee*). CCITT is the former name of the international regulatory body that defines international telecommunications and data communications standards. It has been renamed the Telecommunications Standards Sector of the International Telecommunications Union.

CDDI (*copper data distributed interface*). A subset of the FDDI standard targeted toward copper wiring.

CDMA (*code division multiple access*). A spread spectrum cellular telephone technology, which digitally modulates signals from all channels in a broad spectrum.

CD-ROM (*compact disc-read only memory*). The use of compact discs to store text, data, and other digitized information instead of (or in addition to) audio. One CD-ROM can store up to 700 megabytes of data.

CD-RW (compact disc-rewritable). A special type of compact disc that allows a user to record and erase data, allowing the disc to be used repeatedly.

Cell. The area served by a single cellular telephone antenna. An area is typically divided into numerous cells so that the same frequencies can be used for many simultaneous calls without interference.

Central office (CO). A telephone company facility that handles the switching of telephone calls on the public switched telephone network (PSTN) for a small regional area.

Central processing unit (CPU). The "brains" of a computer, which uses a stored program to manipulate information.

Churn. The percentage of subscribers to a service that are lost and must be replaced in a given period of time, usually a year.

Circuit-switched network. A type of network whereby a continuous link is established between a source and a receiver. Circuit switching is used for voice and video to ensure that individual parts of a signal are received in the correct order by the destination site.

CLEC. See *Competitive local exchange carrier.*

CO. See *Central office.*

Coaxial cable. A type of "pipe" for electronic signals. An inner conductor is surrounded by a neutral material, which is then covered by a metal "shield" that prevents the signal from escaping the cable.

CODEC (*COmpression/DECompression).* A device used to compress and decompress digital video signals.

COFDM (*coded orthogonal frequency division multiplexing*). A flexible protocol for advanced television signals that allows simultaneous transmission of multiple signals at the same time.

Common carrier. A business, including telephone companies and railroads, which is required to provide service to any paying customer on a first-come/first-served basis.

Competitive local exchange carrier (CLEC). An American term for a telephone company that was created after the Telecommunications Act of 1996 made it legal for companies to compete with the ILECs. Contrast with *ILEC.*

Compression. The process of reducing the amount of information necessary to transmit a specific audio, video, or data signal.

Cookies. A file used by a Web browser to record information about a user's computer, including Web sites visited, which is stored on the computer's hard drive.

Core network. The combination of telephone switching offices and transmission plant connecting switching offices together. In the U.S. local exchange network, core networks are linked by several competing interexchange networks; in the rest of the world, the core network extends to national boundaries.

CPE. See *Customer premises equipment.*

CPU. See *Central processing unit.*

Crosstalk. Interference from an adjacent channel.

Customer premises equipment (CPE). Any piece of equipment in a communication system that resides within the home or office. Examples include modems, television set-top boxes, telephones, and televisions.

Cyberspace. The artificial worlds created within computer programs.

DBS. See *Direct broadcast satellite.*

DCC (digital compact cassette). A digital audio format resembling an audiocassette tape. DCC was introduced by Philips (the creator of the common audio "compact cassette" format) in 1992, but never established a foothold in the market.

Dedicated connection. A communications link that operates constantly.

Dial-up connection. A data communication link that is established when the communication equipment dials a phone number and negotiates a connection with the equipment on the other end of the link.

Digital audio broadcasting (DAB). Radio broadcasting that uses digital signals instead of analog to provide improved sound quality.

Digital audiotape (DAT). An audio recording format that stores digital information on 4mm tape.

Digital signal. A signal that takes on only two values, off or on, typically represented by a "0" or "1." Digital signals require less power but (typically) more bandwidth than analog, and copies of digital signals are exactly like the original.

Digital subscriber line (DSL). A data communications technology that transmits information over the copper wires that make up the local loop of the public switched telephone network (see Local loop). It bypasses the circuit-switched lines that make up that network and yields much faster data transmission rates than analog modem technologies. Common varieties of DSL include ADSL, HDSL, IDSL, SDSL, and RADSL. The generic term xDSL is used to represent all forms of DSL.

Digital subscriber line access multiplexer (DSLAM). A device found in telephone company central offices that takes a number of DSL subscriber lines and concentrates them onto a single ATM line.

Digital video compression. The process of eliminating redundancy or reducing the level of detail in a video signal in order to reduce the amount of information that must be transmitted or stored.

Direct broadcast satellites (DBS). High-powered satellites designed to beam television signals directly to viewers with special receiving equipment.

Discrete multitone modulation (DMT). A method of transmitting data on copper phone wires that divides the available frequency range into 256 subchannels or tones, and which is used for some types of DSL.

Domain name system (DNS). The protocol used for assigning addresses for specific computers and computer accounts on the Internet.

Downlink. Any antenna designed to receive a signal from a communication satellite.

DSL. See *Digital subscriber line.*

DSLAM. See *Digital subscriber line access multiplexer.*

DVD (digital video or versatile disc). A plastic disc similar in size to a compact disc that has the capacity to store up to 20 times as much information as a CD and uses both sides of the disc.

DVD-RAM. A type of DVD that can be used to record and play back digital data (including audio and video signals).

DVD-ROM. A type of DVD that is designed to store computer programs and data for playback only.

E

E-1. The European analogue to T1, a standard, high-capacity telephone circuit capable of transmitting approximately 2 Mb/s or the equivalent of 30 voice channels.

Echo cancellation. The elimination of reflected signals ("echoes") in a two-way transmission created by some types of telephone equipment; used in data transmission to improve the bandwidth of the line.

Electromagnetic spectrum. The set of electromagnetic frequencies that includes radio waves, microwave, infrared, visible light, ultraviolet rays, and gamma rays. Communication is possible through the electromagnetic spectrum by radiation and reception of radio waves at a specific frequency.

E-mail. Electronic mail or textual messages sent and received through electronic means.

EV-DO. Short for "Evolution Data Optimized," an emerging standard for the transmission of data over cellular telephone networks at speeds up to 500 Kb/s.

Extranet. A special type of Intranet that allows selected users outside an organization to access information on a company's Intranet.

F

FDDI-I (fiber data distributed interface I). A standard for 100 Mb/s LANs using fiber optics as the network medium linking devices.

FDDI-II (fiber data distributed interface II). Designed to accommodate the same speeds as FDDI-I, but over a twisted-pair copper cable.

Federal Communications Commission (FCC). The U.S. federal government organization responsible for the regulation of broadcasting, cable, telephony, satellites, and other communications media.

Fiber-in-the-loop (FITL). The deployment of fiber optic cable in the local loop, which is the area between the telephone company's central office and the subscriber.

Fiber optics. Thin strands of ultrapure glass or plastic that can be used to carry light waves from one location to another.

Fiber-to-the-cabinet (FTTCab). Network architecture where an optical fiber connects the telephone switch to a street-side cabinet. The signal is converted to feed the subscriber over a twisted copper pair.

Fiber-to-the-curb (FTTC). The deployment of fiber optic cable from a central office to a platform serving numerous homes. The home is linked to this platform with coaxial cable or twisted pair (copper wire). Each fiber carries signals for more than one residence, lowering the cost of installing the network versus fiber-to-the-home.

Fiber-to-the-home (FTTH). The deployment of fiber optic cable from a central office to an individual home. This is the most expensive broadband network design, with every home needing a separate fiber optic cable to link it with the central office.

Fixed satellite services (FSS). The use of geosynchronous satellites to relay information to and from two or more fixed points on the earth's surface.

FM (frequency modulation). A method of superimposing a signal on a carrier wave in which the frequency on the carrier wave is continuously varied. FM radio and the audio portion of an NTSC television signal use frequency modulation.

Footprint. The coverage area of a satellite signal, which can be focused to cover specific geographical areas.

Frame. One complete still image that makes up a part of a video signal.

Frame relay. A high-speed packet switching protocol used in wide area networks (WANs), often to connect local area networks (LANs) to each other, with a maximum bandwidth of 44.725 Mb/s.

Frequency. The number of oscillations in an alternating current that occur within one second, measured in Hertz (Hz).

Frequency division multiplexing (FDM). The transmission of multiple signals simultaneously over a single transmission path by dividing the available bandwidth into multiple channels that each cover a different range of frequencies.

FTP (file transfer protocol). An Internet application that allows a user to download and upload programs, documents, and pictures to and from databases virtually anywhere on the Internet.

FTTC. See *Fiber-to-the-curb*.

FTTH. See *Fiber-to-the-home*.

Full-motion video. The projection of 20 or more frames (or still images) per second to give the eye the perception of movement. Broadcast video in the United States uses 30 frames per second, and most film technologies use 24 frames per second.

Fuzzy logic. A method of design that allows a device to undergo a gradual transition from on to off, instead of the traditional protocol of all-or-nothing.

G.dmt. A kind of asymmetric DSL technology, based on DMT modulation, offering up to 8 Mb/s downstream bandwidth, 1.544 Mb/s upstream bandwidth. "G.dmt" is actually a nickname for the standard officially known as ITU-T Recommendation G.992.1. (See *International Telecommunications Union*.)

G.lite [pronounced "G-dot-light"]. A kind of asymmetric DSL technology, based on DMT modulation, that offers up to 1.5 Mb/s downstream bandwidth, 384 Kb/s upstream, does not usually require a splitter, and is easier to install than other types of DSL. "G.lite" is a nickname for the standard officially known as G.992.2. (See *International Telecommunications Union*.)

Geosynchronous orbit (GEO, also known as geostationary orbit). A satellite orbit directly above the equator at 22,300 miles. At that distance, a satellite orbits at a speed that matches the revolution of the earth so that, from the earth, the satellite appears to remain in a fixed position.

Gigabyte. 1,000,000,000 bytes or 1,000 megabytes (see *Byte*).

Global positioning system (GPS). Satellite-based services that allow a receiver to determine its location within a few meters anywhere on earth.

Gopher. An early Internet application that assisted users in finding information on and accessing the resources of remote computers.

Graphical user interface (GUI). A computer operating system that is based on icons and visual relationships rather than text. Windows and the Macintosh computer use GUIs because they are more user friendly.

Groupware. A set of computer software applications that facilitates intraorganizational communication, allowing multiple users to access and change files, send and receive e-mail, and keep track of progress on group projects.

GSM (Group standard mobile). A type of digital cellular telephony used in Europe that uses time-division multiplexing to carry multiple signals in a single frequency.

GUI. See *Graphical user interface*.

H

Hardware. The physical equipment related to a technology.

HDSL. See *High bit-rate digital subscriber line.*

Hertz. See *Frequency.*

HFC (hybrid fiber/coax). A type of network that includes a fiber optic "backbone" to connect individual nodes and coaxial cable to distribute signals from an optical network interface to the individual users (up to 500 or more) within each node.

High bit-rate digital subscriber line (HDSL). A symmetric DSL technology that provides a maximum bandwidth of 1.5 Mb/s in each direction over two phone lines, or 2 Mb/s over three phone lines.

High bit-rate digital subscriber line II (HDSL II). A descendant of HDSL that offers the same performance over a single phone line.

High-definition television (HDTV). Any television system that provides a significant improvement in existing television systems. Most HDTV systems offer more than 1,000 scan lines, in a wider aspect ratio, with superior color and sound fidelity.

Hologram. A three-dimensional photographic image made by a reflected laser beam of light on a photographic film.

Home networking. Connecting the different electronic devices in a household by way of a local area network (LAN).

Hotspot. An area in which wireless, public Internet access is available using one or more of the Wi-Fi protocols.

HTML. See *Hypertext markup language.*

http (hypertext transfer protocol). The first part of an address (URL) of a site on the Internet, signifying a document written in hypertext markup language (HTML).

Hybrid fiber/coax (HFC). A type of network that includes coaxial cables to distribute signals to a group of individual locations (typically 500 or more) and a fiber optic backbone to connect these groups.

Hypertext. Documents or other information with embedded links that enable a reader to access tangential information at programmed points in the text.

Hypertext markup language (HTML). The computer language used to create hypertext documents, allowing connections from one document or Internet page to numerous others.

Hz. An abbreviation for Hertz. See *Frequency.*

I

ILEC. See *Incumbent local exchange carrier.*

Image stabilizer. A feature in camcorders that lessens the shakiness of the picture either optically or digitally.

Incumbent local exchange carrier (ILEC). A large telephone company that has been providing local telephone service in the United States since the divestiture of the AT&T telephone monopoly in 1982.

Institute of Electrical & Electronics Engineers (IEEE). A membership organization comprised of engineers, scientists, and students that sets standards for computers and communications.

Instructional Television Fixed Service (ITFS). A microwave television service designed to provide closed-circuit educational programming. Underutilization of these frequencies led wireless cable (MMDS) operators to obtain FCC approval to lease these channels to deliver television programming to subscribers.

Interactive TV (ITV). A television system in which the user interacts with the program in such a manner that the program sequence will change for each user.

Interexchange carrier. Any company that provides interLATA (long distance) telephone service.

Interlaced scanning. The process of displaying an image using two scans of a screen, with the first providing all the even-numbered lines and the second providing the odd-numbered lines.

International Organization of Standardization (ISO). Develops, coordinates, and promulgates international standards that facilitate world trade.

Internet Engineering Task Force (IETF). The standards organization that standardizes most Internet communication protocols, including Internet protocol (IP) and hypertext transfer protocol (HTTP).

Internet service provider (ISP). An organization offering and providing Internet access to the public using computer servers connected directly to the Internet.

Intranet. A network serving a single organization or site that is modeled after the Internet, allowing users access to almost any information available on the network. Unlike the Internet, Intranets are typically limited to one organization or one site, with little or no access to outside users.

IP (Internet protocol). The standard for adding "address" information to data packets to facilitate the transmission of these packets over the Internet.

ISDN (Integrated Services Digital Network). A planned hierarchy of digital switching and transmission systems synchronized to transmit all signals in digital form, offering greatly increased capacity over analog networks.

ISDN digital subscriber line (IDSL). A type of DSL that uses ISDN transmission technology to deliver data at 128 Kb/s into an IDSL "modem bank" connected to a router.

ISO. See *International Organization of Standardization*.

ISP. See *Internet service provider*.

ITU (International Telecommunications Union). A U.N. organization that coordinates use of the spectrum and creation of technical standards for communication equipment.

J

JPEG (*Joint Photographic Experts Group*). A committee formed by the ISO to create a digital compression standard for still images. Also refers to the digital compression standard for still images created by this group.

K

Killer app. Short for "killer application," this is a function of a new technology that is so strongly desired by users that it results in adoption of the technology.

Kilobit. One thousand bits (see *Bit*).

Kilobyte. 1,000 bytes (see *Byte*).

Ku-band. A set of microwave frequencies (12 GHz to 14 GHz) used exclusively for satellite communication. Compared to C-band, the higher frequencies produce shorter waves and require smaller receiving dishes.

L

LAN. See *Local area network*.

Laser. From the acronym for "light amplification by stimulated emission of radiation." A laser usually consists of a light-amplifying medium placed between two mirrors. Light not perfectly aligned with the mirrors escapes out the sides, but light perfectly aligned will be amplified. One mirror is made partially transparent. The result is an amplified beam of light that emerges through the partially transparent mirror.

Last mile. See *Local loop*.

LATA (local access transport area). The geographical areas defining local telephone service. Any call within a LATA is handled by the local telephone company, but calls between LATAs must be handled by long distance companies, even if the same local telephone company provides service in both LATAs.

LEO (low earth orbit). A satellite orbit between 400 and 800 miles above the earth's surface. The close proximity of the satellite reduces the power needed to reach the satellite, but the fact that these satellites complete an entire orbit in a few hours means that a large number of satellites must be used in a LEO satellite system in order to have one overhead at all times.

Liner notes. The printed material that accompanies a CD or record album, including authors, identification of musicians, lyrics, pictures, and commentary.

LMDS (local multipoint distribution service). A new form of wireless technology similar to MMDS that uses frequencies above 28 GHz.

Local area network (LAN). A network connecting a number of computers to each other or to a central server so that computers can share programs and files.

Local exchange carrier (LEC). A local telephone company.

Local loop. The copper lines between a customer's premises and a telephone company's central office (see *Central office*).

Mb/s. Megabits per second.

Megabit. One million bits.

Megabyte. 1,000,000 bytes or 1,000 kilobytes (see *Byte*).

Megapixel. Refers to one million pixels, and is used as a measure of the quality of digital cameras. The more pixels used for a picture, the higher the quality.

Microcell. The area, typically a few hundred yards across, served by a single transmitter in a PCS network. The use of microcells allows the reuse of the same frequencies many times in an area, allowing more simultaneous users.

MIDI (*musical instrument digital interface*). An international standard for representing music in digital form. Music can be directly input from a computer keypad and stored to disc or RAM, then played back through a connected instrument. Conversely, a song can be played by the performer on an instrument interfaced with the computer.

MIPS (*millions of instructions per second*). This is a common measure of the speed of a computer processor.

MMDS (*multichannel multipoint distribution systems*). A service similar to cable television that uses microwaves to distribute the signals instead of coaxial cable. MMDS is therefore better suited to sparsely-populated areas than cable.

Mobile satellite services (MSS). The use of satellites to provide navigation services and to connect vehicles and remote regions with other mobile or stationary units.

Modem (*MOdulator/DEModulator).* Enables transmission of a digital signal, such as that generated by a computer, over an analog network, such as the telephone network.

Monochromatic. Light or other radiation with one single frequency or wavelength. Since no light is perfectly monochromatic, the term is used loosely to describe any light of a single color over a very narrow band of wavelengths.

MPEG (*Moving Picture Experts Group*). A committee formed by the ISO to set standards for digital compression of full-motion video. Also stands for the digital compression standard created by the committee that produces VHS-quality video.

MPEG-1. An international standard for the digital compression of VHS-quality, full-motion video.

MPEG-2. An international standard for the digital compression of broadcast-quality, full-motion video.

MTSO (mobile telephone switching office). The "heart" of a cellular telephone network, containing switching equipment and computers to manage the use of cellular frequencies and connect cellular telephone users to the landline network.

Multicast. The transmission of information over the Internet to two or more users at the same time.

Multimedia. The combination of video, audio, and text in a single platform or presentation.

Multiple system operator (MSO). A cable company that owns and operates many local cable systems.

Multiplexing. Transmitting several messages or signals simultaneously over the same circuit or frequency.

Must-carry. A set of rules requiring cable operators to carry all local broadcast television stations.

N

Nanometer. One billionth of a meter. Did you know that "nano" comes from the Greek word "dwarf?"

Narrowband. A designation of bandwidth less than 56 Kb/s.

National information infrastructure (NII). An initiative to support the private sector construction and maintenance of a "seamless web" of communication networks, computers, databases, and consumer electronics that will put vast amounts of information at users' fingertips.

Near video on demand (NVOD). A pay-per-view service offering movies on up to eight channels, each with staggered start times so that a viewer can watch one at any time. NVOD is much more flexible than PPV, and costs far less than VOD to implement.

Network access provider (NAP). Another name for a provider of networked telephone and associated services, usually in the United States.

Network service provider (NSP). A high-level Internet provider that offers high-speed backbone services.

Newbies. New users of an interactive technology, usually identified because they have not yet learned the etiquette of communication in a system.

Nodes. Routers or switches on a broadband network that provide a possible link from point A to point B across a network.

NTSC (*National Television Standards Committee*). The group responsible for setting the U.S. standards for color television in the 1950s.

O

OCR (*optical character recognition*). Refers to computer programs that can convert images of text to text that can be edited.

Octet. A byte, more specifically, an eight-bit byte. The origins of the octet trace back to when different networks had different byte sizes. Octet was coined to identify the eight-bit byte size.

Operating system. The program embedded in most computers that controls the manner in which data are read, processed, and stored.

Optical carrier 3 (OC3). A fiber optic line carrying 155 Mb/s, a U.S. designation generally recognized throughout the telecommunications community worldwide.

Optical network unit (ONU). A form of access node that converts optical signals transmitted via fiber to electrical signals that can be transmitted via coaxial cable or twisted pair copper wiring to individual subscribers. See *Hybrid fiber/coax*.

Overlay. The process of combining a graphic with an existing video image.

Packet switched networks. A network that allows a message to be broken into small "packets" of data that are sent separately by a source to the destination. The packets may travel different paths and arrive at different times, with the destination site reassembling them into the original message. Packet switching is used in most computer networks because it allows a very large amount of information to be transmitted through a limited bandwidth.

Passive optical network (PON). A fiber-based transmission network with no active electronics.

PCN (*personal communications network*). Similar to PCS, but incorporating a wider variety of applications including voice, data, and facsimile.

PCS (*personal communications services*). A new category of digital cellular telephone service which uses much smaller service areas (microcells) than ordinary cellular telephony.

Peripheral. An external device that increases the capabilities of a communications system.

Personal digital assistants (PDAs). Extremely small computers (usually about half the size of a notebook) designed to facilitate communication and organization. A typical PDA accepts input from a special pen instead of a keyboard, and includes appointment and memo applications. Some PDAs also include fax software and a cellular telephone modem to allow faxing of messages almost anywhere.

Photovoltaic cells. A device that converts light energy to electricity.

Pixel. The smallest element of a computer display. The more pixels in a display, the greater the resolution.

Point of presence (POP). The physical point of connection between a data network and a telephone network.

Point-to-multipoint service. A communication technology designed for broadcast communication, where one sender simultaneously sends a message to an unlimited number of receivers.

Point-to-point service. A communication technology designed for closed-circuit communication between two points such as in a telephone circuit.

POTS (*plain old telephone service*). An acronym identifying the traditional function of a telephone network to allow voice communication between two people across a distance.

POTS splitter. A device that uses filters to separate voice from data signals when they are to be carried on the same phone line. Required for several types of DSL service.

PPV (*pay-per-view*). A television service in which the subscriber is billed for individual programs or events.

Progressive scanning. A video display system that sequentially scans all the lines in a video display.

PTT. A government organization that offers telecommunications services within a country. (The initials refer to the antecedents of modern communication: the post [mail], telephone, and telegraph.)

QuickTime. A computer video playback system that enables a computer to automatically adjust video frame rates and image resolution so that sound and motion are synchronized during playback.

RBOC (*regional Bell operating company*). One of the seven local telephone companies formed upon the divestiture of AT&T in 1984. The original seven were Bell Atlantic, NYNEX, Pacific Telesis, BellSouth, SBC Corporation, U S WEST, and Ameritech.

Retransmission consent. The right of a television station to prohibit retransmission of its signal by a cable company. Under the 1992 Cable Act, U.S. television stations may choose between must-carry and retransmission consent.

RF (*radio frequency*). Electromagnetic carrier waves upon which audio, video, or data signals can be superimposed for transmission.

Roaming. Movement of a wireless node between two microcells.

Router. The central switching device in a computer network that directs and controls the flow of data through the network.

S

SCMS (Serial Copy Management System). A method of protecting media content from piracy that allows copies to be made of a specific piece of content, but will not allow copies of copies.

SCSI (*small computer system interface*) [pronounced "scuzzy"]. An outmoded type of interface between computers and peripherals.

SDTV (standard-definition television). Digital television transmissions that deliver approximately the same resolution and aspect ratio of traditional television broadcasts, but do so in a fraction of the bandwidth through the use of digital video compression.

SMPTE (*Society of Motion Picture and Television Engineers*). The industry group responsible for setting technical standards in most areas of film and television production.

Software. The messages transmitted or processed through a communications medium. This term also refers to the instructions (programs) written for programmable computers.

SONET (*Synchronous Optical Network*). A standard for data transfer over fiber optic networks used in the United States that can be used with a wide range of packet- and circuit-switched technologies.

Spot beam. A satellite signal targeted at a small area, or footprint. By concentrating the signal in a smaller area, the signal strength increases in the reception area.

SVOD (subscription video on demand). A pay television service that offers a range of VOD (video on demand) programming for a monthly fee.

Symmetric digital subscriber line (SDSL). A DSL technology that provides a maximum bandwidth of 1.5 Mb/s using one phone line, with a downstream transmission rate that equals the upstream transmission rate and allows use of POTS service on the same phone line. SDSL also refers to single-line digital subscriber line.

Synchronous transmission. The transmission of data at a fixed rate, based on a master clock, between the transmitter and receiver.

T

T1. A standard for physical wire cabling used in networks. A T-1 line has the bandwidth of 1.54 Mb/s.

T3. A standard for physical wire cabling that has the bandwidth of 44.75 Mb/s.

TCP/IP (transmission control protocol/Internet protocol). A method of packet-switched data transmission used on the Internet. The protocol specifies the manner in which a signal is divided into parts, as well as the manner in which "address" information is added to each packet to ensure that it reaches its destination and can be reassembled into the original message.

TDMA (*time division multiple access*). A cellular telephone technology that sends several digital signals over a single channel by assigning each signal a periodic slice of time on the channel. Different TDMA technologies include North America's Interim Standard (IS) 54, Europe's global system for mobile communications, and a version developed by InterDigital Corporation. These systems differ in circuits per channel, timing, and channel width.

Telecommuting. The practice of using telecommunications technologies to facilitate work at a site away from the traditional office location and environment.

Teleconference. Interactive, electronic communication among three or more people at two or more sites. Includes audio-only, audio and graphics, and videoconferencing.

Teleport. A site containing multiple satellite uplinks and downlinks, along with microwave, fiber optic, and other technologies to facilitate the distribution of satellite signals.

Terabyte. 1,000,000,000,000 bytes or 1,000 gigabytes (see *Byte*).

Time division multiplexing (TDM). The method of multiplexing where each device on the network is provided with a set amount of link time.

Transponder. The part of a satellite that receives an incoming signal from an uplink and retransmits it on a different frequency to a downlink.

TVRO (*television receive only*). A satellite dish used to receive television signals from a satellite.

Twisted pair. The set of two copper wires used to connect a telephone customer with a switching office. The bandwidth of twisted pair is extremely small compared with coaxial cable or fiber optics.

U

UHF (*ultra high frequency*). Television channels numbered 14 through 83.

Universal ADSL Working Group (UAWG). An organization composed of leading personal computer industry, networking, and telecommunications companies with the goal of creating an interoperable, consumer-friendly ADSL standard entitled the G.992.2 standard, and commonly referred to as the G.lite standard.

Universal serial bus (USB). A computer interface with a maximum bandwidth of 1.5 Mb/s used for connecting computer peripherals such as printers, keyboards, and scanners.

Universal service provider (USP). A company that sells access to phone, data, and entertainment services and networks.

Universal service. In telecommunications policy, the principle that an interactive telecommunications service must be available to everyone within a community in order to increase the utility and value of the network for all users.

Uplink. An antenna that transmits a signal to a satellite for relay back to earth.

URL (*uniform resource locator)*. An "address" for a specific page on the Internet. Every page has a URL that specifies its server and file name.

V

Variable bit rate (VBR). Data transmission that can be represented by an irregular grouping of bits or cell payloads followed by unused bits or cell payloads. Most applications other than voice circuits generate VBR traffic patterns.

VDSL (Very high bit-rate digital subscriber line). An asymmetric DSL that delivers from 13 Mb/s to 52 Mb/s downstream bandwidth and 1.5 Mb/s to 2.3 Mb/s upstream.

Vertical blanking interval (VBI). In an NTSC television signal, the portion of the signal that is not displayed on a television receiver. Some of the lines in the VBI contain "sync" information that is used to identify the beginning of a new picture. Some of the blank lines in the VBI can be used to carry data such as closed captions.

Vertical integration. The ownership of more than one function of production or distribution by a single company, so that the company, in effect, becomes its own customer.

VHF (very high frequency). Television channels numbered 2 through 13.

Videoconference. Interactive audio/visual communication among three or more people at two or more sites.

Video on demand (VOD). A pay-per-view television service in which a viewer can order a program from a menu and have it delivered instantly to the television, typically with the ability to pause, rewind, etc.

Videophone. A telephone that provides both sound (audio) and picture (video).

Videotext (also known as *videotex*). An interactive computer system using text and/or graphics that allows access to a central computer using a terminal or personal computer to engage in data retrieval, communication, transactions, and/or games.

Virtual reality (VR). A cluster of interactive technologies that gives users a compelling sense of being inside a circumambient environment created by a computer.

VoIP (Voice over Internet Protocol). A set of standards designed to enable telephone communications over data networks using packet-based switching (as opposed to circuit-based switching used in traditional telephone networks.

VRML (*virtual reality markup language).* A computer language that provides a three-dimensional environment for traditional Internet browsers, resulting in a simple form of virtual reality available over the Internet.

VSAT (*very small aperture terminal*). A satellite system that uses relatively small satellite dishes to send and receive one- or two-way data, voice, or even video signals.

Web Services. The name given to the set of protocols including XML, SOAP, and J2EE that facilitate sharing of information across programs and platforms, allowing interconnection of disparate databases.

Wide area network (WAN). A network that interconnects geographically distributed computers or LANs.

Wi-Fi. A collective term for a set of wireless data standards offering speeds up to 54 Mb/s, including 802.11a, 802.11b, and 802.11g.

Wi-MAX. Another name for the IEEE standard 802.16, which consists of a set of standards for fixed and mobile wireless data communication offering speeds of up to 120 Mb/s in the 10-66 GHz frequency range.

Wireless cable. See *MMDS.*

Wireless node. A user computer with a wireless network interface card.

WORM (write once, read many). A technique that allows recording of information on a medium only once, with unlimited playback.

X.25 data protocol. A packet switching standard developed in the mid-1970s for transmission of data over twisted-pair copper wire.

xDSL. See *Digital subscriber line.*

XML. (*Extensible markup language).* A protocol used to code information in a computer program so that information can be shared across numerous programs and platforms.